Responsive Systems for Active Vibration Control

T0183946

NATO Science Series

A Series presenting the results of scientific meetings supported under the NATO Science Programme.

The Series is published by IOS Press, Amsterdam, and Kluwer Academic Publishers in conjunction with the NATO Scientific Affairs Division

Sub-Series

I. **Life and Behavioural Sciences**	IOS Press
II. **Mathematics, Physics and Chemistry**	Kluwer Academic Publishers
III. **Computer and Systems Science**	IOS Press
IV. **Earth and Environmental Sciences**	Kluwer Academic Publishers
V. **Science and Technology Policy**	IOS Press

The NATO Science Series continues the series of books published formerly as the NATO ASI Series.

The NATO Science Programme offers support for collaboration in civil science between scientists of countries of the Euro-Atlantic Partnership Council. The types of scientific meeting generally supported are "Advanced Study Institutes" and "Advanced Research Workshops", although other types of meeting are supported from time to time. The NATO Science Series collects together the results of these meetings. The meetings are co-organized bij scientists from NATO countries and scientists from NATO's Partner countries – countries of the CIS and Central and Eastern Europe.

Advanced Study Institutes are high-level tutorial courses offering in-depth study of latest advances in a field.
Advanced Research Workshops are expert meetings aimed at critical assessment of a field, and identification of directions for future action.

As a consequence of the restructuring of the NATO Science Programme in 1999, the NATO Science Series has been re-organised and there are currently Five Sub-series as noted above. Please consult the following web sites for information on previous volumes published in the Series, as well as details of earlier Sub-series.

http://www.nato.int/science
http://www.wkap.nl
http://www.iospress.nl
http://www.wtv-books.de/nato-pco.htm

Series II: Mathematics, Physics and Chemistry – Vol. 85

Responsive Systems for Active Vibration Control

edited by

André Preumont

Université Libre de Bruxelles,
Bruxelles, Belgium

Kluwer Academic Publishers

Dordrecht / Boston / London

Published in cooperation with NATO Scientific Affairs Division

Proceedings of the NATO Advanced Study Institute on
Responsive Systems for Active Vibration Control
Brussels, Belgium
10–19 September 2001

A C.I.P. Catalogue record for this book is available from the Library of Congress.

ISBN 1-4020-0897-X (HB)
ISBN 1-4020-0898-8 (PB)

Published by Kluwer Academic Publishers,
P.O. Box 17, 3300 AA Dordrecht, The Netherlands.

Sold and distributed in North, Central and South America
by Kluwer Academic Publishers,
101 Philip Drive, Norwell, MA 02061, U.S.A.

In all other countries, sold and distributed
by Kluwer Academic Publishers,
P.O. Box 322, 3300 AH Dordrecht, The Netherlands.

Printed on acid-free paper

Contents

Preface

Structural vibrations have become critical in limiting the performances of many engineering systems, with typical amplitudes ranging from meters (e.g. for a bridge under severe wind or earthquake excitation) to a few nanometers (e.g. a space interferometer); many acoustic nuisances in transportation systems and in residential and office buildings are also related to structural vibrations. This has brought the subject of active vibration control to the forefront of research. Active vibration control involves nine orders of magnitude in terms of vibration amplitude, and this has a strong influence on the technology, which varies considerably from one application to the other; the actuators are the key components for large amplitudes while the sensors tend to dominate the sub-micron vibrations.

Active vibration control is strongly multidisciplinary, because it involves structural vibrations, acoustics, control, signal processing, material sciences, and actuator and sensor technology. All these topics have experienced tremendous improvements over the past few years, The idea of setting up a NATO Advanced Study Institute (ASI) on active vibration control came up during a discussion with Ben Wada in Barga, in 1999, at the unforgettable conference on "Engineering Adaptive Structures for Noise and Vibration Control". We thought that the ASI format would give a unique opportunity to offer a deep and comprehensive coverage of the interdisciplinary aspects, by a group of international experts, and would provide a forum for discussing the technical challenges of the next decade, to meet the needs of the various engineering areas, including civil engineering, aeronautics, space, ground transportation, optical instruments and consumer products. We were able to assemble a group of 14 international experts from universities, research labs and industry, who were selected for their complementarity as well as teaching skills.

The Advanced Study Institute took place at the university of Brussels (ULB) on September 10-19, 2001. We received more than 150 applications, but the number of participants was limited to 93 (from 25 countries), to comply with the ASI rules. The course was very successful in spite of the tragic events of September 11, which prevented only Prof. I.Chopra from participating. The interest of the course was attested by the responsiveness of the audience, which was still asking questions at six o'clock on the ninth day of the course!

I take this opportunity to thank warmly the co-director of the course, Rimas Bansevicius, the members of the organizing committee, Steve Elliott and Ben Wada, and the lecturers, Inderjit Chopra, Paolo Gaudenzi, Lothar Gaul, Wodek Gawronski, Hartmut Janocha, Ronan LeLetty, Kiong Lim, Patrice Minotti, Vytautas Ostasevicius and Jan Viba, for accepting to give their time and share their experience, and delivering excellent and always well attended lectures. I also want to express our gratitude to the NATO for its generous support, and to the program director, Dr F.Pedrazzini, who was always helpful during the preparation of the course. Finally, I extend my thanks to Prof. Y.Baudoin from the Royal Military Academy, and to my young colleagues from the Active Structures Laboratory at ULB, Ahmed, Arnaud, David, Frederic, Mihaita, Pierre, Valerie, Vincent, for their help in organizing all the practical details and assisting the participants.

The chapters of this book contain a written version of most of the lectures presented during the ASI. It was not possible to include all the lectures, because some lecturers were not able to produce the written material within the time frame allowed for preparing this volume. I express my thanks to the contributors for their efforts in this matter.

André Preumont
Brussels, May 2002.

AN INTRODUCTION TO ACTIVE VIBRATION CONTROL

A. PREUMONT
Université Libre de Bruxelles
Brussels, Belgium

1. Introduction

In order to motivate the use of active vibration control, consider the future interferometric missions planned by NASA or ESA (one such a mission, called "Terrestrial Planet Finder" aims at detecting earth-sized planets outside the solar system; other missions include the mapping of the sky with an accuracy one order better than that achieved by *Hypparcos*).

The purpose is to use a number of smaller telescopes as an interferometer to achieve a resolution which could only be achieved with a much larger monolithic telescope. One possible spacecraft architecture for such an interferometric mission is represented in Fig.1; it consists of a main truss supporting a set of independently pointing telescopes.

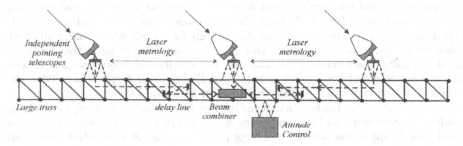

Figure 1: Schematic view of a future interferometric mission.

The relative position of the telescopes is monitored by a sophisticated metrology and the optical paths between the individual telescopes and the beam combiner are accurately controlled with optical delay lines, based on the information coming from a wave front sensor. Typically, the distance between the telescopes could be 50 m or more, and the order of magnitude of the error allowed on the optical path length is a few nanometers; the pointing error of the individual telescopes is as low as a few nanoradians (i.e. one order of magnitude better than the Hubble space telescope). Clearly, such stringent geometrical requirements cannot be achieved with a precision monolithic structure, but rather by active means as suggested in Fig.1. Let us first consider the supporting truss: given its size and environment, the main requirement on the supporting truss is not

1

A. Preumont (ed.), Responsive Systems for Active Vibration Control, 1–41.
© *2002 Kluwer Academic Publishers. Printed in the Netherlands.*

precision but *stability*, the accuracy of the optical path being taken care of by the wide-band vibration isolation/steering control system of individual telescopes and the optical delay lines (described below). Geometric stability includes thermal stability, vibration damping and prestressing the gaps in deployable structures (this is a critical issue for deployable trusses). In addition to the geometric requirements mentioned above, this spacecraft would be sent in deep space (perhaps as far as the orbit of Jupiter) to ensure maximum sensitivity; this makes the weight issue particularly important.

Another interesting subsystem necessary to achieve the stringent specifications is the six d.o.f. vibration isolator at the interface between the attitude control module and the supporting truss; this isolator allows the low frequency attitude control torque to be transmitted while filtering out the high frequency disturbances generated by the unbalanced centrifugal forces in the reaction wheels. The same general purpose vibration isolator may be used at the interface between the truss and the independent telescopes; in this case however, its vibration isolation capability is combined with the steering (pointing) of the telescopes. The third component relevant of active control is the optical delay line; it consists of a high precision single degree of freedom translational mechanism supporting a mirror, whose function is to control the path length between every telescope and the beam combiner, so that these distances are kept identical to a fraction of the wavelength (e.g. $\lambda/20$).

Performance and weight savings are the prime motivations of the foregoing example. However, as technology develops and with the availability of low cost electronic components, it is likely that there will be a growing number of applications where active solutions will become cheaper than passive ones, for the same level of performance.

The reader should not conclude that *active* will always be better and that a control system can compensate for a bad design. In most cases, a bad design will remain bad, active or not, and an active solution should normally be considered only after all other passive means have been exhausted. One should always bear in mind that feedback control can compensate external disturbances only in a limited frequency band that is called the *bandwidth* of the control system. One should never forget that outside the bandwidth, the disturbance is actually amplified by the control system.

In recent years, there has been a growing interest for *semi-active* control, particularly for vehicle suspensions; this has been driven by the reduced cost as compared to active control, due mainly to the absence of a large power actuator. A semi-active device can be broadly defined as a passive device in which the properties (stiffness, damping, ...) can be varied in real time with a low power input. Although they behave in a strongly nonlinear way, semi-active devices are inherently passive and, unlike active devices, cannot destabilize the system; they are also less vulnerable to power failure. Semi-active suspension devices may be based on classical viscous dampers with a variable orifice, or on magneto-rheological (MR) fluids.

2. Smart materials and structures

An active structure consists of a structure provided with a set of actuators and sensors coupled by a controller; if the bandwidth of the controller includes some vibration modes of the structure, its dynamic response must be considered. If the set of actuators and sensors are located at discrete points of the structure, they can be treated separately. The distinctive feature of *smart* structures is that the actuators and sensors are often distributed and have a high degree of integration inside the structure, which makes a

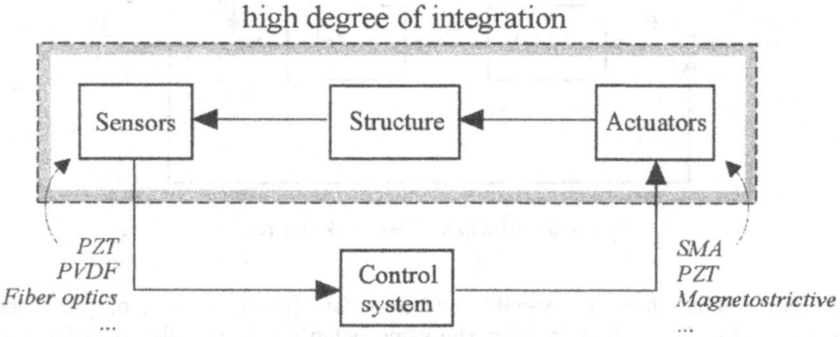

high degree of integration

Figure 2: Smart structure.

separate modelling impossible (Fig.2). Moreover, in some applications like vibroacoustics, the behaviour of the structure itself is highly coupled with the surrounding medium; this also requires a coupled modelling.

From a mechanical point of view, classical structural materials are entirely described by their elastic constants relating stress and strain, and their thermal expansion coefficient relating the strain to the temperature. *Smart materials* are materials where strain can also be generated by different mechanisms involving temperature, electric field or magnetic field, etc... as a result of some coupling in their constitutive equations. The most celebrated smart materials are briefly described below:

- *Shape Memory Alloys (SMA)* allow one to recover up to 5 % strain from the phase change induced by temperature. Although two-way applications are possible after education, *SMAs* are best suited for one-way tasks such as deployment. In any case, they can be used only at low frequency and for low precision applications, mainly because of the difficulty of cooling. Fatigue under thermal cycling is also a problem. The best known *SMA* is the *NITINOL*; *SMAs* are little used in vibration control and will not be discussed in this book.

- *Piezoelectric materials* have a recoverable strain of 0.1 % under electric field; they can be used as actuators as well as sensors. There are two broad classes of piezoelectric materials used in vibration control: ceramics and polymers. The piezopolymers are used mostly as sensors, because they require extremely high voltages and they have a limited control authority; the best known is the *polyvinylidene fluoride* (*PVDF* or PVF_2). Piezoceramics are used extensively as actuators and sensors, for a wide range of frequency including ultrasonic applications; they are well suited for high precision in the nanometer range ($1nm = 10^{-9}m$). The best known piezoceramic is the *Lead Zirconate Titanate (PZT)*.

- *Magnetostrictive materials* have a recoverable strain of 0.15 % under magnetic field; the maximum response is obtained when the material is subjected to compressive loads. Magnetostrictive actuators can be used as load carrying elements (in compression alone) and they have a long lifetime. They can also be used in high precision applications. The best known is the *TERFENOL-D*.

- *Magneto-rheological (MR)* fluids consists of viscous fluids containing micron-sized particles of magnetic material. When the fluid is subjected to a magnetic field, the particles create columnar structures requiring a minimum shear stress to initiate

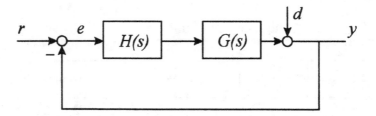

Figure 3: Principle of feedback control.

the flow. This effect is reversible and very fast (response time of the order of millisecond). Some fluids exhibit the same behaviour under electrical field; they are called *electro-rheological* (*ER*) fluids; however, their performances (limited by the electric field breakdown) are significantly inferior to *MR* fluids. *MR* and *ER* fluids are used in semi-active devices.

This brief list of commercially available smart materials is just a flavor of what is to come: *phase change materials* are currently under development and are likely to become available in a few years time; they will offer a recoverable strain of the order of 1 % under an electric field, one order of magnitude more than the piezoceramics.

The range of available devices to measure position, velocity, acceleration and strain is extremely wide, and there are more to come, particularly in optomechanics. Displacements can be measured with inductive, capacitive and optical means (laser interferometer); the latter two have a resolution in the nanometer range. Piezoelectric accelerometers are very popular but they cannot measure a d.c. component. Strain can be measured with strain gages, piezoceramics, piezopolymers and fiber optics. The latter can be embedded in a structure and give a global average measure of the deformation; they offer a great potential for health monitoring as well. We will see that piezopolymers can be shaped to react only to a limited set of vibration modes (modal filters).

3. Control strategies

There are two radically different approaches to disturbance rejection: feedback and feedforward. Although this text is entirely devoted to feedback control, it is important to point out the salient features of both approaches, in order to enable the user to select the most appropriate one for a given application.

The objective of active damping is to reduce the effect of the resonant peaks on the response of the structure. From

$$\frac{y(s)}{d(s)} = \frac{1}{1 + GH} \tag{1}$$

This requires $GH \gg 1$ near the resonances. Active damping can generally be achieved with moderate gains; another nice property is that it can be achieved without a model of the structure, and with guaranteed stability, provided that the actuator and sensor are collocated and have perfect dynamics. Of course actuators and sensors always have finite dynamics and any active damping system has a finite bandwidth.

The control objectives can be more ambitious, and we may wish to keep a control variable y (a position, or the pointing of an antenna) to a desired value r in spite of

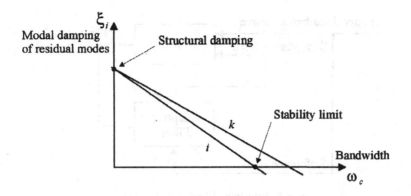

Figure 4: Effect of the control bandwidth on the net damping of the residual modes.

external disturbances d in some frequency range. From the previous formula and

$$F(s) = \frac{y(s)}{r(s)} = \frac{GH}{1 + GH} \qquad (2)$$

we readily see that this requires large values of GH in the frequency range where $y \simeq r$ is sought. $GH \gg 1$ implies that the closed-loop transfer function $F(s)$ is close to 1, which means that the output y tracks the input r accurately. From Equ.(1), this also ensures disturbance rejection within the bandwidth of the control system. In general, to achieve this, we need a more elaborate strategy involving a mathematical model of the system which, at best, can only be a low-dimensional approximation of the actual system $G(s)$. There are many techniques available to find the appropriate compensator, and only the simplest and the best established will be reviewed in this text. They all have a number of common features:

- The bandwidth ω_c of the control system is limited by the accuracy of the model; there is always some destabilization of the flexible modes outside ω_c (residual modes). The phenomenon whereby the net damping of the residual modes actually decreases when the bandwidth increases is known as *spillover* (Fig.4).

- The disturbance rejection within the bandwidth of the control system is always compensated by an amplification of the disturbances outside the bandwidth.

- When implemented digitally, the sampling frequency ω_s must always be two orders of magnitude larger than ω_c to preserve reasonably the behaviour of the continuous system. This puts some hardware restrictions on the bandwidth of the control system.

3.1. FEEDFORWARD

When a signal correlated to the disturbance is available, feedforward adaptive filtering constitutes an attractive alternative to feedback for disturbance rejection; it was originally developed for noise control [23], but it is very efficient for vibration control too [14]. Its principle is explained in Fig.5. The method relies on the availability of a reference signal correlated to the primary disturbance; this signal is passed through an adaptive filter, the output of which is applied to the system by secondary sources. The

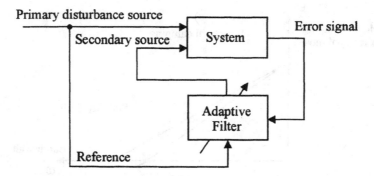

Figure 5: Principle of feedforward control.

filter coefficients are adapted in such a way that the error signal at one or several critical points is minimized. The idea is to produce a secondary disturbance such that it cancels the effect of the primary disturbance at the location of the error sensor. Of course, there is no guarantee that the global response is also reduced at other locations and, unless the response is dominated by a single mode, there are places where the response can be amplified; the method can therefore be considered as a local one, in contrast to feedback which is global. Unlike active damping which can only attenuate the disturbances near the resonances, feedforward works for any frequency and attempts to cancel the disturbance completely by generating a secondary signal of opposite phase.

The method does not need a model of the system, but the adaption procedure relies on the measured impulse response. The approach works better for narrow-band disturbances, but wide-band applications have also been reported. Because it is less sensitive to phase lag than feedback, feedforward control can be used at higher frequency (a good rule of thumb is $\omega_c \simeq \omega_s/10$); this is why it has been so successful in acoustics.

The main limitation of feedforward adaptive filtering is the availability of a reference signal correlated to the disturbance. There are many applications where such a signal can be readily available from a sensor located on the propagation path of the perturbation. For disturbances induced by rotating machinery, an impulse train generated by the rotation of the main shaft can be used as reference. Table 1 summarizes the main features of the two approaches.

4. Open-loop frequency response

Consider a lightly damped flexible structure provided with a point force actuator and a displacement sensor. The open-loop frequency response function (FRF) can be expanded in modal coordinates as

$$G(\omega) = \sum_{i=1}^{n} \frac{\phi_i(a)\phi_i(s)}{\mu_i(\omega_i^2 - \omega^2 + 2j\xi_i\omega_i\omega)} \tag{3}$$

where ω_i is the natural frequency of mode i, μ_i its modal mass, ξ_i its modal damping, and $\phi_i(a)$ and $\phi_i(s)$ are the modal amplitudes at the actuator and sensor locations, respectively; in principle, the sum extends to all the modes of the structure. If one wish to truncate the modal expansion above the frequency range of interest, it is very

TABLE 1: Comparison of control strategies.

Type of control	Advantages	Disadvantages
Feedback		
Active damping	• no model needed • guaranteed stability when collocated	• effective only near resonances
Model based ($LQG, H_\infty ...$)	• global method • attenuates all disturbances within ω_c	• limited bandwidth ($\omega_c \ll \omega_s$) • disturbances outside ω_c are amplified • spillover
Feedforward		
Adaptive filtering of reference (*x-filtered LMS*)	• no model necessary • wider bandwidth ($\omega_c \simeq \omega_s/10$) • works better for narrow-band disturb.	• reference needed • local method (response may be amplified in some part of the system) • large amount of real time computations

important to keep the static contribution of the high frequency modes:

$$G(\omega) \simeq \sum_{i=1}^{m} \frac{\phi_i(a)\phi_i(s)}{\mu_i(\omega_i^2 - \omega^2 + 2j\xi_i\omega_i\omega)} + \sum_{i=m+1}^{n} \frac{\phi_i(a)\phi_i(s)}{\mu_i\omega_i^2} \qquad (4)$$

The second sum is often called *residual mode*; it is independent of ω and introduces a *feedthrough* component in the FRF. It can be shown that this term plays an important role in the location of the open-loop zeros of the system. Upon writing Equ.(4) for $\omega = 0$, it is readily obtained that the residual mode can be written alternatively

$$R = \sum_{i=m+1}^{n} \frac{\phi_i(a)\phi_i(s)}{\mu_i\omega_i^2} = G(0) - \sum_{i=1}^{m} \frac{\phi_i(a)\phi_i(s)}{\mu_i\omega_i^2} \qquad (5)$$

4.1. COLLOCATED SYSTEMS

Consider an undamped system with collocated actuator and sensor. Since $\phi_i(a) = \phi_i(s)$, Equ.(4) becomes

$$G(\omega) \simeq \sum_{i=1}^{m} \frac{\phi_i^2(a)}{\mu_i(\omega_i^2 - \omega^2)} + R \qquad (6)$$

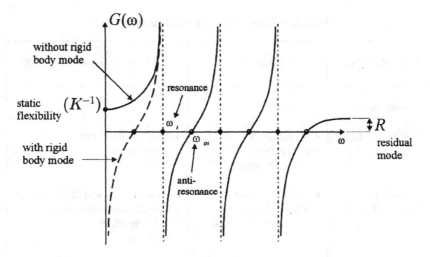

Figure 6: FRF of an undamped structure with collocated actuator and sensor.

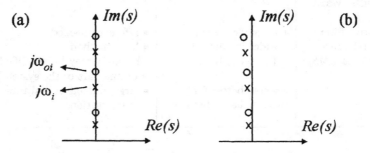

Figure 7: Pole/zero pattern of a structure with collocated actuator and sensor. (a) Undamped. (b) Lightly damped. (Only the upper half of the complex plane is shown, the diagram is symmetrical with respect to the real axis)

We note that the residues are all positive. The behaviour of $G(\omega)$ is represented in Fig.6; the amplitude of $G(\omega)$ goes to $\pm\infty$ at the resonance frequencies ω_i (corresponding to a pair of imaginary poles in the system). Besides, as $G(\omega)$ is an increasing function of ω^2, in every interval between consecutive resonance frequencies, there is a frequency ω_{0i} where the amplitude of the FRF vanishes; these frequencies are known in structural dynamics as *anti-resonance*; they correspond to purely imaginary zeros. Thus an undamped structure with collocated actuator and sensor has *alternating poles and zeros on the imaginary axis* (Figure 7.a). The transfer function can be written alternatively

$$G(s) = k \frac{\prod_{\text{zeros}}(s^2 + \omega_{0i}^2)}{\prod_{\text{poles}}(s^2 + \omega_i^2)} \tag{7}$$

If some damping is added, the poles and zeros are slightly moved into the left half plane as indicated in Fig.7.b, without changing the dominant feature of interlacing. A collocated system always exhibits Bode and Nyquist plots similar to those represented in Fig.8.

Each flexible mode introduces a circle in the Nyquist diagram; it is more or less centered on the imaginary axis which is intersected at $\omega = \omega_i$ and $\omega = \omega_{0i}$; the radius of each circle is proportional to the inverse of the modal damping, ξ_i^{-1}. In the Bode plots, a 180^0 phase lag occurs at every natural frequency, and is compensated by a 180^0 phase lead at every imaginary zero; the phase always oscillates between 0 and $-\pi$, as a result of the interlacing property of the poles and zeros. It is worth pointing out that the zeros (anti-resonance) of a collocated system are identical to the resonance frequencies of the system with an additional restraint at the actuator/sensor location.

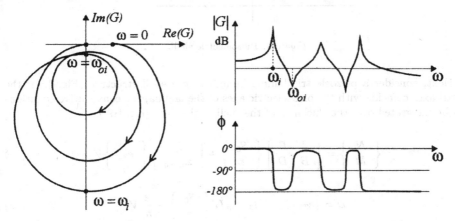

Figure 8: Nyquist diagram and Bode plots of a lightly damped structure with collocated actuator and sensor.

5. Laminar piezoelectric actuator

5.1. SMART PIEZOELECTRIC SHELL

Consider a two-dimensional piezoelectric lamina in a plane (x, y); the poling direction is z (normal to the lamina) and the electric field is also applied along z. In the piezoelectric principal axes, the constitutive equations read

$$\{T\} = [C]\{S\} - \left\{ \begin{array}{c} e_{31} \\ e_{32} \\ 0 \end{array} \right\} E \tag{8}$$

$$D = \{e_{31}\ e_{32}\ 0\}\{S\} + \varepsilon E \tag{9}$$

where

$$\{T\} = \left\{ \begin{array}{c} \sigma_x \\ \sigma_y \\ \sigma_{xy} \end{array} \right\} \qquad \{S\} = \left\{ \begin{array}{c} \varepsilon_x = \dfrac{\partial u}{\partial x} \\ \varepsilon_y = \dfrac{\partial v}{\partial y} \\ \gamma_{xy} = \dfrac{\partial u}{\partial y} + \dfrac{\partial v}{\partial x} \end{array} \right\} \tag{10}$$

are the stress and strain vector, respectively, $[C]$ is the matrix of elastic constant, E is the component of the electric field along z, D is the z-component of the electric displacement and ε the dielectric constant and e_{31} and e_{32} are the piezoelectric constants.

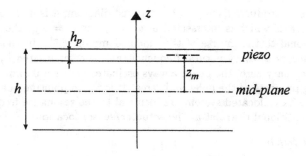

Figure 9: Piezoelectric shell.

Next, consider a piezoelectric lamina bounded on a shell structure (Fig.9). If the global axes coincide with the piezoelectric axes of the lamina, the constitutive equations can be integrated over the thickness of the shell in the form (e.g. [19])

$$\left\{ \begin{array}{c} N \\ M \end{array} \right\} = \left[\begin{array}{cc} A & B \\ B & D \end{array} \right] \left\{ \begin{array}{c} S_0 \\ \kappa \end{array} \right\} + \left[\begin{array}{cc} I_3 & \\ & z_m I_3 \end{array} \right] \left\{ \begin{array}{c} e_{31} \\ e_{32} \\ 0 \end{array} \right\} V \tag{11}$$

$$D = \{ e_{31} \ e_{32} \ 0 \} [I_3 \ z_m I_3] \left\{ \begin{array}{c} S_0 \\ \kappa \end{array} \right\} - \frac{\varepsilon}{h_p} V \tag{12}$$

where $\{N\}$ is the vector of in-plane resultant forces and $\{M\}$ the vector of bending moments;

$$\{N\} = \int_{-h/2}^{h/2} \{T\} dz \qquad \{M\} = \int_{-h/2}^{h/2} \{T\} z \ dz \tag{13}$$

$\{S_0\}$ is the deformation vector of the mid-plane and $\{\kappa\}$ is the vector of curvatures:

$$\{S_0\} = \left\{ \begin{array}{c} \frac{\partial u_0}{\partial x} \\ \frac{\partial v_0}{\partial y} \\ \frac{\partial u_0}{\partial y} + \frac{\partial v_0}{\partial x} \end{array} \right\} \qquad \{\kappa\} = \left\{ \begin{array}{c} \frac{\partial^2 w}{\partial x^2} \\ \frac{\partial^2 w}{\partial y^2} \\ \frac{\partial^2 w}{\partial x \partial y} \end{array} \right\} \tag{14}$$

The matrices A,B,D are the classical stiffness matrices of the shell theory (e.g. [3]); h_p is the thickness of the piezoelectric lamina and z_m is the distance between its mid-plane and the mid-plane of the shell.

If the piezoelectric lamina is connected to a charge amplifier, the voltage between the electrodes is set to $V = 0$ and the sensor equation (12) can be integrated over the electrode to produce the sensor output

$$Q = \int_\Omega [e_{31} \frac{\partial u_0}{\partial x} + e_{32} \frac{\partial v_0}{\partial y} + z_m (e_{31} \frac{\partial^2 w}{\partial x^2} + e_{32} \frac{\partial^2 w}{\partial y^2})] dS \tag{15}$$

where the integral extends over the surface of the electrode (the part of the piezo not covered by the electrode does not contribute to the signal). The first part of the integral is the contribution of the membrane strain while the second one is due to bending. If

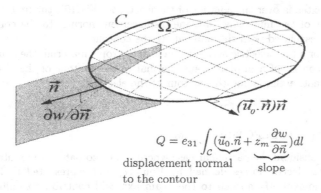

$$Q = e_{31} \int_C (\vec{u}_0.\vec{n} + \underbrace{z_m \frac{\partial w}{\partial \vec{n}}}_{\text{slope}}) dl$$

displacement normal
to the contour

Figure 10: Sensor equation for an isotropic piezo.

$$\underbrace{N_p = -e_{31} V}_{\text{normal force}} \quad \underbrace{M_p = -e_{31} z_m V}_{\text{bending moment}}$$

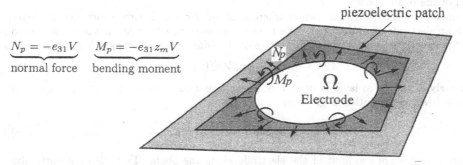

Figure 11: Equivalent piezoelectric forces for an isotropic piezo.

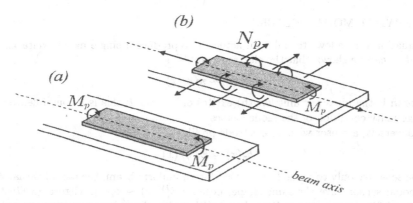

Figure 12: Equivalent piezoelectric loads of a rectangular piezoceramic patch on a beam. (a) Beam theory. (b) Shell theory.

the piezoelectric properties are isotropic ($e_{31} = e_{32}$), the surface integral can be further transformed into a contour integral using one of the Green integrals:

$$Q = e_{31} \int_C (\vec{u}_0.\vec{n} + z_m \frac{\partial w}{\partial \vec{n}}) dl \qquad (16)$$

This integral extends over the contour of the electrode (Fig.10); the first contribution is the component of the mid-plane in-plane displacement normal to the contour and the second one is associated with the slope along the contour.

Similarly, for a piezoelectric actuator made of isotropic material, the equivalent piezo-electric loads consist of a in-plane force, normal to the contour of the electrode, and a constant moment, acting along the contour of the electrode (Fig.11):

$$N_p = -e_{31}V \qquad\qquad M_p = -e_{31}z_mV \qquad\qquad (17)$$

5.2. SMART PIEZOELECTRIC BEAM

Figure 12 considers the particular case where the piezo patch is mounted on a beam. Of all the piezoelectric forces defined by Equ.(17) and represented in Fig.12.b, only the bending moment M_p normal to the beam axis will contribute significantly to the transverse displacements of the beam (Fig.12.a); this is the corresponding equivalent load of the beam theory.

In a more general configuration where a beam is covered with a piezoelectric layer with an electrode of width $b_p(x)$, the equivalent piezoelectric load consists of a distributed load proportional to the second derivative of the width of the electrode (e.g. [25]):

$$p = -e_{31}z_mV\, b_p''(x) \qquad\qquad (18)$$

Similarly, if the piezo layer is used as a sensor, the amount of electric charge generated by the beam deformation is given by

$$Q = -e_{31}z_m \int_a^b \frac{\partial^2 w}{\partial x^2} b_p(x)\, dx \qquad\qquad (19)$$

where a and b are the limit of the electrode along the beam. Equ.(19) is a particular case of Equ.(15), with the assumptions of the beam theory.

5.3. SPATIAL MODAL FILTERS

Equation (18) allows to tailor an actuator to produce a single mode excitation [20]. Indeed, it can be shown that the electrode profile

$$b_p''(x) \sim m\, \phi_l(x) \qquad\qquad (20)$$

(where m is the mass per unit length) excites only mode l; this is a consequence of the orthogonality condition of the mode shapes.

Conversely, a sensor with an electrode profile

$$b_p(x) \sim EI\, \phi_l''(x) \qquad\qquad (21)$$

will be sensitive only to mode l. Note that for an uniform beam, the modal actuator and the modal sensor have the same shape, because $\phi_i^{IV}(x) \sim \phi_i(x)$. Figure 13 illustrates the modal filters used for a uniform beam with various boundary conditions; the change of sign indicates a change in the polarity of the strip, which is equivalent to negative values of $b_p(x)$. As an alternative, the part of the sensor with negative polarity can be bonded on the opposite side of the beam. The reader will notice that the electrode shape of the simply supported beam is the same as the mode shape, while for the cantilever beam, the electrode shape is that of the mode shape of a beam supported at the opposite end. The extension of distributed modal filters to two-dimensional structures (plates) has been considered in [31].

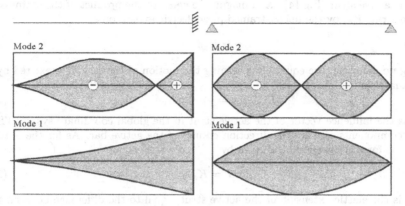

Figure 13: Modal filters for the first two modes of a beam for various boundary conditions: (a) cantilever; (b) simply supported.

6. Active truss

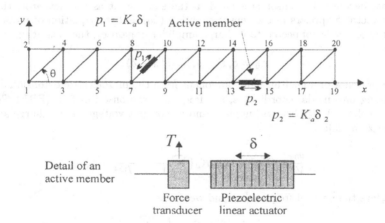

Figure 14: Active truss. The active struts consist of a piezoelectric linear actuator colinear with a force transducer.

Consider the active truss of Fig.14; when a voltage V is applied to an unconstrained linear piezoelectric actuator, it produces an expansion δ.

$$\delta = d_{33}nV = g_a V \tag{22}$$

where d_{33} is the piezoelectric coefficient, n is the number of piezoelectric ceramic elements in the actuator; g_a is the actuator gain. This equation neglects the hysteresis of the piezoelectric expansion. If the actuator is placed in a truss, its effect on the structure can be represented by equivalent piezoelectric loads acting on the passive structure. As for thermal loads, the pair of self equilibrating piezoelectric loads applied axially to both ends of the active strut (Fig.14) has a magnitude equal to the product of the stiffness of the active strut, K_a, by the unconstrained piezoelectric expansion δ:

$$p = K_a \delta \tag{23}$$

Assuming no damping, the equation governing the motion of the structure excited by a single actuator is

$$M\ddot{x} + Kx = bp = bK_a\delta \tag{24}$$

where b is the influence vector of the active strut in the global coordinate system. The non-zero components of b are the direction cosines of the active bar. As for the output signal of the force transducer, it is given by

$$y = T = K_a \delta_e \tag{25}$$

where δ_e is the elastic extension of the active strut, equal to the difference between the total extension of the strut and its piezoelectric component δ. The total extension is the projection of the displacements of the end nodes on the active strut, $\Delta = b^T x$. Introducing this into Equ.(25), we get

$$y = T = K_a(b^T x - \delta) \tag{26}$$

Note that, because the sensor is located in the same strut as the actuator, the same influence vector b appears in the sensor equation (26) and the equation of motion (24). If the force sensor is connected to a charge amplifier of gain g_s, the output voltage v_0 is given by

$$v_0 = g_s T = g_s K_a(b^T x - \delta) \tag{27}$$

Note the presence of a feedthrough component from the piezoelectric extension δ. Upon transforming into modal coordinates, the frequency response function (FRF) $G(\omega)$ between the voltage V applied to the piezo and the output voltage of the charge amplifier can be written [25]:

$$\frac{v_0}{V} = G(\omega) = g_s g_a K_a \{ \sum_{i=1}^{n} \frac{\nu_i}{1 - \omega^2/\Omega_i^2} - 1 \} \tag{28}$$

where Ω_i are the natural frequencies, and we define

$$\nu_i = \frac{K_a(b^T \phi_i)^2}{\mu_i \Omega_i^2} = \frac{K_a(b^T \phi_i)^2}{\phi_i^T K \phi_i} \tag{29}$$

The numerator and the denominator of this expression represent respectively twice the strain energy in the active strut and twice the total strain energy when the structure vibrates according to mode i; $\nu_i (\geq 0)$ is therefore called the *modal fraction of strain energy* in the active strut. From Equ.(28), we see that ν_i determines the residue of mode i, that is the amplitude of the contribution of mode i in the transfer function between the piezo actuator and the force sensor; it can therefore be regarded as a compound index of controllability and observability of mode i. ν_i is readily available from commercial finite element programs; it can be used to select the proper location of the active strut in the structure: the best location is that with the highest ν_i for the modes that we wish to control [29]. The FRF (28) has alternating poles and zeros on the imaginary axis (or near, if the structural damping is taken into account)(Fig.15).

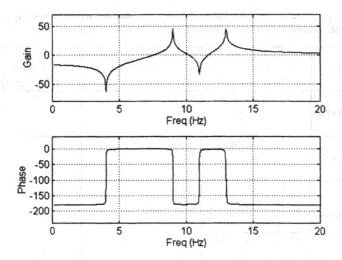

Figure 15: Open-loop FRF $G(\omega)$ of the active truss (a small damping is assumed).

7. Active damping with collocated pairs

7.1. INTRODUCTION

The role of damping is to reduce the settling time of the transient response to impulsive loads, and the resonant response to broad band stationary excitations. In this section, we examine various ways of achieving active damping augmentation with collocated actuator/sensor pairs. As we have seen in section '4.', this special configuration leads to alternating poles and zeros near the imaginary axis; thanks to this property, a number of active damping schemes with guaranteed stability have been developed and tested with various types of actuators and sensors (Table 2); they can be implemented in a decentralized manner, each actuator interacting only with its collocated sensor.

7.2. DIRECT VELOCITY FEEDBACK

Consider an undamped structure controlled with a set of point force actuators u collocated with a set of velocity sensors $\dot{y}$; the governing equations are
structure:

$$M\ddot{x} + Kx = f + Bu \tag{30}$$

sensor:

$$\dot{y} = B^T \dot{x} \tag{31}$$

control:

$$u = -G\dot{y} \tag{32}$$

where B is the control influence matrix and G is the positive definite matrix of control gains [5, 6]. The fact that B^T appears in the sensor equation is due to collocation.

TABLE 2: Collocated active damping compensators for various actuator/sensor pairs. The column indicates the type of actuator, and the row the type of sensor.

$gD(s)$	Force	Strain (d_{31} piezo)	Linear (d_{33} piezo)
Displacement	*Lead* $g\dfrac{s}{s+a}$		
Velocity	*Direct Velocity F.* g		
Acceleration	*DVF:* g/s $\dfrac{g}{s^2 + 2\xi_f\omega_f s + \omega_f^2}$		
Strain (d_{31} piezo)		*Positive Position F.* $\dfrac{-g\omega_f^2}{s^2 + 2\xi_f\omega_f s + \omega_f^2}$	
Force			*Integral Force F.* $-g/s$

Combining the three equations, we find the closed-loop equation

$$M\ddot{x} + BGB^T\dot{x} + Kx = f \tag{33}$$

Therefore, the control forces appear as a viscous damping (electrodynamic damping). The damping matrix $C = BGB^T$ is positive semi definite, because the actuators and sensors are collocated.

7.2.1. *Lead compensator*

Let us examine the *SISO* case a little closer. In this case, the matrix B degenerates into a control influence vector b . The open-loop transfer function between the control force u and the collocated displacement y is

$$G_0(s) = \frac{Y(s)}{U(s)} = \sum_i \frac{b^T\phi_i\phi_i^T b}{\mu_i(\omega_i^2 + s^2)} \tag{34}$$

where the sum extends to all the modes. We know that the corresponding poles and zeros alternate on the imaginary axis.

Because the amplitude of the derivative compensation increases linearly with the frequency, which would lead to noise amplification at high frequency, it is not desirable to implement the compensator as in Equ.(32), but rather to supplement it by a low-pass filter to produce:

$$D(s) = g\frac{s}{s+a} \tag{35}$$

A pole has been added at some distance a along the negative real axis. This compensator behaves like a derivator at low frequency ($\omega \ll a$). The block diagram of the control system is shown in Fig.16; a displacement sensor is now assumed and the structural damping is again omitted for simplicity. Typical root locus plots are shown in Fig.17 for two values of the low-pass filter corner frequency a. The closed-loop pole trajectories go from the open-loop poles to the open-loop zeros following branches which are entirely contained in the left half plane. Since there are two poles more than zeros, there are two asymptotes at $\pm 90^0$. The system is always stable, and this property is not sensitive to parameter variations, because the alternating pole-zero pattern is preserved under parameter variations.

7.3. ACCELERATION FEEDBACK

The easiest way to use the acceleration is to integrate it to obtain the absolute velocity; the direct velocity feedback can then be used. In practice, however, piezoelectric

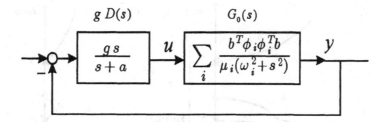

Figure 16: Block diagram of the modified direct velocity feedback.

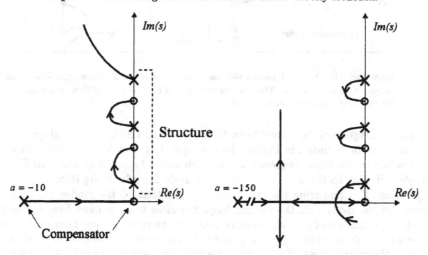

Figure 17: Root locus plots for two values of the low-pass filter corner frequency a (only the upper half is shown).

18

accelerometers use charge amplifiers which behave as high-pass filters; this does not affect significantly the results if the corner frequency of the charge amplifier is well below the vibration mode of the structure. Next, we consider an alternative controller which also enjoys guaranteed stability and exhibits a larger roll-off at high frequency [32, 34].

7.3.1. Second order filter

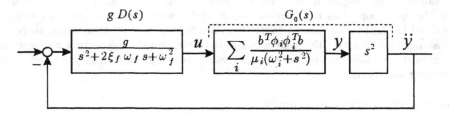

Figure 18: Acceleration feedback for a SISO collocated system.

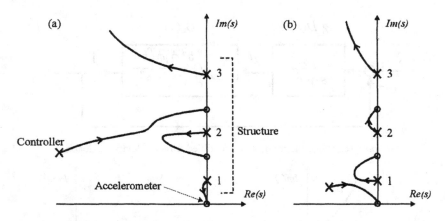

Figure 19: Root locus of the acceleration feedback for a m.d.o.f. structure. (a) The control is targeted at mode 2. (b) The control is targeted at mode 1. (Different scales are used for the real and imaginary parts)

The block diagram of the second order filter applied to a SISO collocated system with acceleration sensor is shown in Fig.18; the corresponding root locus is shown in Fig.19 for two values of the filter frequency ω_f; in both cases, $\xi_f = 0.5$ is used. In Fig.19.a, ω_f is selected close to the natural frequency of mode 2 while in Fig.19.b, it is selected close to mode 1. Comparing the two figures, we see that all the modes are positively damped, but the mode with the natural frequency close to ω_f is more heavily damped. Thus, the performance of the compensator relies on the tuning of the filter on the mode that we wish to damp (this aspect may become problematic if the system is subject to changes in the parameters). The maximum achievable damping ratio increases with ξ_f; a value of ξ_f between 0.5 and 0.7 is recommended. For closely spaced modes, stability is still guaranteed, but a large damping ratio cannot be achieved simultaneously for the two modes; besides, small variations of the filter frequency may significantly change the root locus and the modal damping.

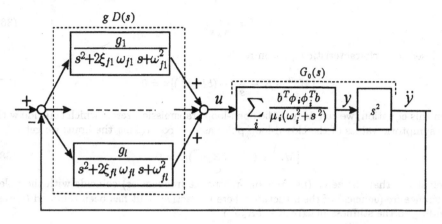

Figure 20: Targeting several modes with a SISO acceleration feedback.

If several modes must be damped, several compensators may be used in parallel as represented in Fig.20, where the ω_{fi} are tuned on the targeted modes.

As compared to the Direct Velocity Feedback, $gD(s) = g/s$, the new compensator has a larger roll-off at high frequency (-40 $dB/decade$ instead of -20 $dB/decade$), which may decrease the risk of destabilizing high frequency modes. The need for tuning the filter frequency ω_f may be a drawback if the natural frequency is not known accurately, or is subject to changes.

7.4. POSITIVE POSITION FEEDBACK

The *Positive Position Feedback* (PPF) [12,16] is appropriate for a structure equipped with strain actuators and sensors; the objective is, once again, to use a second order filter to improve the roll-off of the control system, allowing high frequency gain stabilization. The block diagram is represented in Fig.21. As compared to Fig.18, the output y is now proportional to the displacements (e.g. strain sensor) and a minus sign appears in the controller block (together with the minus sign in the feedback loop, this produces a positive feedback.). Figure 22.a and .b show the root locus when the controller is tuned on mode 1 and mode 2, respectively. We see that the tuning property of the controller is very similar to that of Fig.18 and, even in presence of a feedthrough component, the open-loop transfer function has a roll-off of $-40dB/decade$. However, there is a stability limit which is reached when the open-loop static gain is equal to 1.

7.5. INTEGRAL FORCE FEEDBACK

Consider the active truss of Fig.14; the open-loop FRF of Equ.(28) has alternating poles and zeros and has no roll-off at high frequencies. This system can be actively damped by a *positive Integral Force Feedback* [29] (Fig.23); the corresponding root locus is shown in Fig.24.

7.5.1. *Modal damping*

Combining the structure equation (24), the sensor equation (26) and the control law

$$\delta = \frac{g}{K_a s} y \tag{36}$$

the closed-loop characteristic equation reads

$$[Ms^2 + K - \frac{g}{s+g}(bK_a b^T)]x = 0 \tag{37}$$

From this equation, we can deduce the open-loop transmission zeros, which coincide with the asymptotic values of the closed-loop poles as $g \to \infty$. Taking the limit, we get

$$[Ms^2 + (K - bK_a b^T)]x = 0 \tag{38}$$

which states that the zeros (i.e. the anti-resonance frequencies) coincide with the poles (resonance frequencies) of the structure where the active strut has been removed (corresponding to the stiffness matrix $K - bK_a b^T$).

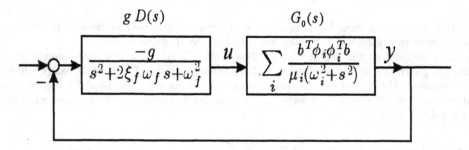

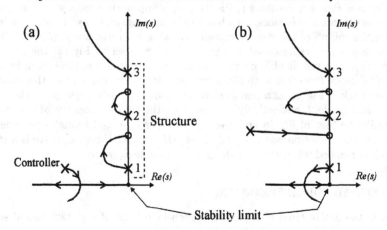

Figure 21: Positive Position Feedback for a SISO collocated system.

Figure 22: Root locus of the PPF. (a) The control is targeted at mode 1. (b) The control is targeted at mode 2.

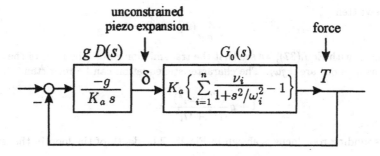

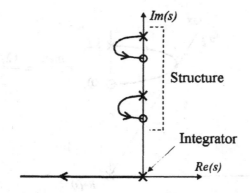

Figure 23: Block diagram of the integral force feedback.

Figure 24: Root locus of the integral force feedback.

To evaluate the modal damping, Equ.(37) must be transformed in modal coordinates with the change of variables $x = \Phi z$. Assuming that the mode shapes have been normalized according to $\Phi^T M \Phi = I$ and taking into account that $\Phi^T K \Phi = \text{diag}(\Omega_i^2) = \Omega^2$, we have

$$[Is^2 + \Omega^2 - \frac{g}{s+g} \Phi^T (bK_ab^T)\Phi]z = 0 \tag{39}$$

The matrix $\Phi^T(bK_ab^T)\Phi$ is, in general, fully populated; if we assume that it is diagonally dominant, and if we neglect the off-diagonal terms, it can be rewritten

$$\Phi^T (bK_ab^T)\Phi \simeq \text{diag}(\nu_i\Omega_i^2) \tag{40}$$

where ν_i is the *fraction of modal strain energy* in the active member when the structure vibrates according to mode i; ν_i is defined by Equ.(29). Substituting Equ.(40) into (39), we find a set of decoupled equations

$$s^2 + \Omega_i^2 - \frac{g}{s+g}\nu_i\Omega_i^2 = 0 \tag{41}$$

and, after introducing

$$\omega_i^2 = \Omega_i^2(1 - \nu_i) \tag{42}$$

it can be rewritten

$$s^2 + \Omega_i^2 - \frac{g}{s+g}(\Omega_i^2 - \omega_i^2) = 0 \tag{43}$$

By comparison with Equ.(37), we see that the transmission zeros (the limit of the closed-loop poles as $g \to \infty$) are $\pm j\omega_i$. The characteristic equation can be rewritten

$$1 + g\frac{(s^2 + \omega_i{}^2)}{s(s^2 + \Omega_i{}^2)} = 0 \tag{44}$$

The corresponding root locus is shown in Fig.25. The depth of the loop in the left half

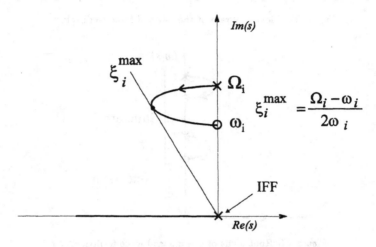

Figure 25: Root locus of the closed-loop pole for the IFF.

plane depends on the frequency difference $\Omega_i - \omega_i$, and the maximum modal damping is given by

$$\xi_i{}^{max} = \frac{\Omega_i - \omega_i}{2\omega_i} \tag{45}$$

it is obtained for $g = \Omega_i\sqrt{\Omega_i/\omega_i}$. For small gains, it can be shown [29] that

$$\xi_i = \frac{g\nu_i}{2\Omega_i} \tag{46}$$

This interesting result tells us that, for small gains, the active damping ratio in a given mode is proportional to the fraction of modal strain energy in the active element. This result is very useful for the design of active trusses; the active struts should be located in order to maximize the fraction of modal strain energy ν_i in the active members for the critical vibration modes. The preceding results have been established for a single active member; if there are several active members operating with the same control law and the same gain g, this result can be generalized under similar assumptions. It can be shown that each closed-loop pole follows a root locus governed by Equ.(44) where the pole Ω_i is the natural frequency of the open-loop structure and the zero ω_i is the natural frequency of the structure where the active members have been removed.

Figure 26: Active truss with piezoelectric struts (ULB).

7.5.2. *Experimental results*

The test structure is shown in Fig.26. Figures 27 and 28 illustrate typical results. The modal damping ratio of the first two modes is larger than 10 %. Note that, in addition to being simple and robust, the control law can be implemented in a analog controller which performs better in microvibrations.

8. Active tendon control

The use of cables to achieve lightweight structures is not new; it can be found in Herman Oberth's early books on astronautics. The use of guy cables is probably the most efficient way to stiffen a structure, in terms of weight. They can also be used to prestress a deployable structure and eliminate the geometric uncertainty due to the gaps. Cables structures are also extensively used in civil engineering. One further step consists of providing the cables with active tendons to achieve active damping in the structure. This approach has been developed in [1, 2, 7, 8, 25–28].

8.1. ACTIVE DAMPING OF CABLE STRUCTURES

When using a displacement actuator (e.g. a piezo) and a force sensor, the (positive) Integral Force Feedback (36) belongs to the class of "energy absorbing" control : indeed, if

$$\delta \sim \int T dt \qquad (47)$$

24

the power flow from the control system is $W = -T\dot{\delta} \sim -T^2 \leq 0$. This means that the control can only extract energy from the system, and this applies to nonlinear structures as well; all the states which are controllable and observable are asymptotically stable for all positive gains (infinite gain margin). The control concept is represented schematically in Fig.29 where the spring-mass system represents an arbitrary structure. Note that the damping introduced in the cable is usually very low, but experimental results have confirmed that it remains always stable, even at the parametric resonance (when the natural frequency of the structure is twice that of the cables).

8.2. MODAL DAMPING

If we assume that the dynamics of the cables can be neglected, that their interaction with the structure is restricted to the tension in the cables, and that the global mode shapes are identical with and without the cables, one can develop an approximate linear theory for the closed-loop system; the following results can be established, which follow closely those obtained for active trusses in the foregoing section (we assume no structural damping):

- The open-loop poles are $\pm j\Omega_i$ where Ω_i are the natural frequencies of the structure including the active cables; the open-loop zeros are $\pm j\omega_i$ where ω_i are the natural frequencies of the structure where the active cables have been removed.
- If the same control gain is used for every local control loop, as g goes from 0 to ∞, the closed-loop poles follow the root locus defined by Equ.(44) (Fig.30). Equ.(45) and (46) also apply in this case.

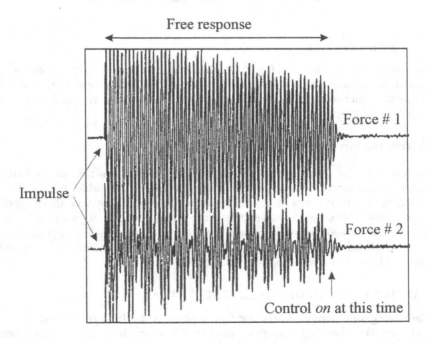

Figure 27: Force signal from the two active struts during the free response after impulsive load.

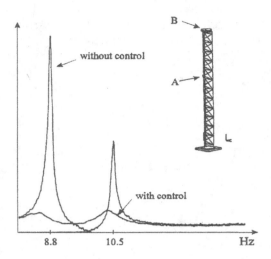

Figure 28: FRF between a force in A and an accelerometer in B, with and without control.

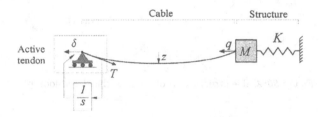

Figure 29: Control strategy for active damping of cable structures.

8.3. ACTIVE TENDON DESIGN

Figure 31 shows two possible designs of the active tendon: the first one (bottom left) is based on a linear piezoactuator from PI and a force sensor from B&K; a lever mechanism (top view) is used to transform the tension in the cable into a compression in the piezo stack, and amplifies the translational motion to achieve about 100 μm. This active element is identical to that in an active strut. In the second design (bottom center and right), the linear actuator is replaced by an amplified actuator from CEDRAT Research, also connected to a B&K force sensor and flexible tips. In addition to being more compact, this design does not require an amplification mechanism, and a tension of the flexible tips produces a compression in the piezo stack, which expands in the transverse direction, at the center of the elliptical structure.

8.4. EXPERIMENTAL RESULTS

Figure 32 shows the test structure; it is representative of a scale model of the JPL-Micro-Precision-Interferometer [22] which consists of a large trihedral passive truss of

26

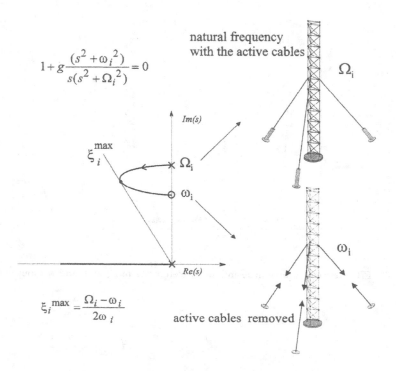

$$1 + g \frac{(s^2 + \omega_i^2)}{s(s^2 + \Omega_i^2)} = 0$$

natural frequency
with the active cables

Ω_i

ξ_i^{max} $\times \ \Omega_i$

$\circ \ \omega_i$

$Im(s)$

$Re(s)$

ω_i

$$\xi_i^{max} = \frac{\Omega_i - \omega_i}{2\omega_i}$$

active cables removed

Figure 30: Cable structure: root locus of the closed-loop poles.

about $9m$. The free-floating condition during the test is simulated by hanging the structure from the ceiling of the lab with soft springs. In this study, two different types of cables have been used: a fairly soft cable of $1mm$ diameter of polyethylene ($EA \approx 4000N$) and a stiffer one of synthetic fiber "Dynema" ($EA \approx 18000N$); in both cases, the tension in the cables was chosen in order to set the first cable mode at $400rad/sec$ or more, far above the first five flexible modes for which active damping is sought. The table inset into Fig.32 gives the measured natural frequencies ω_i (without cables) and Ω_i (with cables), for the two sets of cables.

Figure 33 compares the experimental closed-loop poles obtained for increasing gain g of the control with the root locus prediction of Equ.(44). The results are consistent with the analytical predictions, although a larger scatter is observed with stiffer cables. Note, however, that the experimental results tend to exceed the root locus predictions. Figure 34 compares typical FRF with and without control. An analytical study was conducted in [27] to investigate the possibility of using three Kevlar cables of $2mm$ diameter connecting the tips of the three trusses of the JPL-MPI. Using the root locus technique of Fig.30, a damping ratio between 14% and 21% was predicted in the first three flexible modes.

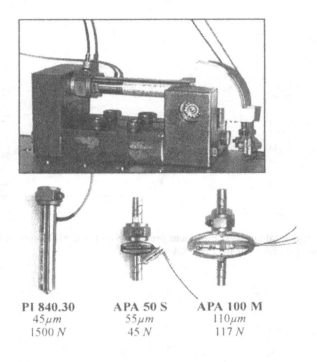

PI 840.30
45μm
1500 N

APA 50 S
55μm
45 N

APA 100 M
110μm
117 N

Figure 31: Various designs of active tendon or active strut (ULB).

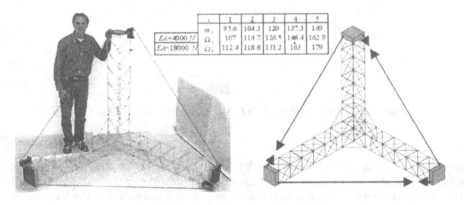

	1	2	3	4	5
EA=4000 N ω_1	95.6	104.3	120	137.3	149
Ω_1	107	114.7	126.5	146.4	162.9
EA=18000 N Ω_1	112.4	118.6	131.2	163	179

Figure 32: Free floating truss with active tendons.

9. Active damping generic interface

The active strut discussed in section '6.' can be developed into a generic 6 d.o.f. interface which can be used to connect arbitrary substructures. Such an interface is

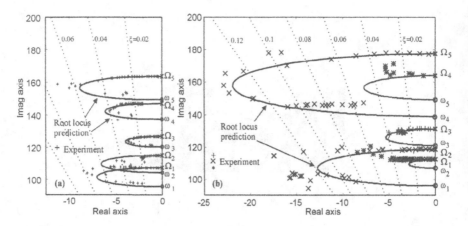

Figure 33: Experimental poles vs. root-locus prediction for the flexible modes of the free floating truss. (a) $EA = 4000N$. (b) $EA = 18000N$.

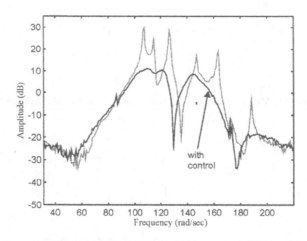

Figure 34: Typical FRF with and without control ($EA = 4000N$).

shown in Fig.35; it consists of a Stewart platform with cubic architecture [15]. Each leg consists of an active strut similar to that shown at the center of Fig.31: piezotranslator of the amplified design collocated with a force sensor, and connected to the base plates by flexible tips acting like spherical joints. The cubic architecture provides a uniform control capability in all directions, a uniform stiffness in all directions, and minimizes the cross-coupling amongst actuators (which are mutually orthogonal). The control is decentralized with the same gain for all loops. Figure 35.c shows the generic interface mounted between a truss and the supporting structure. Figure 36 shows the evolution of the first two closed-loop poles of this system when we increase the gain of the decentralized controller; the continuous line shows the root locus prediction of Equ.(44); Ω_i are the

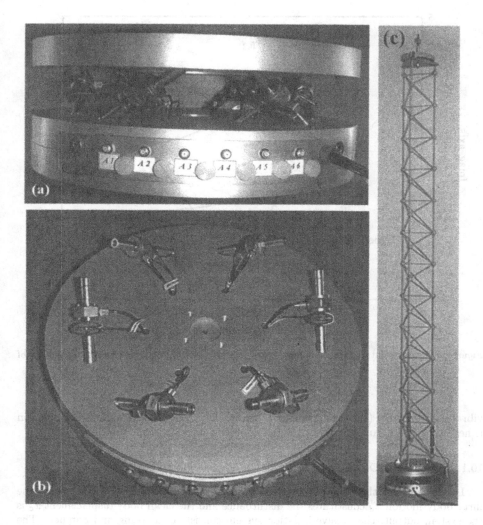

Figure 35: Stewart platform with piezoelectric legs as generic active damping interface. (a) General view. (b) With the upper base plate removed. (c) Interface acting as a support of a truss.

open-loop natural frequencies while ω_i are the high-gain asymptotes of the closed-loop poles.

10. Active vibration isolation

Many operating equipments (e.g. a car engine or an attitude control reaction wheel assembly in a spacecraft) generate oscillatory forces which can propagate in the supporting structure. Conversely, sensitive equipments may be supported by a structure which

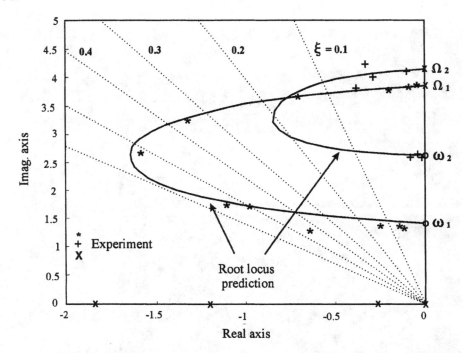

Figure 96: Experimental poles and root locus prediction from Equ.(44) for the structure of Fig.35.c.

vibrates appreciably (e.g. a telescope in a spacecraft). In both cases, a vibration isolation is necessary and it turns out that the two problems have the same solution.

10.1. PASSIVE ISOLATION

Let us consider the "dirty body/clean body" isolation problem (Fig.37), where the dirty body motion x_d constitutes the disturbance and the clean body displacement x_c is the system output; the passive isolation system consists of a spring and damper. The transmissibility of the isolation system is defined as

$$\frac{X_c(s)}{X_d(s)} = \frac{1 + 2\xi s/\omega_n}{1 + 2\xi s/\omega_n + s^2/\omega_n^2} \tag{48}$$

The amplitude diagram is represented in Fig.37 for various values of the damping ratio. We observe that

- All the curves are larger than 1 for $\omega < \sqrt{2}\,\omega_n$ and become smaller than 1 for $\omega > \sqrt{2}\,\omega_n$. Thus the critical frequency $\sqrt{2}\,\omega_n$ separates the domains of amplification and attenuation of the isolator.
- When $\xi = 0$, the high frequency decay rate is $1/s^2$, that is -40 dB/decade, while very large amplitudes occur near the corner frequency ω_n (the natural frequency of the spring-mass system).

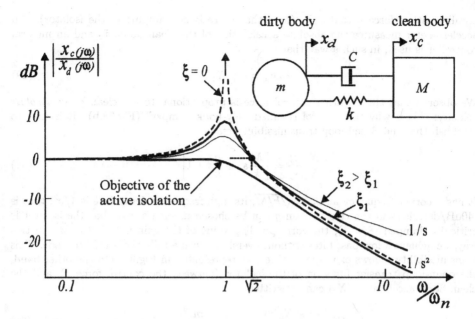

Figure 37: Passive isolator transmissibility FRF for various values of the damping ξ.

- The damping reduces the amplitude at resonance, but also tends to reduce the effectiveness at high frequency; the high frequency decay rate becomes $1/s$ (-20dB/decade).

The design of a passive isolator involves a trade-off between the resonance amplification and the high frequency attenuation; the ideal isolator should have a frequency dependent damping, with high damping below the critical frequency $\sqrt{2}\,\omega_n$ to reduce the amplification peak, and low damping above $\sqrt{2}\,\omega_n$ to improve the decay rate. The objective in designing an active isolation system is to achieve no amplification below ω_n and a decay rate of -40dB/decade at high frequency, as represented in Fig.37.

10.2. THE "SKY-HOOK" DAMPER

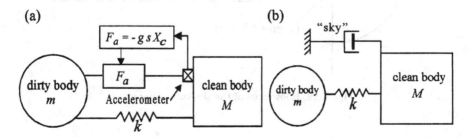

Figure 38: (a) Soft isolator with acceleration feedback. (b) Equivalent "sky-hook" damper.

Consider the single axis isolator of Fig.38.a; it consists of a soft spring k acting in parallel with a force actuator F_a (note that there is no damping in the isolator). An accelerometer measures the absolute acceleration of the clean body, $\ddot{x}_c$ and an integral controller is used, in such a way that

$$F_a = -gsX_c \tag{49}$$

We observe that the resulting control force is proportional to the clean body *absolute* velocity; this is why this control is called "sky-hook damper" (Fig.38.b). It is easy to establish that the closed-loop transmissibility reads

$$\frac{X_c(s)}{X_d(s)} = \left[\frac{M}{k}s^2 + \frac{g}{k}s + 1\right]^{-1} \tag{50}$$

It has a corner frequency at $\omega_n = \sqrt{k/M}$, its high frequency decay rate is $1/s^2$, that is -40dB/decade, and the control gain g can be chosen in such a way that the isolator is critically damped ($\xi = 1$); the corresponding value of the gain is $g = 2\sqrt{kM}$. In this way, we achieve a low-pass filter without overshoot with a roll-off of -40dB/decade. This transmissibility follows exactly the objective represented in Fig.37. On the other hand, the open-loop transfer function of the isolator (between the control force F_a and the clean body acceleration $\ddot{X}_c$) can be written

$$G(s) = \frac{s^2 X_c(s)}{F_a(s)} = \frac{ms^2}{Mms^2 + k(M+m)} \tag{51}$$

The open-loop poles are the natural frequencies of the system without control. The rigid body modes do not appear in the transfer function (51) because they are not controllable from F_a. The root locus of the closed-loop poles as the gain g of the controller increases is shown in Fig.39.

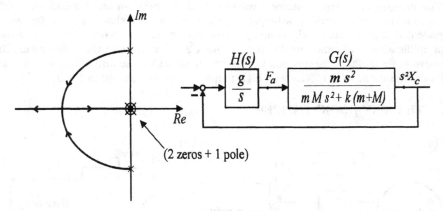

Figure 39: Root locus of the sky-hook damper.

10.3. FORCE FEEDBACK

If the clean body is rigid, its acceleration is proportional to the total force transmitted by the interface, $F = F_a + F_k$. As a result, the sky-hook damper can be obtained

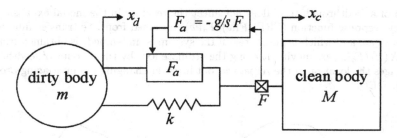

Figure 40: Force feedback isolator.

alternatively with the control configuration of Fig.40, where a force sensor has been substituted to the acceleration sensor.

The control strategies based on acceleration feedback and on force feedback appear as totally equivalent for the isolation of rigid bodies. However, the force feedback has two advantages. The first one is related to sensitivity: force sensors with a sensitivity of $10^{-3}N$ are commonplace and commercially available; if we consider a clean body with a mass of 1000 kg (e.g. a telescope), the corresponding acceleration is $10^{-6}m/s^2$. Accelerometers with such a sensitivity are more difficult to find; for example, the most sensitive accelerometer available in the Bruel & Kjaer catalogue is $2.10^{-5}m/s^2$(model 8318). The second advantage is stability when the clean body is flexible. In this case, the sky-hook damper appears to be only conditionally stable (for small gain) when the clean body becomes very flexible, so that the corner frequency of the isolator overlaps with the natural frequencies of the clean body. On the contrary, the stability of the force feedback remains guaranteed, because the following result can be established [30]:

If two arbitrary flexible, undamped structures are connected with a single axis soft isolator with force feedback (Fig.41), the poles and zeros in the open loop transfer function $F(s)/F_a(s)$ alternate on the imaginary axis.

The proof stems from the property of the collocated systems with *energetically conjugated* input and output variables (e.g. force input and displacement output, or torque input and angle output): For such a system, all the residues in the modal expansion of the transfer function have the same sign and this results in alternating poles and zeros on the imaginary axis [25].

If we now examine the transfer function between the control force F_a and the output of the force sensor F (Fig.41), although the actuator and sensor are collocated, F and F_a are not energetically conjugated and the preceding property does not apply. However, the total force F transmitted by the isolator is the sum of the control force F_a and the spring force, $k\Delta x$, where Δx is the relative displacement of the two structures along the isolator axis,

$$F = k\Delta x - F_a$$

or

$$\frac{F(s)}{F_a(s)} = k\frac{\Delta X(s)}{F_a(s)} - 1 \qquad (52)$$

Thus, the open-loop transfer function F/F_a is the sum of $k\Delta X/F_a$ and a negative unit feedthrough. The input F_a and the output Δx involved in the transfer function $\Delta X/F_a$ are energetically conjugated and, as a result, the transfer function $\Delta X/F_a$ has all its residues positive and possesses alternating poles and zeros along the imaginary axis. The

34

addition of a feedthrough term does not affect the residues in the modal expansion; the frequency response function (FRF) $F(\omega)/F_a(\omega)$ (obtained from the transfer function by setting $s = j\omega$, and which is purely real if the system is undamped) is obtained from the FRF $\Delta X(\omega)/F_a(\omega)$ by moving it along the ordinate axis by the amount of feedthrough; this changes the location of the zeros, without however changing the interlacing property (Fig.42). QED.

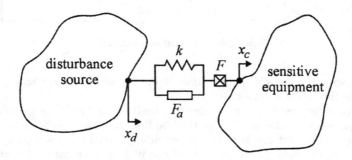

Figure 41: Two arbitrary flexible structures connected with a single axis soft isolator with force feedback.

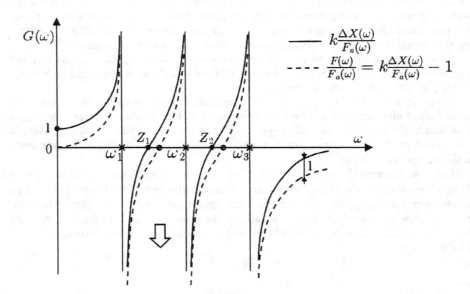

Figure 42: FRFs $k\frac{\Delta X(\omega)}{F_a(\omega)}$ and $\frac{F(\omega)}{F_a(\omega)}$ for an undamped structure (they are purely real). ω_i are the resonance frequencies and Z_i the transmission zeros. The unit feedthrough component which appears in F/F_a alters the location of the zeros (from • to ○) without changing the interlacing property.

Figure 43: Multi-purpose soft isolator based on a Stewart platform (JPL).

10.4. 6 D.O.F. ISOLATOR

The foregoing section describes a single axis active isolator which combines a $-40dB/decade$ attenuation rate in the roll-off region with no overshoot at the corner frequency. To fully isolate two rigid bodies with respect to each other, we need six such isolators, judiciously placed, that could be controlled either in a centralized or (more likely) in a decentralized manner. For a number of space applications, generic multi-purpose 6 d.o.f. isolators have been developed with a standard *Stewart platform* architecture (Fig.43) [21, 35]. The Stewart platform uses 6 identical active struts arranged in a mutually orthogonal configuration connecting the corner of a cube (same cubic architecture than in section '9.'). In addition to the properties discussed in the foregoing section, this cubic architecture also tends to minimize the spread of the modal frequencies of the isolator.

10.5. DECENTRALIZED CONTROL OF THE 6 D.O.F. ISOLATOR

Assuming that the base is fixed and that the payload attached to the upper part of the isolator is a rigid body, the dynamic equation (for small rotations) of the isolator is

$$M\ddot{x} + Kx = Bu \tag{53}$$

where $x = (x_r, y_r, z_r, \theta_x, \theta_y, \theta_z)^T$ is the vector describing the small displacements and rotations in the payload frame, $u = (u_1, ..., u_6)^T$ is the vector of active control forces in strut 1 to 6, and B is their influence matrix in the payload frame. M is the mass matrix and $K = kBB^T$ is the stiffness matrix.

If each leg is equipped with a force sensor as in Fig.40, the output equation reads

$$y = -kB^T x + u \tag{54}$$

This equation expresses the fact that the total force is the sum of the spring force and the control force. Once again, we note that the same matrix B appears in Equ.(53) and (54) because the sensors and actuators are collocated. Using a decentralized integral force feedback with constant gain, the controller equation reads

$$u = -\frac{g}{s}y \tag{55}$$

Combining Equ.(53), (54) and (55), the closed-loop equation reads

$$Ms^2x + Kx = \frac{g}{s+g}kBB^Tx$$

and, taking into account that $K = kBB^T$,

$$[Ms^2 + K\frac{s}{s+g}]x = 0 \tag{56}$$

If we transform into modal coordinates, $x = \Phi z$, and take into account the orthogonality relationships of the mode shapes, the characteristic equation is reduced to a set of uncoupled equations

$$\left(s^2 + \Omega_i^2\frac{s}{s+g}\right)z_i = 0 \tag{57}$$

Thus, every mode follows the characteristic equation

$$s^2 + \Omega_i^2\frac{s}{s+g} = 0$$

or

$$1 + g\frac{s}{s^2 + \Omega_i^2} = 0 \tag{58}$$

The corresponding root locus is shown in Fig.44. It is identical to Fig.39 for a single-axis isolator; however, unless the 6 natural frequencies are identical, a given value of the gain g will lead to different pole locations for the various modes and it will not be possible to achieve the same damping for all modes. This is why it is recommanded to locate the payload in such a way that the spread of the modal frequencies is minimized.

It is worth noting that:

(i) The foregoing model is based on the assumption that there is no structural damping and that the only contribution to the stiffness matrix is the axial stiffness of the struts, so that $K = kBB^T$. In practice, however, the spherical joints at the connection between the legs and the base plates of the Stewart plateform are replaced by flexible connections with high longitudinal stiffness and low bending stiffness, in such a way that the stiffness matrix has an additional contribution, K_e, from the elastic joints. Thus,

$$K = kBB^T + K_e \tag{59}$$

and the closed loop equation becomes

$$[Ms^2 + K_e + kBB^T\frac{s}{s+g}]x = 0 \tag{60}$$

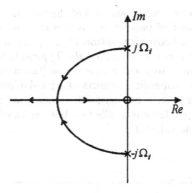

Figure 44: Root locus of the modes of the six-axis isolator with integral force feedback.

According to this equation, the transmission zeros, which are the asymptotic solutions of Equ.(60) as $g \to \infty$, are no longer at the origin ($s = 0$), but are solutions of the eigenvalue problem

$$[Ms^2 + K_e]x = 0 \tag{61}$$

This shift of the zeros may have a substantial influence on the practical performances of actual Stewart platforms.

(ii) The foregoing model assumes that the legs of the Stewart platform have no mass. In fact, the magnetic circuit of the voice coil actuator is fairly heavy and the local dynamics of the legs may interfere with that of the Stewart platform, resulting in a reduced attenuation of the isolation system.

A deeper discussion of the technological aspects can be found in [33, 36]

10.6. VEHICLE SUSPENSION

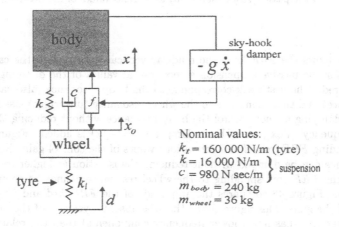

Nominal values:
$k_t = 160\,000$ N/m (tyre)
$k = 16\,000$ N/m
$c = 980$ N sec/m $\Big\}$ suspension
$m_{body} = 240$ kg
$m_{wheel} = 36$ kg

Figure 45: Quarter-car model and sky-hook damper.

Figure 45 shows a quarter-car model of a vehicle. Although this 2 d.o.f. model is too

38

simple for performing a comprehensive analysis of the ride motion, it is sufficient to gain some insight in the behaviour of passive and active suspensions in terms of vibration isolation (represented by the body acceleration $\ddot{x}$), suspension travel $(x - x_0)$ and road holding (represented by the tyre deflexion $x_0 - d$). Typical numerical values used in the simulation reported later are also given in the figure (taken from [9]). The stiffness k_t corresponds to the tyre; the suspension consists of a passive part (spring k + damper c) and an active one, assumed to be a perfect force actuator acting as a sky-hook damper in this case (the active control force is applied on both sides of the active device, to the body and to the wheel of the vehicle).

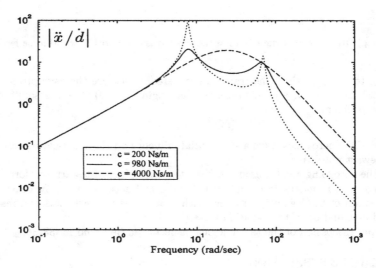

Figure 46: FRF of the passive suspension $(\ddot{x}/\dot{d})$ for various values of the damping coefficient.

Figure 46 shows the *FRF* from the roadway vertical velocity $\dot{d}$ to the car body acceleration $\ddot{x}$ for the passive suspension alone; several values of the damping coefficient c are considered. The first peak corresponds to the body resonance (also called sprung mass resonance) and the second one to the wheel resonance (unsprung mass resonance). The passive damping cannot control the body resonance without reducing the isolation at higher frequency. Next, a sky-hook damper $(f = -g\dot{x})$ is added. Figure 47 shows the corresponding *FRF* from $\dot{d}$ to $\ddot{x}$ for various values of the control gain. Note that the body resonance can be damped without reducing the isolation at higher frequency but the peak in the *FRF* corresponding to the wheel resonance cannot be changed by the active control. Figure 48 compares the amplitude of the *FRF* $\dot{x}/\dot{d}$ and $(\dot{x} - \dot{x}_0)/\dot{d}$ for two values of the gain. This figure shows that the absolute velocity of the body $\dot{x}$ rolls-off much faster (i.e. has much lower frequency components) than the relative velocity $(\dot{x} - \dot{x}_0)$. This point is important in the design of semi-active suspension devices which try to emulate the sky-hook damper by acting on the flow parameters of the damper acting on the *relative* velocity.

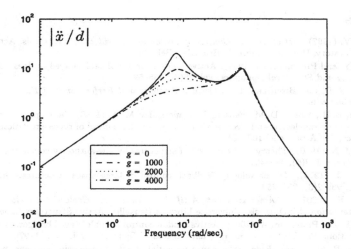

Figure 47: FRF of the active suspension ($\ddot{x}/\dot{d}$) for various values of the gain g of the sky-hook damper (all the other parameters have the nominal values listed in Fig.45).

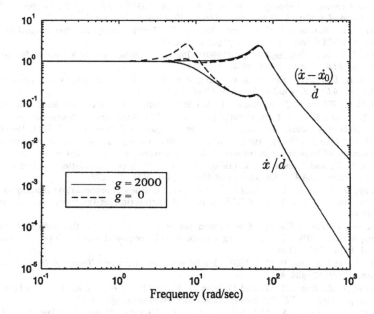

Figure 48: Comparison of $|\dot{x}/\dot{d}|$ and $|(\dot{x} - \dot{x}_0)/\dot{d}|$.

11. Acknowledgment

This contribution is based on the work of Y. Achkire, A. Abu Hanieh, F. Bossens, P. De Man, A. Francois, N. Loix and V. Piefort. This work was partly supported by the Inter University Attraction Pole IUAP-5 on Advanced Mechatronic Systems.

References

1. Achkire, Y. (1997). *Active Tendon Control of Cable-Stayed Bridges*. Ph.D. thesis, Active Structures Laboratory, Université Libre de Bruxelles, Belgium.

2. Achkire, Y. and Preumont, A. (1996). Active tendon control of cable-stayed bridges. *Earthquake Engineering and Structural Dynamics*, vol. 25(6), 585–597.

3. Agarwal, B. D. and Broutman, L. J. (1990). *Analysis and Performance of Fiber Composites*. Wiley, second edn.

4. Anderson, E. H., Moore, D. M., Fanson, J. L. and Ealey, M. A. (1990). Development of an active member using piezoelectric and electrostrictive actuation for control of precision structures. SDM Conference, AIAA paper 90-1085-CP.

5. Aubrun, J. N. (1980). Theory of the control of structures by low-authority controllers. *AIAA J. of Guidance*, vol. 3(5), 444–451.

6. Balas, M. J. (1979). Direct velocity feedback control of large space structures. *AIAA J. of Guidance*, vol. 2(3), 252–253.

7. Bossens, F. (2001). *Amortissement Actif Des Structures Cablees: De la Theorie a L'implementation*. Ph.D. thesis, Universite Libre de Bruxelles, Active Structures Laboratory.

8. Bossens, F. and Preumont, A. (2001). Active tendon control of cable-stayed bridges: A large-scale demonstration. *Earthquake Engineering and Structural Dynamics*, vol. 30, 961–979.

9. Chalasani, R. M. (1984). Ride performance potential of active suspension systems, part1: Simplified analysis based on a quarter-car model. ASME Symposium on Simulation and Control of Ground vehicles and Transportation systems, *Anaheim, CA*.

10. Chen, G., Lurie, B. and Wada, B. (1989). Experimental studies of adaptive structure for precision performance. *Proceedings of the 30th AIAA/ASME/ASCE/AHS Structures, Structural Dynamics, and Materials Conference, AIAA, Washington DC*, pp. 1462–1472.

11. Fanson, J. L., Blackwood, G. H. and Chen, C. C. (1989). Active member control of precision structures. SDM Conference, AIAA paper 89-1329-CP.

12. Fanson, J. L. and Caughey, T. K. (1990). Positive position feedback control for large space structures. *AIAA Journal*, vol. 28(4), 717–724.

13. Forward, R. L. (1981). Electronic damping of orthogonal bending modes in a cylindrical mast experiment. *AIAA Journal of Spacecraft*, vol. 18(1), 11–17.

14. Fuller, C. R., Elliott, S. J. and Nelson, P. A. (1996). *Active Control of Vibration*. Academic Press.

15. Geng, Z. J. and Haynes, L. S. (1994). Six degree-of-freedom active vibration control using the stewart platforms. *IEEE Transactions on Control Systems Technology*, vol. 2(1), 45–53.

16. Goh, C. and Caughey, T. K. (1985). On the stability problem caused by finite actuator dynamics in the control of large space structures. *Int. J. of Control*, vol. 41(3), 787–802.

17. Kaplow, C. E. and Velman, J. R. (1980). Active local vibration isolation applied to a flexible space telescope. *AIAA J. Guidance and Control*, vol. 3(3), 227–233.

18. Karnopp, D. C. and Trikha, A. K. (1969). Comparative study of optimization techniques for shock and vibration isolation. Trans. ASME, Journal of Engineering for Industry, *series B*, vol. 91(4), 1128–1132.

19. Lee, C.-K. (1990). Theory of laminated piezoelectric plates for the design of distributed Sensors/Actuators-Part I : Governing equations and reciprocal relationships. *J. Acoust.-Soc. Am*, vol. 87(3), 1144–1158.

20. Lee, C.-K. and Moon, F. C. (1990). Modal sensors/actuators. *Trans. ASME, J. of Applied Mechanics*, vol. 57, 434–441.

21. Mc Inroy, J. E., Neat, G. W. and O'Brien, J. F. (1999). A robotic approach to fault-tolerant, precision pointing. *IEEE Robotics & Automation Magazine*, pp. 24–31.

22. Neat, G., Abramovici, A., Melody, J., Calvet, R., Nerheim, N. and O'Brien, J. (1997). Control technology readiness for spaceborne optical interferometer missions. *Proceedings SMACS-2, Toulouse*, pp. 13–32.

23. Nelson, P. A. and Elliott, S. J. (1992). *Active Control of Sound*. Academic Press.

24. Piefort, V. (2001). *Finite Element Modeling of Piezoelectric Active Structures*. Ph.D. thesis, Université Libre de Bruxelles, Active Structures Laboratory.

25. Preumont, A. (1997). *Vibration Control of Active Structures, An Introduction*. Kluwer Academic Publishers.

26. Preumont, A. and Achkire, Y. (1997). Active damping of structures with guy cables. *AIAA, J. of Guidance, Control, and Dynamics*, vol. 20(2), 320–326.

27. Preumont, A., Achkire, Y. and Bossens, F. (2000). Active tendon control of large trusses. *AIAA Journal*, vol. 38(3), 493–498.

28. Preumont, A. and Bossens, F. (2000). Active tendon control of vibration of truss structures : Theory and experiments. *Journal of Intelligent Material Systems and Structures*, vol. 11(2), 91–99.

29. Preumont, A., Dufour, J. P. and Malekian, C. (1992). Active damping by a local force feedback with piezoelectric actuators. *AIAA J. of Guidance*, vol. 15(2), 390–395.

30. Preumont, A., Francois, A., Bossens, F. and Abu-Hanieh, A. (Accepted for publication in 2001). Force feedback versus acceleration feedback implementation in active vibration isolation. *Journal of Sound and Vibration*.

31. Preumont, A., Francois, A., De Man, P. and Loix, N. (2002). A novel electrode concept for spatial filtering with piezoelectric films. In *Active 2002*. Southampton.

32. Preumont, A., Loix, N., Malaise, D. and Lecrenier, O. (1993). Active damping of optical test benches with acceleration feedback. *Machine Vibration*, vol. 2, 119–124.

33. Rahman, Z., Spanos, J. and Laskin, R. (1998). Multi-axis vibration isolation, suppression and steering system for space observational applications. In *SPIE International Symposium on Astronomical Telescopes and Instrumentation*. Kona, Hawaii. Paper no. 3351-44.

34. Sim, E. and Lee, S. W. (1993). Active vibration control of flexible structures with acceleration or combined feedback. *AIAA J. of Guidance*, vol. 16(2), 413–415.

35. Spanos, J., Rahman, Z. and Blackwood, G. (1995). A soft 6-axis active vibration isolator. *Proceedings of the American Control Conference, Seattle, WA*, pp. 412–416.

36. Thayer, D., Campell, M., Vagner, J. and Von Flotow, Á. (2002). Six-axis vibration isolation system using soft actuators and multiple sensors. *Journal of Spacecraft and Rockets*, vol. 39(2).

37. Thayer, D., Vagners, J., Von Flotow, A., Hardham, C. and Scribner, K. (1998). Six-axis vibration isolation system using soft actuators and multiple sensors. *Proc. of Annual American Astronautical Society Rocky Mountain Guidance and Control Conference (AAS-98-064)*, pp. 497–506.

ACTIVE SOUND CONTROL

S.J.ELLIOTT
University of Southampton
Highfield, Southampton SO17 1BJ, UK

1. Introduction

One of the reasons for actively controlling the vibration of a structure may be to reduce its sound radiation. Before considering such active vibroacoustic control systems in a later chapter, we use this chapter to review the physical limitations of purely acoustical active control systems, i.e. those in which a soundfield is controlled using acoustical sources. The nature of sound propagation also means that a time-advanced reference signal is available in some applications and this allows feedforward control systems to be implemented, which do not suffer from the limitations of the Bode sensitivity integral in the same way that feedback control systems do. In order to be robust to changes in the disturbance and plant such feedforward controllers are often made adaptive and are also widely used to control tonal acoustic disturbances, in particular those experienced in the passenger cabins of propeller aircraft, with typical commercial systems having 36 loudspeakers and 72 microphones.

Although the principle of active sound control has been known since the 1930's [1], Figure 1, and single-channel analogue control systems were developed in the 1950's [2],[3] it was not until the development of modern digital signal processing, DSP, devices in the 1980's that adaptive digital controllers enabled the technique to be used in many practical problems [4],[5]. Since then there has been considerable interest in the commercial application of active sound control and this has led to a more detailed investigation of its fundamental acoustic limitations [6],[7].

2. Control of Wave Transmission

In this section, we consider the active control of sound that is transmitted as one-dimensional propagating waves in a duct. For sufficiently low frequency excitation only *plane waves* of sound can propagate in such a duct. In that case the waves have a uniform pressure distribution across any section of the duct and obey the one-dimensional wave equation [8].

We will assume that an incident tonal wave of sound, travelling in the positive x direction along a duct, is controlled by a single acoustic secondary source such as a loudspeaker mounted in the wall of the duct, as illustrated in Figure 2. The duct is assumed to be infinite in length, to have rigid walls and to have a constant cross section. The complex pressure due to the incident primary wave is assumed to be

43

A. Preumont (ed.), Responsive Systems for Active Vibration Control, 43–58.
© 2002 *Kluwer Academic Publishers. Printed in the Netherlands.*

June 9, 1936.   P. LUEG   2,043,416
PROCESS OF SILENCING SOUND OSCILLATIONS
Filed March 8, 1934

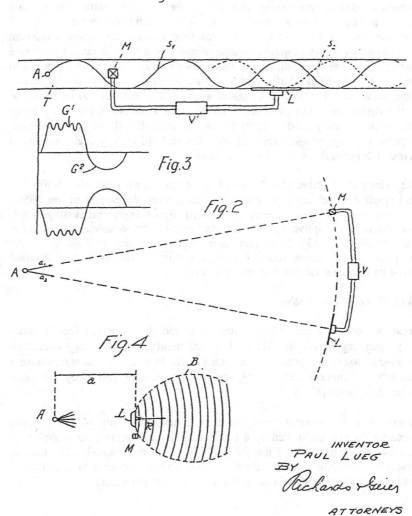

INVENTOR
PAUL LUEG
BY

ATTORNEYS

Figure 1. Illustrations page from an early active noise control patent by Lueg (1936).

$$p_{p+}(x)=Ae^{-jkx} \quad , \text{ for all } x, \qquad (2.1)$$

where the subscript $_{p+}$ denotes the primary wave travelling in a positive x direction or *downstream*. An acoustic source such as a loudspeaker driven at the same frequency as that of the incident wave will produce acoustic waves travelling both in the downstream direction and in the *upstream* or negative-x direction, whose complex pressures can be written as

$$p_{s+}(x) = Be^{-jkx} \text{ , for } x > 0, \quad p_{s-}(x) = Be^{+jkx} \text{ , for } x < 0 , \quad (2.2\text{a,b})$$

where the secondary source has been assumed to be at the position corresponding to $x = 0$, and B is a complex amplitude, which is linearly dependent on the electrical input to the secondary source, u, in Figure 2. If this electrical input is adjusted in amplitude and phase so that $B = -A$, then the total downstream pressure will be

$$p_{p+}(x) + p_{s+}(x) = 0 \quad \text{, for } x > 0 , \qquad (2.3)$$

i.e. the pressure will be perfectly cancelled at all points downstream of the secondary source. This suggests that a practical way in which the control input could be adapted is by monitoring the tonal pressure at any point downstream of the secondary source and adjusting the amplitude and phase of the control input until this pressure is zero. We are mainly interested here in the physical consequences of such a control strategy, however, and so we calculate the total pressure to the left, i.e. on the upstream side, of the secondary source, which in general will be

$$p_{p+}(x) + p_{s-}(x) = Ae^{-jkx} + Be^{+jkx} \quad , x < 0 . \qquad (2.4)$$

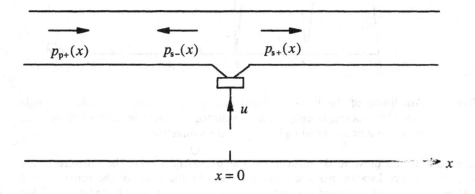

Figure 2.   Active control of plane sound waves in an infinite duct using a single secondary source.

46

If the secondary source is adjusted to cancel the pressure on the downstream side, then $B = -A$, and the pressure on the upstream side becomes

$$p_{p+}(x) + p_{s-}(x) = -2jA \sin kx \quad , x < 0, \tag{2.5}$$

since $e^{jkl} - e^{-jkl} = 2j\sin kl$. Thus a perfect acoustic *standing wave* is generated by interference between the positive-going primary wave and the negative-going wave generated by the secondary source. Notice that this standing wave has nodes of pressure at $x = 0$, i.e. at the position of the secondary source, and $x = -\lambda/2$, $x = -\lambda$ etc, where λ is the *acoustic wavelength*, and that when $x = -\lambda/4$, $x = -3\lambda/4$ etc, its amplitude is exactly twice the amplitude of the incident primary wave. The distribution of the pressure amplitude in the duct under these circumstances is shown in Figure 3.

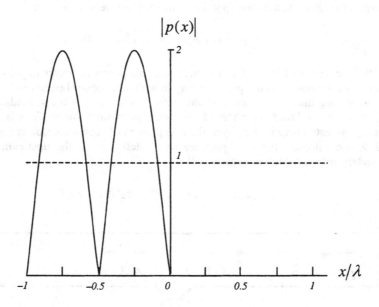

Figure 3. Amplitude of the pressure distribution in an infinite duct after a single secondary source, at x=0, has been adjusted to cancel an incident plane wave of unit amplitude traveling in the positive x direction.

In cancelling the pressure downstream of the secondary source, the pressure at the secondary source location has been driven to zero by the effect of the active control system. The secondary source thus acts to create a pressure-release boundary condition as far as the incident wave is concerned, and effectively *reflects* this wave back up the duct with equal amplitude and inverted phase, which gives rise to the standing wave observed in Figure 3. The acoustic power generated by a loudspeaker is equal to the time-averaged product of its volume velocity and the acoustic pressure in front of it. The fact that the acoustic pressure at the secondary source location is zero, means that the secondary source can generate no acoustic power when operating to cancel the incident wave, and it acts as a purely reactive element.

Feedforward control systems are generally used to implement such active sound controllers in ducts, using an upstream microphone as a reference signal, as illustrated in the upper illustration of Figure 1. If either the reference microphone (M) or the secondary loudspeaker system (L) are directional, the controller (V) required to reduce broadband disturbances becomes a pure delay, because of the non-dispersive nature of acoustic propagation. A very similar feedforward control strategy can also be used to control flexural waves propagating on a beam [9], except that such waves are dispersive and the controller must have a more complicated frequency response than a pure delay in this case.

3. Control of Sound in Enclosures

The soundfield in long ducts can be accurately represented as the superposition of two plane propagating waves travelling in opposite directions. In larger enclosures, for which the width and height are not small compared with the acoustic wavelength, sound waves can propagate in any direction and the soundfield becomes diffuse. At very high frequencies such a diffuse field model is a useful way of describing enclosed soundfields, but if the active control of the entire soundfield is being contemplated, a more helpful model of an enclosed soundfield is in terms of a superposition of its acoustic modes.

4. Control of Enclosed Soundfields

The steady state complex pressure distribution at a point x, y, z in an enclosure, when excited at a frequency ω, can be written in terms of its acoustic modes as

$$p(x, y, z, \omega) = \sum_{n=0}^{\infty} a_n(\omega)\psi_n(x, y, z) \ , \qquad (4.1)$$

where $a_n(\omega)$ is amplitude of the n-th acoustic mode, which has a mode shape given by $\psi_n(x,y,z)$. The mode shapes are orthogonal and will be assumed here to be entirely real, and are normalised so that

$$\frac{1}{V}\int_v \psi_n(x, y, z)\psi_m(x, y, z)\mathrm{d}x\,\mathrm{d}y\,dz = \begin{cases} 1 \text{ if } n = m \\ 0 \text{ if } n \neq m \end{cases}, \qquad (4.2)$$

where V is the volume of the enclosure. The acoustic mode shapes for a rigid rectangular enclosure of dimensions L_x by L_y by L_z for example, are given by

$$\psi_n(x,y,z) = \sqrt{\varepsilon_{n1}\,\varepsilon_{n2}\,\varepsilon_{n3}}\,\cos\!\left(\frac{n_1\pi\,x}{L_x}\right)\cos\!\left(\frac{n_2\pi\,y}{L_y}\right)\cos\!\left(\frac{n_3\pi\,z}{L_z}\right), \qquad (4.3)$$

where $\varepsilon_{ni} = 1$ if $n_{i=0}$ and $\varepsilon_{ni} = 2$ if $n_i > 0$ and n_1, n_2 and n_3 are the three modal integers denoted by the single index, n, in equation (4.1).

If the enclosure is excited by a point monopole of source strength, q, applied at position (x_0, y_0, z_0) then the mode amplitude can be written as

$$a_n(\omega) = \frac{\rho_0\,c_0^2\,A_n(\omega)\psi_n(x_0, y_0, z_0)}{V}\,q, \qquad (4.4)$$

where ρ_0 and c_0 are the density and speed of sound. The term $A_n(\omega)$ in equation (4.4) denotes the modal resonance term which can be written as

$$A_n(\omega) = \frac{\omega}{B_n\omega + j\!\left(\omega^2 - \omega_n^2\right)}, \qquad (4.5)$$

where B_n is the modal bandwidth, and ω_n is the natural frequency of the n-th mode. If viscous damping is assumed then $B_n = 2\,\omega_n\zeta_n$, where ζ_n is the modal damping ratio.

The *total potential energy* stored in the panel is proportional to the volume integral of the mean-square pressure, and can be written as [6]

$$E_p(\omega) = \frac{1}{4\rho_0\,c_0^2}\int_v |p(x,y,z)|^2\,dv. \qquad (4.6)$$

If the pressure distribution is expressed in terms of the modal expansion, equation (4.1), then the orthonormal properties of the modes, equation (4.2), can be used to show that the total potential energy can also be written as

$$E_p(\omega) = \frac{V}{4\rho_0\,c_0^2}\sum_{n=0}^{\infty}|a_n|^2 \qquad (4.7)$$

and so is proportional to the sum of the squared mode amplitudes.

The total acoustic potential energy provides a convenient cost function for evaluating the effect of global active control of sound in an enclosure. Each of the mode amplitudes in equation (4.7) for E_p can be expressed in terms of the contribution from the primary and secondary sources, as in equation (4.4). Thus the total acoustic potential energy in the enclosure at a single frequency is a Hermitian quadratic function of the complex strengths of the secondary acoustic sources, which can be minimised analytically [6].

A simulation has been carried out of minimising the total acoustic potential energy in an enclosure of dimensions 1.9×1.1×1.0 m, as illustrated in Figure 4, in which the acoustic modes have an assumed damping ratio of 10%, which is fairly typical for a reasonably well damped acoustic enclosure such as a car interior at low frequencies. The acoustic mode shapes in a rigid-walled rectangular enclosure are proportional to the product of three cosine functions in the three dimensions. The lowest order mode, for the zero-th order cosine functions, has an equal mode amplitude throughout the enclosure and corresponds to a uniform compression of the air at all points. The mode with the next highest natural frequency corresponds to fitting a half wavelength into the longest dimension of the enclosure, and this first longitudinal mode has a natural frequency of about 90 Hz for the enclosure shown in Figure 4, whose size is similar to the interior of a small car.

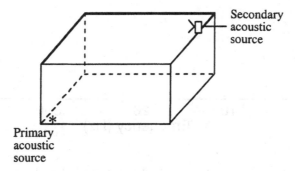

Figure 4. Physical arrangement for the simulation of the active control of tonal sound in a rectangular enclosure, which is about the size of a car interior, excited by a primary acoustic source in one corner and a secondary acoustic source in the opposite corner.

Figure 5 shows the total acoustic potential energy in the enclosure when driven only by the primary source placed in one corner of the enclosure and when the total acoustic potential energy is minimised by a single secondary acoustic source, as indicated as a loudspeaker in Figure 4, in the opposite corner (dashed line) or by seven secondary acoustic sources positioned at each of the corners of the enclosure not occupied by the primary source (dash-dot line). The enclosure is assumed to be driven at one frequency at a time and the secondary sources are adjusted to minimise the total acoustic potential energy at that frequency. Figure 5 shows the result of a series of such calculations carried out at a set of discrete frequencies up to 400 Hz. Significant attenuations in the total acoustic potential energy are achieved with a single secondary source for excitation frequencies below about 20 Hz, where only the zero-th order acoustic mode is significantly excited, and for excitation frequencies close to the natural frequency of the first longitudinal mode at about 90 Hz.

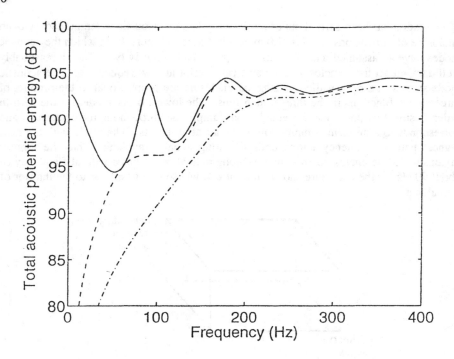

Figure 5. Total acoustic potential energy in the enclosure when driven by the primary
acoustic source alone at discrete frequencies (solid line) and when the total
potential energy has been minimized using either the single secondary source
shown in Figure 4, optimally adjusted at each excitation frequency (dashed
line), or seven secondary acoustic sources placed in all the corners of the
enclosure not occupied by the primary source (dash-dot line).

The response of the system does not , however, show clear modal behaviour for
excitation frequencies above about 150 Hz, and very little attenuation can be achieved
with a single secondary source above this frequency. This is because the spacing of the
natural frequencies for the acoustic modes in a three-dimensional enclosure becomes
closer together with higher mode order and a single secondary source cannot control the
multiple modes which are excited by even a pure tone source at these frequencies.
There is a more specific problem in the enclosure used to generate the results in
Figure 5, and in the interior of most cars, which is that the length is about twice the
width and the height and so the second longitudinal mode has a similar natural
frequency to the first transverse and the first vertical mode. This clump of modes
occurs at about 180 Hz. Although introducing seven secondary sources into the
enclosure does provide some attenuation at 180 Hz global control still cannot be
maintained at frequencies above about 250 Hz.

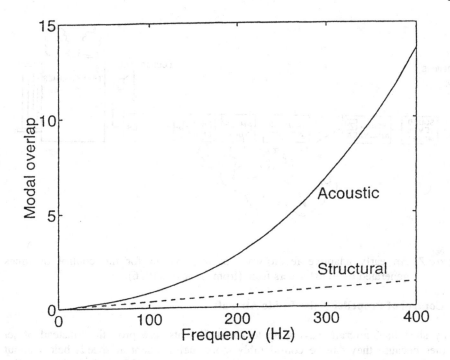

Figure 6. The modal overlap, $M(\omega)$, for the acoustic modes in the enclosure used for the simulations in Figure 5.

The total acoustic potential energy in an enclosure is proportional to the volume integral of the mean-square pressure. In a practical active control system, the total potential energy can be estimated using the sum of the squared outputs of a number of pressure microphones. Practical systems which adjust the amplitude and phase of the tonal signals driving the loudspeakers to minimise the sum of the squared microphones have developed rapidly over the past decade [11], but are in principle exactly the same as the controller originally put forward by Conover [3] and illustrated in Figure 7. This is an example of an adaptive feedforward system in which the amplitude and phase of a reference signal at the frequency to be controlled is adjusted to minimise the pressure at a remote reference microphone. Although the adjustment is being performed manually and for a single loudspeaker and microphone in Figure 7, modern digital systems, which can control many actuators to minimise the response at many sensors, employ essentially the same philosophy of control. The number of microphones required to obtain an accurate estimate of the total acoustic potential energy is proportional to the number of significantly excited acoustic modes within the enclosure and thus increases sharply at higher excitation frequencies.

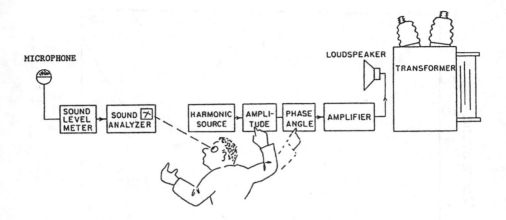

Figure 7. An early adaptive feedforward control system for the control of tones generated by transformers from (from Conover, 1956).

5. Control of Propeller Noise Inside Aircraft

Many short-haul aircraft, having up to about 50 seats, use propellers instead of jet engines because they can be considerably more fuel efficient at speeds below about 300 mph. An example of a modern propeller aircraft, the Q400, made by Bombardier Aerospace is shown in Figure 8. The development of prop-fan engines to power larger faster aircraft has also been suggested because of their improved fuel efficiency. The spectrum of the sound pressure in the passenger cabins of such aircraft contain strong tonal components at the *blade passage frequency* (BPF) of the propellers as shown in Figure 9, which are difficult to attenuate using passive absorption [12],[13]. Active control of these tones has been considered since the early 1980's [4],[14],[15] and is a good solution to this problem since active sound control works particularly well at low frequencies, where the acoustic modal overlap is relatively small, as described in Section 4. Also, with lightweight loudspeakers, active control potentially carries a significantly smaller weight penalty than passive methods of noise control. The original flight trials of a practical active control system operating in the passenger cabin of a turboprop aircraft were conducted in early 1988 [16],[17],[18]. Attenuation of 14 dB in the sum of the mean square pressures measured with 32 microphones were achieved at the BPF, 88 Hz in this case, using 16 loudspeakers.

Figure 8.   The Bombardier Q400, an example of a quiet modern turboprop aircraft which uses active control (picture courtesy of Bombardier Aerospace).

At the second and third harmonics, the attenuations in the sum of squared microphone signals are somewhat smaller than at the blade passage frequency. This reflects the greater complexity of the acoustic field within the aircraft cabin at these higher frequencies, which makes the sound field more difficult to control, as discussed in Section 4. By moving some of the loudspeakers that were previously uniformly distributed in the cabin to a circumferential array in the plane of the propellers, somewhat greater reductions were measured at the second and third harmonics. This improvement is achieved by the ability of the secondary loudspeakers to more closely match the fuselage vibration caused by the propellers, and thus control the sound at source, without it being radiated into the cabin. Some care must be taken in the interpretation of the results from the error microphones at these higher frequencies, however, since the greater complexity in the sound field makes it more likely that the pressure level has been substantially reduced only at the locations of the error microphones, and that the average pressure level in the rest of the cabin has remained substantially unaltered. More recent flight trials have included a separate array of monitoring microphones, which are not used in the cost function minimised by the active control system, so that an independent measure of attenuation can be achieved [19].

54

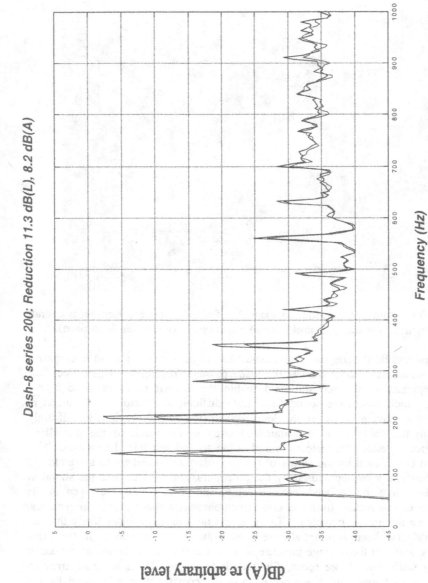

Figure 9. Typical spectrum of the sound pressure inside a propeller aircraft, showing the high levels at harmonics of the blade passing frequency, BPF, which is about 70 Hz in this case, and the spectrum after control with an active noise control system (figure courtesy of Ultra Electronics).

Although the aircraft control system did not have to respond very quickly under straight and level cruise conditions if the speeds of the two propellers were locked together, this was not possible on the aircraft used for the original flight trials. As the speeds of the two engines are brought closer together, the sound fields due to the two propellers beat together and the amplitude tracking properties of the adaptive algorithm allow the control system to follow these beats, provided they are not quicker than about 2 beats per second. Reductions in overall A-weighted sound pressure level, of up to 7 dB(A) were measured with the active control system operating at all 3 harmonics [18]. The ability of the control system to track rapid changes in the amplitude and phase of various tonal components in the disturbance were also found to be very important in the implementation of an active system to control sound inside a helicopter cabin [20].

Since this early work, which demonstrated the feasibility of actively controlling the internal propeller noise using loudspeakers, a number of commercial systems have been developed [21],[22] and are now in service on a number of aircraft. Instead of using loudspeakers as secondary sources it is also possible to use structural actuators attached to the fuselage to generate the secondary sound field. These have a number of potential advantages over loudspeakers, since they are capable of reducing both the cabin vibration as well as the sound inside the aircraft. It is also possible to integrate structural actuators more easily into the aircraft manufacturing process, since no loudspeakers would need to be mounted on the trim panels. Early work in this area using piezoceramic actuators on the aircraft fuselage [23] demonstrated that relatively few secondary actuators may be needed for efficient acoustic control. More recent systems have used inertial electromagnetic actuators mounted on the aircraft frame [24] or actively-tuned resonant mechanical systems [25],[26] to achieve a more efficient mechanical excitation of the fuselage at the blade passage frequency and its harmonics.

Such a system, with 42 electromagnetic shakers and 88 microphones, is used on the Q400 aircraft shown in Figure 8 and significant reductions in the internal pressure are achieved at the blade passing frequency and several of its harmonics, as shown in Figure 9. Detailed measurements of the pressure distribution throughout the passenger cabin show that the high sound pressure levels experienced at the front of the aircraft, particularly in the plane of the propellers, have been reduced to the significantly lower levels that were originally present at the rear of the aircraft, and the distribution of sound pressure level is much more uniform throughout the cabin after control.

6. Summary and Conclusions

The original idea of controlling sound with sound was developed in the 1930's and 1950's, but the practical implementation of multichannel systems only became feasible with advances in digital signal processors in the 1980's. This generated a surge of interest and a re-examination of the fundamental limitations of active sound control.

Active noise control is most effective at low frequencies, for which the wavelengths are large compared with the spatial difference between the soundfields due to the primary and secondary sources. The physical principles of active sound control are most easily illustrated for the one-dimensional problem of plane waves propagating in a duct. At higher frequencies multiple waves can propagate in the duct and multiple actuators then need to be used to control the transmission of acoustic energy.

The problem of spatial matching becomes more severe when secondary acoustic sources are used to attenuate the acoustic output of a primary source in a three-dimensional space. When sound in a three-dimensional enclosure is considered, the number of modes which are significantly excited at a given driving frequency can increase with the cube of that frequency and a single secondary source is not sufficient to control such an enclosed soundfield above the first resonant frequency. Multiple secondary sources can be used to achieve control in this case, provided there are at least as many secondary sources as significantly excited acoustic modes. Minimising the total acoustic potential energy in the enclosure has almost the same effect as minimising the total acoustic power input. By using an array of microphones, a practical approximation to the total acoustic potential energy can be measured in an enclosure, which can be used to automatically adjust the amplitude and phase of the multiple secondary sources to minimise this quantity.

Such a control system with 16 loudspeakers and 32 microphones was used in the original demonstration of the control of propeller noise in a passenger aircraft. Commercial systems are now available for this purpose, although more recent systems use electromagnetic shakers attached to the fuselage as actuators instead of loudspeakers, which can control the vibration in the passenger cabin as well as the noise, and typically use 42 shakers, 88 microphones and a fully-coupled adaptive feedforward control system.

Active control techniques are thus seen to complement conventional passive control methods since they are most effective at low frequencies. Current applications are mainly in the aerospace field, where the reduced weight of an active system for low frequency noise control, compared with a passive one, is worth the additional cost and complexity of a digital electronic system, or in markets such as the active headset where inexpensive analogue feedback controllers can give good performance.

Although active control has been demonstrated to be an effective way of reducing both low frequency engine noise and road noise in cars some time ago [27],[28], the cost of the controller has been an important factor in limiting the widespread use of this technology in the automotive market. Recently, however, active sound control systems have begun to find their way into cars by integrating them with the audio system [29]. With the continuing reductions in the price of devices which can be used for the real-time implementation of active controllers, other useful applications of active sound control using loudspeakers will, no doubt, be found.

Another technique for reducing the sound generated by a structure is to actively control the vibration of the structure, and thus reduce the noise at source. It is this technique, called active structural acoustic control or active vibroacoustic control, that is the subject of a separate chapter.

7. References

1. Lueg P. (1936) Process of silencing sound oscillations. *U.S. Patent*, No. 2,043,416.
2. Olson H.F. and May E.G., 1953, Electronic sound absorber. *Journal of the Acoustical Society of America*, 25, 1130-1136.
3. Conover W.B., 1956, Fighting noise with noise. *Noise Control.* 2, p.78-82.
4. Chaplin G.B.B. (1983) Anti-sound – The Essex breakthrough. *Chartered Mechanical Engineer.* 30, .41-47.
5. Roure A. (1985) Self adaptive broadband active noise control system. *Journal of Sound and Vibration.* 101, 429-441.
6. Nelson P.A. and Elliott S.J. (1992) *Active Control of Sound* (Academic Press).
7. Hansen C.H. and Snyder S.D. (1997) *Active Control of Noise and Vibration*, E&FN Spon.
8. Kinsler L.E., Frey A.R., Coppens A.B. and Sanders J.V. (1982) *Fundamentals of Acoustics*, 3rd Edition, John Wiley.
9. Elliott S.J. and Billet L., 1993, Adaptive control of flexural waves propagating in a beam. *Journal of Sound and Vibration.* **163,** 295-310.
10. Morse P.M. (1948) *Vibration and Sound* (2nd ed.) (New York, McGraw-Hill) (reprinted in 1981 by the Acoustical Society of America)
11. Elliott S.J. (2001) *Signal Processing for Active Control*, Academic Press.
12. Wilby J.F., Rennison D.C., Wilby E.G. and Marsh A.H. (1980) Noise control prediction for high speed propeller-driven aircraft. *American Institute of Aeronautics and Astronautics 6th Aeroacoustic Conference*, Paper AIAA-80-0999.
13. Metzger F.B. (1981) Strategies for reducing propeller aircraft cabin noise. *Automotive Engineering*, 89, 107-113.
14. Ffowcs-Williams J.E. (1984) Review Lecture: Anti-Sound. *Royal Society of London*, A395, 63-88.
15. Bullmore A.J., Nelson P.A., Elliott S.J., Evers J.F. and Chidley B. (1987) Models for evaluating the performance of propeller aircraft active noise control systems. *AIAA 11th Aeroacoustics Conference*, Palo Alto, C.A., Paper AIAA-87-2704.
16. Elliott S.J., Nelson P.A., Stothers I.M. and Boucher C.C. (1989) Preliminary results of in-flight experiments on the active control of propeller-induced cabin noise. *Journal of Sound and Vibration*, 128, 355-357.
17. Dorling C.M., Eatwell G.P., Hutchins S.M., Ross C.F. and Sutcliffe S.G.C. (1989) A demonstration of active noise reduction in an aircraft cabin. *Journal of Sound and Vibration*, 128, 358-360.
18. Elliott S.J., Nelson P.A., Stothers I.M. and Boucher C.C. (1990) In-flight experiments on the active control of propeller-induced cabin noise. *Journal of Sound and Vibration*, 140, 219-238.
19. Borchers I.U. et al. (1994) Advanced study of active noise control in aircraft (ASANCA), in *Advances in Acoustics Technology*, J.M.M.Hernandez (Ed.) (John Wiley and Sons).
20. Elliott S.J., Boucher C.C. and Sutton T.J. (1997) Active control of rotorcraft interior noise. *Proceedings of the Conference on Innovations in Rotorcraft Technology*, Royal Aeronautical Society, London. 15.1-15.6.
21. Emborg U. and Ross C.F. (1993) Active control in the SAAB 340, *Recent Advances in the Active Control of Sound and Vibration*, 567-573.

22. Billout G., Norris M.A., and Rossetti D.J. (1995) System de controle actif de bruit Lord NVX pour avions d'affaire Beechcraft Kingair, un concept devanu produit. *Active Control Conference*, Cenlis.
23. Fuller C.R. (1985) Experiments on reduction of aircraft interior noise using active control of fuselage vibration. *Journal of the Acoustical Society of America*, 78 (S1), S88.
24. Ross C.F. and Purver M.R.J. (1997) Active cabin noise control. *ACTIVE 97*, xxxix-xlvi.
25. Fuller C.R., Maillard J.P., Meradal M. and von Flotow A.H. (1995) Control of aircraft interior noise using globally detuned vibration absorbers. *The First Joint CEAS/AIAA Aeroacoustics Conference*, Munich, Germany, Paper CEAS/AIAA-95-082, 615-623.
26. Fuller C.R. (1997) Active control of cabin noise - lessons learned? *Fifth International Congress on Sound and Vibration*, Adelaide.
27. Elliott S.J., Stothers I.M., Nelson P.A., McDonald A.M., Quinn D.C. and Saunders T. (1988) The active control of engine noise inside cars. *InterNoise '88*, Avignon, 987-990.
28. Sutton T.J., Elliott S.J. and McDonald A.M. (1994) Active control of road noise inside vehicles. *Noise Control Engineering Journal*, 42, 137-147.
29. Sano, H. et al. (2001) Active control system for low-frequency road noise combined with an audio system. *IEEE Transactions on Speech and Audio Processing*, 9(7), 755-763.

ACTIVE VIBROACOUSTIC CONTROL

S.J.ELLIOTT and P.GARDONIO
Institute of Sound and Vibration Research
University of Southampton
Highfield, Southampton SO17 1BJ, UK

1. Introduction

In many practical applications the reason for controlling the vibration of a structure is to reduce its radiated sound. In a previous chapter we saw that sound could be controlled using acoustic actuators such as loudspeakers, and in this chapter we will discuss how vibration actuators, acting directly on the structure, can be used to reduce its sound radiation. This may be termed active structural acoustic control [1] or active vibroacoustic control.

If sensors integrated into the structure can also be used to estimate the sound radiation, and a feedback controller can be implemented to control these sensors using integrated actuators, then a truly smart vibroacoustic structure can be implemented. Feedback controllers are used in this application, in contrast to the feedforward control employed for active sound control in ducts, because no time-advanced reference signal is generally available, and the waveform of the disturbances is generally unknown.

We begin by discussing the formulation of the sound radiation problem in terms of both structural modes and radiation modes. Using the latter formulation it becomes clear that at frequencies for which the wavelength is large compared with the size of the structure, it is the overall volume velocity of the structure which makes the greatest contribution to the sound power radiated. This has led to the development of sensors and actuators which predominantly detect and drive this component of a structure's motion, and single channel control systems for the control of volume velocity. When these transducers are made using a distribution of piezoelectric film, however, the plant response is severely disrupted at high frequencies by in-plane coupling effects.

This had led to the investigation of arrays of smaller actuators and sensors for use in this application, and further consideration as to whether these should be connected together using a single or multichannel controller. Finally, some recent work will be described in which arrays of piezoelectric actuators and inertial velocity sensors are connected only using local feedback loops, to form a modular active structure for active vibroacoustic control.

2. Vibration and Sound Radiation of a Panel

We begin by discussing the vibration and sound radiation of a lightly-loaded, baffled, panel, radiating into an infinite space, in terms of its structural modes. The steady state complex velocity distribution on a panel in the x, y plane when excited at a frequency ω can thus be written, using the notation of Fuller et al. [1], as

A. Preumont (ed.), Responsive Systems for Active Vibration Control, 59–82.
© 2002 *Kluwer Academic Publishers. Printed in the Netherlands.*

$$v(x, y, \omega) = \sum_{n=0}^{\infty} a_n(\omega)\psi_n(x, y) \ , \tag{2.1}$$

where $a_n(\omega)$ is amplitude of the n-th mode of vibration, which has a mode shape given by $\psi_n(x,y)$. The mode shapes are orthogonal and will be assumed here to be entirely real and normalised so that

$$\frac{1}{S}\int_s \psi_n(x, y)\psi_m(x, y)\mathrm{d}x\mathrm{d}y = \begin{cases} 1 & \text{if } n = m \\ 0 & \text{if } n \neq m \end{cases}, \tag{2.2}$$

where S is the surface area of the panel. The structural mode shapes for a uniform panel of dimensions L_x by L_y for example, which is constrained not to have any linear motion at its edges but whose edges can have angular motion, i.e. it is *simply supported*, are given by

$$\psi_n(x, y) = 4\sin\left(n_1\pi x/L_x\right)\sin\left(n_2\pi y/L_y\right) \tag{2.3}$$

where n_1 and n_2 are the two modal integers denoted by the index, n, in equation (2.1).

If the panel is excited by a point force, f, applied at position (x_0, y_0) then the mode amplitude can be written as

$$a_n(\omega) = \frac{A_n(\omega)\psi_n(x_0, y_0)}{M}f \tag{2.4}$$

where M is the total mass of the panel. The term $A_n(\omega)$ in equation (2.4) denotes the modal resonance term which can be written as

$$A_n(\omega) = \frac{\omega}{B_n\omega + j\left(\omega^2 - \omega_n^2\right)} \ , \tag{2.5}$$

where B_n is the modal bandwidth, and ω_n is the natural frequency of the n-th mode. If viscous damping is assumed then $B_n = 2\,\omega_n\zeta_n$, where ζ_n is the modal damping ratio.

In order to illustrate the difference between controlling vibration and sound power radiation, we can express the total kinetic energy of the panel in terms of the vector of amplitudes of the N significant structural modes [2],

$$E_k = \frac{M}{4}\mathbf{a}^H\mathbf{a} \ , \tag{2.6}$$

where M is the mass of the panel. If the panel is excited by a primary excitation which generates the vector of mode amplitudes $\mathbf{a}_p$ and these modes are also driven by a set of secondary forces $\mathbf{f}_s$ via modal coupling matrix $\mathbf{B}$, so that

$$a = a_p + Bf_s \ , \tag{2.7}$$

then the total kinetic energy can be expressed as a standard Hermitian quadratic function of f_s, and is minimised by the set of secondary forces given by [1]

$$f_{s,opt:E_k} = -\left[B^H B\right]^{-1} B^H a_p \ . \tag{2.8}$$

The sound power radiated by the panel is a more complicated function of the structural mode amplitudes than the total kinetic energy, however. If one mode were vibrating on its own, the ratio of the radiated sound power to the mean-square velocity of the vibration on the panel, which is proportional to the *self radiation efficiency* of the mode [3,1], is a function of both the mode shape and the excitation frequency. When the velocity of the panel is due to the excitation of more than a single mode, however, the sound fields due to each of the modes generally interact and the radiated sound power will not just be proportional to the mean-square amplitude of each of the individual modes. The radiated sound power is then related to the amplitudes of the structural modes by the expression

$$\Pi_R = \sum_{i=1}^{N} \sum_{j=1}^{N} M_{ij} a_i^* a_j \tag{2.9}$$

where M_{ij} are entirely real frequency-dependant numbers [1], and so the radiated power can be written in matrix form as

$$\Pi_R = a^H M a \ . \tag{2.10}$$

The diagonal terms in the matrix M are proportional to the *self radiation efficiencies* of the structural modes, and the off-diagonal terms are proportional to the *mutual radiation efficiencies*, or combined mode radiation efficiency, which account for the interaction of the sound fields generated by different structural modes. The radiation efficiency is a dimensionless quantity which is equal to the ratio of radiated sound power of the panel to the sound power radiated by an equal area of infinite rigid panel having the same mean-square velocity.

If we consider the modes of a simply supported panel, for example, the mode shapes have the particularly simple form

$$\psi_n(x, y) = 4 \sin(n_1 \pi x / L_x) \sin(n_2 \pi y / L_y) \tag{2.11}$$

where the factor of 4 is chosen so that the mode satisfies the orthonormality condition, equation (2.2), L_x and L_y are the dimensions of the panel in the x and y directions and n_1 and n_2 are the modal integers. The structural mode with modal integers $n_1 = 2$ and $n_2 = 1$, for example, is commonly referred to as the (2,1) mode. The shapes of some of these modes on a simply supported panel are shown in Figure 1.

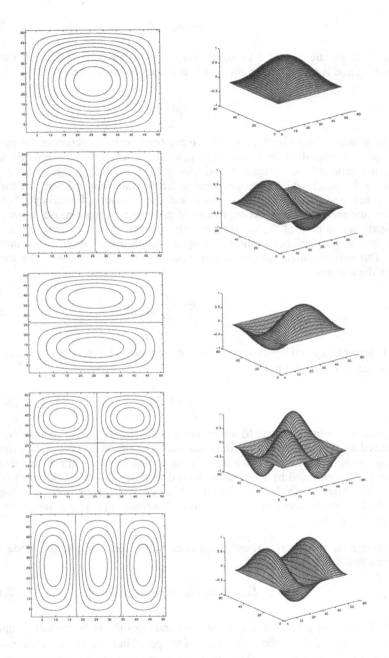

Figure 1. Mode shapes for the first five structural modes of a simply supported rectangular panel.

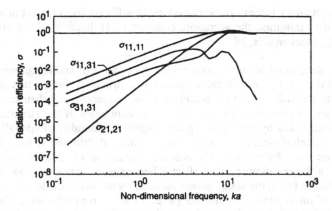

Figure 2. Self and mutual radiation efficiencies of the (1,1), (2,1) and (3,1)/structural modes. $\sigma_{11,11}$, $\sigma_{21,21}$ and $\sigma_{31,31}$ are the self radiation efficiencies, $\sigma_{11,31}$ is a mutual radiation efficiency and in this case $\sigma_{11,21}$ etc. are zero.

The self and mutual radiation efficiencies of the (1,1), (2,1) and (3,1) structural modes, for a panel with $L_y/L_x \approx 0.79$, mounted in an infinite baffle, are shown in Figure 2 as a function of normalised excitation frequency, L_x/λ. The self radiation efficiency of the (1,1) structural mode, for example, is denoted $\sigma_{11,11}$ and the mutual radiation efficiency of the (1,1) structural mode and the (3,1) structural mode is denoted $\sigma_{11,31}$. It can be seen that all the modes considered radiate with similar efficiency if $L_x \bullet \lambda$, and there is a decreasing amount of interaction between the modes as the excitation frequency is increased. In the low frequency region, however, when the size of the panel is small compared with the acoustic wavelength, the (1,1) mode and the (3,1) mode radiate considerably more efficiently than the (2,1) mode, but also there is considerable interaction between the sound radiation due to the (1,1) and the (3,1) modes. The important difference between the (1,1) and (3,1) modes and the (2,1) mode is that the latter does not have any net volumetric component. When the panel is vibrating in the (2,1) mode at low frequencies the air in front it is transferred from one side of the panel to the other as the mode vibrates, generating very little sound. The modes with an (odd, odd) mode number, however, have a space-average volumetric component and so displace fluid into the medium as they vibrate. It is this volumetric component that creates sound at low frequencies, and the way the volumetric contributions from the (odd, odd) modes add together cause their interaction.

To return to the active control of such sound radiation, we can see that if equation (2.7) is substituted into the matrix expression for the radiated sound power, equation (2.10), then another Hermitian quadratic function is obtained that has a unique minimum given when the complex amplitudes of the secondary sources are now equal to

$$\mathbf{f}_{s,opt:\Pi_R} = -\left[\mathbf{B}^H\mathbf{M}\mathbf{B}\right]^{-1}\mathbf{M}\mathbf{B}^H\mathbf{a}_p. \qquad (2.12)$$

64

This vector of secondary forces can be considerably different from the set of secondary sources required to minimise the vibration, equation (2.8), because of the presence of the radiation efficiency matrix, $\mathbf{M}$.

One way of implementing an active system for the control of sound radiation from a structure would thus be to measure the amplitudes of structural modes, to use equation (2.9) to estimate the radiated sound power at each frequency, and then to adjust the secondary sources to minimise this quantity. The amplitudes of the structural modes could be measured either by manipulating the outputs of a number of spatially discrete sensors such as accelerometers [1] or by using spatially distributed sensors integrated into the structure [4]. Either way, the accurate estimation of radiated sound power would require a large number of well-matched sensors, even at low excitation frequencies. Alternatively, the sound power radiated by the panel could be estimated by using an array of microphones around the panel. The microphones would have to be positioned some distance from the panel to avoid near-field effects and a large number of microphones would potentially be required for accurate estimation of the sound power. A more efficient implementation would involve controlling the output of a number of structural sensors that measure only the velocity distributions on the panel that are efficient at radiating sound.

3. Radiation Modes and the Control of Volume Velocity

It is possible to transform the problem of actively controlling sound radiation into a form that gives a clearer physical insight into the mechanisms of control, and also suggests a particularly efficient implementation for an active control system, particularly at low frequencies. This transformation involves an eigenvalue/eigenvector decomposition of the matrix of radiation efficiencies, which can be written as

$$\mathbf{M} = \mathbf{P}\underline{\Omega}\mathbf{P}^{\mathrm{T}} \qquad (3.1)$$

where $\mathbf{P}$ is an orthogonal matrix of real eigenvectors, since $\mathbf{M}$ has real elements and is symmetric, and $\underline{\Omega}$ is a diagonal matrix of eigenvalues, which are all real and positive, since $\mathbf{M}$ is positive definite. The radiated sound power, equation (2.10) can thus be written as

$$\Pi_{\mathrm{R}} = \mathbf{a}^{\mathrm{H}}\mathbf{M}\mathbf{a} = \mathbf{a}^{\mathrm{H}}\mathbf{P}\underline{\Omega}\mathbf{P}^{\mathrm{T}}\mathbf{a} , \qquad (3.2)$$

and if we define a vector of transformed mode amplitudes to be $\mathbf{b} = \mathbf{P}^{\mathrm{T}}\mathbf{a}$, the radiated sound power can be written as

$$\Pi_{\mathrm{R}} = \mathbf{b}^{\mathrm{H}}\underline{\Omega}\mathbf{b} = \sum_{n-1}^{N} \Omega_n |b_n|^2 . \qquad (3.3)$$

These transformed modes thus have no mutual radiation terms and hence radiate sound *independently* of each other [5,6,7,8,9].

250 Hz

1000 Hz

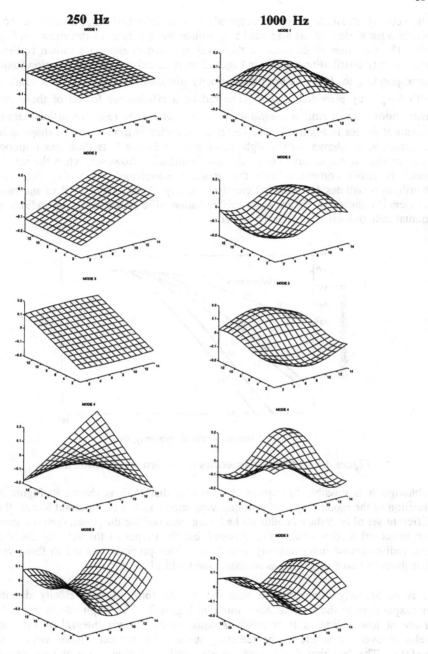

Figure 3. First five radiation modes of the rectangular plate considered in Figure 1 at 250 (left) and 1000 (right) Hz.

66

The velocity distributions that correspond to this transformed set of modes are shown in Figure 3 for a panel of the size used to calculate the structural modes shown in Figure 1 [9]. The radiation efficiencies of these velocity distributions are shown in Figure 4. The velocity distributions shown in Figure 3 were calculated for an excitation frequency corresponding to $L_x \approx \lambda/5$, but these velocity distributions only vary to a small extent with frequency provided $L_x \bullet \lambda$. The radiation efficiencies for all of these velocity distributions are of similar magnitudes if $L_x \bullet \lambda$, as are the radiation efficiencies of the structural modes in Figure 2. The fact that the radiation modes change shape at higher frequencies, as shown on the right hand side of Figure 3, is thus less important in practice than it might have been. At low frequencies, however, when the size of the panel is small compared with the acoustic wavelength, then the first velocity distribution radiates far more efficiently than any of the others. All of the radiation efficiencies shown in Figure 4 are self radiation efficiencies since, by definition, the mutual radiation efficiency terms are zero.

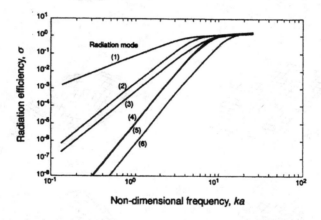

Figure 4. Radiation efficiencies of the first six radiation modes.

Although it is possible to express the velocity distributions shown in Figure 3 as a function of the mode shapes of a simply-supported panel, it is important to note that if a different set of boundary conditions had been assumed for the panel, then the shapes of the structural modes would have changed, but the shapes of the velocity distributions that radiate sound independently would not. This property has led to these velocity distributions being described as *radiation modes* [5,9].

It is particularly interesting to note the simple form of the velocity distribution corresponding to the first radiation mode in Figure 3. The amplitude of this radiation mode at low excitation frequencies is equal to the surface integral of the complex velocity over the whole of the radiating surface, i.e. the net *volume velocity* of the surface. The fact that this is such an efficiently radiating mode at low frequencies suggests that significant reductions in the sound power radiated from a surface can be achieved by the active control of this single quantity. The net volume velocity of a clamped panel can be directly measured using a single distributed sensor [10,11,12], so that it is not necessary to measure the amplitudes of all the individual structural modes.

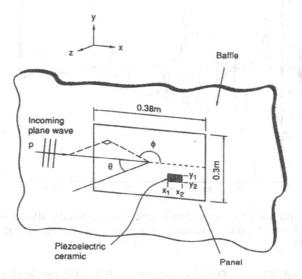

Figure 5. Baffled panel with a secondary piezoelectric actuator patch considered in the analysis.

As an example of the effect of controlling the net volume velocity on the sound radiation from a panel, Johnson and Elliott [13] presented simulation results for the arrangement shown in Figure 5, in which an incident plane acoustic wave excites the vibration of a panel which then radiates sound on the other side. The panel was made of aluminium, had dimensions 380mm × 300mm × 1mm with a damping ratio of 0.2%, and was excited by a plane acoustic wave incident at an angle of $\theta = 45°$, $\phi = 45°$. Figure 6 shows the ratio of the sound power radiated by the panel to the incident sound power, which is equal to the *power transmission ratio*, and is the inverse of the transmission loss. The power transmission ratio with no control is shown as the solid line in Figure 6. This is large when the panel is excited near the natural frequencies of its (odd, odd) modes, i.e. about 40 Hz, 180 Hz, 280 Hz, etc, since the panel then vibrates strongly and these modes radiate sound efficiently. The volume velocity of the panel is then cancelled at each frequency by the action of a piezoceramic actuator mounted on the panel as a secondary source, and the power transmission ratio is again calculated, as shown by the dashed-dotted line in Figure 6. The sound power radiated by the panel has been significantly reduced by the cancellation of its volume velocity at excitation frequencies below about 400 Hz, for which the largest dimension of the panel is less than half an acoustic wavelength. The residual levels of sound power that are radiated after control are due to the weakly radiating structural modes, which do not contribute to the volume velocity. At some higher frequencies, however, the sound radiation is slightly increased by volume velocity cancellation because the amplitude of these weakly radiating modes is increased.

68

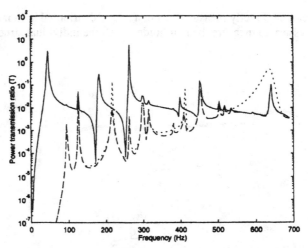

Figure 6. The sound power transmission before control (solid), after minimisation of radiated sound power (dashed), and after cancellation of volume velocity (dotted).

The effect of using the piezoceramic actuator to minimise the total sound power radiated by the panel is also plotted as the dashed line in Figure 6. This would be a very difficult control problem to implement in practice since it would involve the use of large numbers of microphones to estimate the radiated sound power, for example. It does, however, provide a benchmark for the best possible results that can be obtained in controlling sound transmission with this actuator. Apart from a few isolated frequencies, then it can be seen that below about 400 Hz, the reduction of sound power by volume velocity cancellation is almost as great as the best possible reduction of sound power that can be obtained using this piezoceramic actuator. The mechanism by which the piezoceramic actuators can generate vibration on a panel is described by Fuller et al. [1] for example, and a more detailed review of both actuators and sensors is provided by Hansen and Snyder [14].

The control of sound radiation using structural actuators can thus be seen to require a rather different strategy than the control of vibration. Relatively simple control systems can be effective in reducing sound radiation from structures at low frequencies, when the size of the structure is small compared with the acoustic wavelength. The control problem becomes progressively harder as the frequency increases, setting an upper limit on the frequency range of a practical control system.

4. Implementation of a collocated Volume Velocity Sensor and Uniform Force Actuator using Shaped Piezoelectric Film

The higher frequencies increases in vibration and sound radiation levels of the panel with a volume velocity sensor and one piezoelectric control actuator discussed above are primarily due to control spillover [15,16]. The secondary piezoelectric actuator excites radiation modes that are weakly excited by the primary disturbance in its attempt to control the volumetric vibration component of the panel. Moreover at higher frequencies the volumetric vibration of the panel does not provide a good estimate of the first radiation mode of the panel so that the spillover effect of the piezoelectric actuator is magnified.

If the sensor–actuator pair are collocated on the panel they will observe and excite the structure in a similar manner [17,18]. This property also carries over to distributed actuators and sensors. In particular, a reciprocal transducer, originally designed as a volume velocity sensor, will generate a uniform force over the surface of the panel when driven by a voltage. If a volume velocity sensor and a uniform–force actuator are used as a sensor–actuator pair to control the sound power radiation, the cancellation of the volume velocity can be achieved without causing increases in the vibration and sound radiation levels. Figure 7 shows the sound power transmission before and after control when using a distributed uniform–actuator. The use of a uniform–force actuator instead of a piezoelectric patch increases the useful frequency range to above 600 Hz and also increases the attenuations possible over the entire frequency range. The control strategies of minimisation of sound power and the cancellation of volume velocity produce extremely similar results.

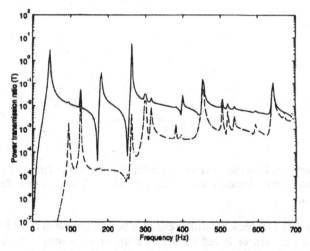

Figure 7. The sound power transmission before control (solid), after minimisation of radiated sound power (dashed), and after cancellation of volume velocity (dotted) when using a uniform force actuator.

Having the matched volume velocity sensor and uniform force actuator transducers also implies that the transfer function between these transducers is minimum phase [13]. The frequency response of a minimum phase system can always, in principle, be exactly compensated for to give an overall response with no delay. In an adaptive feedforward control system the speed of convergence is always limited by the delays in the system under control [2] and so this arrangement of actuator and sensor could result in a very fast–acting controller.

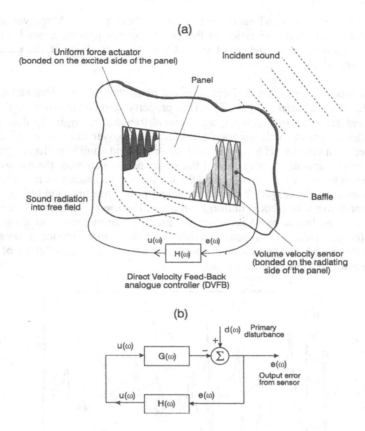

(a)

Uniform force actuator
(bonded on the excited side of the panel)

Incident sound

Panel

Sound radiation
into free field

Baffle

u(ω)   e(ω)

H(ω)

Volume velocity sensor
(bonded on the radiating
side of the panel)

Direct Velocity Feed-Back
analogue controller (DVFB)

(b)

d(ω)   Primary
disturbance

u(ω)

G(ω)   Σ

e(ω)

Output error
from sensor

u(ω)   e(ω)

H(ω)

Figure 8.   Smart panel for direct velocity feedback control. (a) The smart panel with two piezoelectric shaped transducers. (b) Block diagram of the feedback control scheme.

The possibility of a feedback controller can also be contemplated in which the output of the volume velocity sensor is fed back to the uniform pressure actuator as shown in Figure 8a and the overall frequency response from the primary disturbance to error signal is given by

$$\frac{e(j\omega)}{d(j\omega)} = \frac{1}{1 + G(j\omega)H(j\omega)} ,\qquad (4.1)$$

whose equivalent block diagram is shown in Figure 8b. If the system under control is minimum phase, then, in principle, its response $G(j\omega)$ can be perfectly compensated for by the controller $H(j\omega)$, which can also incorporate a large feedback gain such that $1 + G(j\omega)H(j\omega)$ is always entirely real and very large. Under these circumstances $e(j\omega)/d(j\omega)$ is very small and so the performance of the feedback system is the same as the feedforward one. The advantage of a feedback system is that no reference signal is necessary and therefore broadband noise as well as harmonic disturbances can be controlled.

There are several ways of measuring the volume velocity of a panel. The most common one uses piezoelectric films with shaped electrodes [10,11,12,19,20,21,22,23]. When such type of sensors are bonded onto a panel, the vibration of the panel deforms the piezoelectric material so that it generates a distribution of charges on the surfaces of the two electrodes by means of the piezoelectric effect [24]. A charge output, $q(t)$, proportional to the plate vibration can therefore be measured by connecting a charge amplifier to the two electrodes. According to Lee's formulation [25], the total charge output is given by two components: the charge generated by the bending vibration of the panel $q_b(t)$ and the charge due to the in-plane vibration of the panel $q_i(t)$ so that:

$$q(t) = q_b(t) + q_i(t) . \tag{4.2}$$

where

$$q_b(t) = -\int_0^{L_y} \int_0^{L_x} h(x,y) S(x,y) \left[e_{31} \frac{\partial^2 w}{\partial x^2} + e_{32} \frac{\partial^2 w}{\partial y^2} + 2e_{36} \frac{\partial^2 w}{\partial x \partial y} \right] dx dy , \tag{4.3}$$

$$q_i(t) = \int_0^{L_y} \int_0^{L_x} S(x,y) \left[e_{31} \frac{\partial u}{\partial x} + e_{32} \frac{\partial v}{\partial y} + e_{36} \left(\frac{\partial u}{\partial y} + \frac{\partial v}{\partial x} \right) \right] dx dy . \tag{4.4}$$

In the above two formulae u, v and w are the displacements of the plate along the two in-plane directions x and y and the out of plane, z, direction, $h = h_s + h_{pe}/2$ represents the distance in z direction from the neutral plane of the plate to the middle plane of the piezoelectric film; h_s is half of the plate thickness and h_{pe} is the thickness of the sensor (piezoelectric material and electrodes). e_{31}, e_{32} and e_{36} are the piezoelectric stress constants and $S(x,y)$ is the spatial sensitivity function of the sensor.

Considering a rectangular panel clamped along the four edges, Rex and Elliott [10], have shown that if the distributed piezoelectric transducer has a quadratic sensitivity along the x-direction and a constant sensitivity along the y-direction, so that

$$S(x,y) = -k(x^2 - L_x x) , \tag{4.5}$$

where k is a constant, then the charge output of equation (4.3) is proportional to the integrated transverse displacement of the whole surface of the panel:

$$q_b(t) = 2e_{31} hk \int_0^{L_y} \int_0^{L_x} w(x,y) dx dy . \tag{4.6}$$

Equation (4.6) shows that, this sensitivity distribution for a piezoelectric transducer bonded on a clamped panel generates a charge output proportional to the transverse displacement integrated over the panel surface. Thus, when the panel is excited in bending and current is measured, $i(t) = dq(t)/dt$, then the sensor output is proportional to the volume velocity of the panel $Q(t) = \int_S \dot{w}(x, y)\, dS$.

There are various methods of achieving the required quadratic weighted sensitivity. The easiest one is to implement a quadratically shaped electrode as shown in Figure 9 [12]. The electrical impedance of the piezoelectric material is very high and therefore the electrical output is due purely to areas that are covered by the metal electrode. The electrode can be etched away to produce shaped strips which define the area over which the film is sensitive. The quadratic sensitivity in the x-direction has been obtained by etching the electrode surface to produce a discrete number of strips whose width varies quadratically along the x-axis, as shown in Figure 9. The constant sensitivity is guaranteed if the strips have a sufficiently small width so that they are not affected by a large change in transverse displacement from one edge of the strip to the other. This assumption is generally valid if the wavelength of a structural bending mode is large compared to the width of the strip [12]. Therefore, for low bending wave-number, the "strip geometry" of the electrodes shown in Figure 9 approximates a quadratically weighted sensitivity in the x-direction and a constant sensitivity in the y-direction.

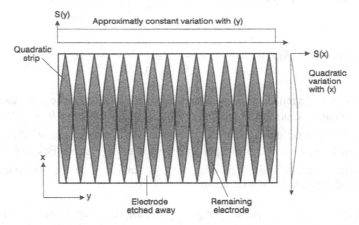

Figure 9. A rectangular plate with a set of quadratically shaped electrodes for the measurement of the plate volume velocity vibration component.

When a voltage is applied between the two electrodes, an electric field is generated across the piezoelectric material which then deforms because of the piezoelectric effect. When such a piezoelectric actuator is bonded to the surface of a plate the piezoelectric effect produces a force field on the surface of the panel, which, in general, will cause it to bend, twist and stretch. Using the formulation presented by Lee [25], the forced equations for the uncoupled transverse (equation 4.7 below) and in-plane (equations 4.8 and 4.9 below) vibration of the panel generated by a layer of piezoelectric material bonded on to one surface of the panel can be written as [26]

$$D_s\left(\frac{\partial^4 w}{\partial x^4}+2\frac{\partial^4 w}{\partial x^2\partial y^2}+\frac{\partial^4 w}{\partial y^4}\right)+m\frac{\partial^2 w(t)}{\partial t^2}=-hV(t)\left(e_{31}\frac{\partial^2 S(x,y)}{\partial x^2}+2e_{36}\frac{\partial^2 S(x,y)}{\partial x\partial y}+e_{32}\frac{\partial^2 S(x,y)}{\partial y^2}\right),\ (4.7)$$

$$\frac{2Y_sh_s}{1-v_s^2}\left(\frac{\partial^2 u}{\partial x^2}+\frac{1-v_s}{2}\frac{\partial^2 u}{\partial y^2}\right)+\frac{2Y_sh_s}{2(1-v_s)}\frac{\partial^2 v}{\partial x\partial y}-m\frac{\partial^2 u(t)}{\partial t^2}=\left\{e_{31}\frac{\partial S(x,y)}{\partial x}+e_{36}\frac{\partial S(x,y)}{\partial y}\right\}V(t),\ (4.8)$$

$$\frac{2Y_sh_s}{2(1-v_s)}\frac{\partial^2 u}{\partial y\partial x}+\frac{2Y_sh_s}{1-v_s^2}\left(\frac{1-v_s}{2}\frac{\partial^2 v}{\partial x^2}+\frac{\partial^2 v}{\partial y^2}\right)-m\frac{\partial^2 v(t)}{\partial t^2}=\left\{e_{36}\frac{\partial S(x,y)}{\partial x}+e_{31}\frac{\partial S(x,y)}{\partial y}\right\}V(t),\ (4.9)$$

where $V(t)$ is the voltage applied across the piezoelectric transducer, D_s and Y_s are the bending and axial stiffness of the panel and piezoelectric transducers and m is the mass per unit surface of the panel and piezoelectric transducers. The condition necessary to generate a transverse uniform force actuator is to have a piezoelectric material with constant sensitivity function along the y-direction and quadratic sensitivity along the x-direction as given by equation (4.5). Indeed, substituting the sensitivity function in equation (4.5) into equation (4.7) and considering the boundary conditions due to the clamping of the panel, the transverse excitation term is found to be [27]:

$$f_n(x,y,t)=\text{cost}=2\left(h_s+\frac{h_{pe}}{2}\right)e_{31}kV(t)\ .\qquad(4.10)$$

A piezoelectric actuator with constant sensitivity along the y-direction and quadratic sensitivity along the x-direction can be again obtained in practice by etching the electrode surface in such a way as to get a discrete number of strips whose width varies quadratically with x as shown in Figure 9.

The volume velocity sensor and the uniform force actuator described above could be used together to form a matched sensor–actuator pair. A piezoelectric film with quadratically shaped electrodes, as shown in Figure 9, could be bonded to each side of a panel for example, so that one film is used to measure the net volume velocity due to the bending vibration of the panel and the other film is used to exert a transverse uniform force over the panel surface. The actuator and sensor are connected such that the phase response for bending vibration is between $\pm 90^o$.

The ratio $G(\omega)=i(\omega)/V(\omega)$, output current per input voltage, of the sensor–actuator transducers can be expressed as the sum of two transfer functions, one related to the bending vibration and the other related to in-plane longitudinal and shear vibration of the panel, so that:

$$G(\omega)=G_b(\omega)+G_i(\omega)\ .\qquad(4.11)$$

Gardonio et al. [26,27] have derived the two frequency response functions components $G_b(\omega)$ and $G_i(\omega)$ due to the coupling of the sensor–actuator transducers via bending and in-plane vibrations of the panel, which are shown in Figure 10 (faint and dashed lines respectively) together with the total frequency response function $G(\omega)$ (solid line).

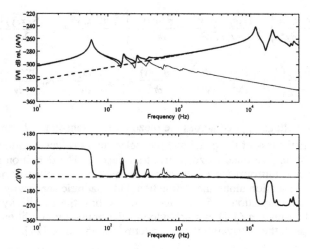

Figure 10. Simulation of the output current from the volume velocity sensor per unit voltage input to the force actuator; solid line both bending and axial vibration effects; dashed line: only axial vibration effects; faint line: only bending vibration effects.

This plot shows that below about 100 Hz the amplitude of the total frequency response function is dominated by the component related to bending vibration of the panel and the (1,1) bending mode resonance frequency at 59 Hz is clearly seen. Above 100 Hz, however, the component related to in-plane vibration of the panel becomes more important and hides the resonances due to the higher order bending modes detected by the volume velocity sensor. The $G_i(\omega)$ frequency response function component shows a constant rising trend, typical of a low frequency stiffness control response, until the first two resonance frequencies for in-plane vibration occurs at about 12 kHz and 21 kHz. The response in correspondence of these two resonance frequencies is about 20 dB higher than that in correspondence of the first bending resonance frequency.

The phase plot in Figure 10 shows a phase shift of $-180°$ at the resonance frequency of the first bending mode of the panel at 59 Hz and a second phase shift of about $-180°$ at the resonance frequencies of the uncoupled in-plane modes of the piezoelectric transducers. The phase shift recovers to about $-90°$ between 5 and 10 kHz, but at the resonance frequencies of 12 kHz and 21 kHz, due to the in-plane longitudinal and shear natural modes of the panel, another phase shift of $-180°$ is introduced.

The sensor–actuator frequency response function is therefore not strictly positive real. If a simple direct velocity feedback control scheme is used, then there is a clear stability problem caused by the extra $-180°$ phase shift at the resonance frequencies due to the uncoupled in-plane modes of the two piezoelectric films and at the resonance frequencies due to the in-plane modes of the panel. Although a compensator circuit could be considered to improve the stability of the control system, the increasing amplitude of the plant response due to in-plane coupling requires a relatively high order low pass characteristic to reduce its effects, which will inevitably lead to large phase shifts and potential stability problems at lower frequencies.

Several options have been considered to avoid the problem of in-plane coupling between the sensor and actuator transducers. A possible solution would be to use two piezoelectric films bonded on each side of the panel for both the volume velocity sensor and transverse uniform force transducers. If the output of the two piezoelectric films is subtracted then only the effect of bending vibration in the panel would be measured. Also, if the two films were driven out of phase, then only bending vibration would be generated on the panel. This type of arrangement would require a four layer smart panel with two piezoelectric films on each side, which however has shown some problems of local coupling between two adjacent piezoelectric films [28]. Alternatively a smart panel with one piezoelectric film on each side could be used provided the two films are operated simultaneously as sensors and actuators. This requires a signal conditioning system which is able to filter out the very little measured output signal from the input control excitation signals [29]. More recently scientists have started to consider the possibility of building a volume velocity sensor with a discrete grid of accelerometers which would then avoid the problem of measuring in-plane vibrations [30,31,32]. A smarter solution has been presented by [33], where they have discretised the distributed piezoelectric film into a shaped grid of small piezoelectric tails.

5. Multiple Local Feedback Loops

The positions of the actuators and sensors used in active vibroacoustic control systems are often chosen so that they can couple into the structural modes that dominate the vibration or the sound radiation [34,35], or detect the radiation modes of the structure, as described above. As the frequency of excitation increases, however, the detailed shape of the structural modes become increasingly sensitive to the boundary conditions and external loads on the structure and hence become more uncertain. It may thus be preferable to use a larger number of actuators and sensors than are strictly required, and arrange them in a regular array so that the structural modes are controlled whatever their shape.

There are considerable advantages in collocating the actuators and the sensors in such a feedback control system. When the actuator and sensor are also dual, in the sense that the product of the actuator input and the sensor response is proportional to the power supplied to the structure, [36], the plant response, from actuator input to sensor output, will have a positive real part, since the uncontrolled structure is passive. If a collocated force actuator and velocity sensor were used, for example, the plant response would be proportional to the input, or point, mobility of the structure, which must have a positive real part. The bandwidth over which this passivity property holds will, in practice, be limited by the dynamics of the transducers used. Provided the frequency response of the feedback controller also has a positive real part, the polar plot of the open loop frequency response function, i.e. the Nyquist plot, must stay in the right-hand half of the complex plane and so the system is unconditionally stable, since the polar plot cannot encircle the Nyquist point. The generalisation of this simple passivity property to multichannel systems is discussed in [17,37], where it is shown that if the collocated actuators and sensors are coupled only by local feedback control loops with positive feedback gains, then the controller is passive and stability is assured for a passive plant. Such an array of locally acting feedback loops is referred to as a decentralised control system. It may also be possible to economically implement such an array of integrated transducers using micro electro-mechanical systems (MEMS) technology. In this section we investigate the consequences of a decentralised feedback control strategy, which uses a set of 16 collocated actuators and sensors on a panel, when the panel is subject to an incident acoustic excitation [37].

The objective is to investigate the effect of this control strategy on the sound radiated by the panel. It will be assumed that each of the sensors measures the panel velocity at the corresponding point, which could be achieved in practice by integrating the output of a small accelerometer for example. Although force actuators have useful theoretical properties, as discussed above, they require an inertial base to react off. It would be more useful in practice to have actuators, that are fully integrated with the panel. In this section the results will be described of using a system with velocity sensors and small strain actuators directly underneath them, such as could be implemented using piezoelectric devices. The similarity between the behaviour of the "ideal" force actuator and velocity sensor control system has been compared with that of the "practical" control system with piezoelectric actuators and velocity sensors by Elliott et al. [37].

Although we can theoretically guarantee the unconditional stability of the sixteen channel decentralised feedback control system in the case of point force actuators and velocity sensors, Balas [17], since this system is passive, Joshi [38], the stability range of this controller must be determined when piezoelectric actuators are used before the control performance can be calculated.

A uniform array of 4×4 actuators and sensors was modelled on a 278×247mm×1mm panel excited by an incident acoustic wave and control systems were investigated for which each pair of the 16 individual actuators and sensors were connected in sixteen control loops, as illustrated in Figure 11 for the case of piezoelectric actuators. The transducers were uniformly arranged on the panel so that their centres were $\frac{1}{8}$, $\frac{3}{8}$, $\frac{5}{8}$ and $\frac{7}{8}$ of the plate length and breadth away from the edges.

Figure 12 shows the ratio of the sound power radiated on one side of the panel to the incident sound power due to the plane wave excitation on the other side, which is termed the sound transmission ratio, T. Before control only the modes whose modal

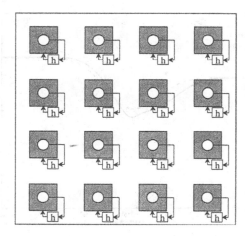

Figure 11. Arrangement of 16 piezoelectric actuators, as shown by the squares, driven locally by the output of 16 velocity sensors, as sown by circles, via individual control loops with gain of *h*.

integers are both odd radiate sound significantly at low frequencies and also anti-resonances appear, due to destructive interference between the sound pressures radiated by adjacent odd-odd modes. As the feedback gains are increased, the sound radiation at the dominant modes is suppressed to a greater and greater extent, but at very high values of the feedback gain the panel is pinned at the sensor locations by the control system and new system resonances begin to appear. The new resonance at about 200 Hz has the greatest prominence, since its velocity distribution has the greatest net volume velocity.

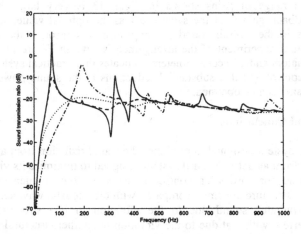

Figure 12. Sound transmission ratio of the plane wave-excited panel with no control (solid line) and with a 16 channel decentralized feedback controller with feedback gains of 1 (dashed line), 10 (dotted line), and 100 (dot-dashed line).

78

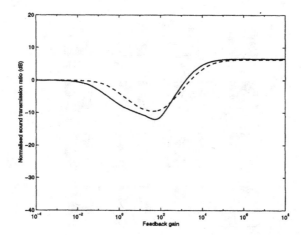

Figure 13. Normalized sound transmission ratio level, integrated from 0 Hz to 1 kHz, plotted against the gain in the decentralized feedback controller, h, for the force actuators (solid lie) and the piezoelectric actuators (dashed line).

The variation of the sound transmission ratio, integrated up to 1 kHz, with the feedback gain has been plotted in Figure 13, together with this variation for force actuators [37]. A clear minimum in the integrated kinetic energy is again seen for one value of feedback gain, for which the overall attenuation is about 10 dB. This value of feedback gain is about the same as it was when using the force actuators, although it should be noted that the piezoelectric responses have been scaled to be similar to force responses at the first resonance [37]. At higher values of feedback gain the sound transmission ratio begins to increase and for very high feedback gains the sound transmission is slightly greater with control than without it, due to the effect of the new system resonances. The range of gains shown in Figure 13 is very large, however, and provided the feedback gain is of the same order as the optimal value, significant and useful reductions in the overall sound power transmitted through the panel can be obtained. An initial experiment of the arrangement shown in Figure 11 [39], with 16 piezoelectric actuators and 16 accelerometers, indicates that practical systems are stable with only local control and that substantial reductions in the sound power transmitted through such a panel can be obtained.

6. Summary and Conclusions

An active control system designed to minimise the sound radiation from a structure can be very different from an active control system designed to minimise its vibration. This is particularly true for low-order control systems at low frequencies, when the dimensions of the structure are small compared with the acoustic wavelength. At these frequencies the radiated sound due to the vibration in one structural mode can destructively interfere with that due to the vibration in another structural mode, and so reducing the vibration by actively controlling only one structural mode can actually increase the structure's sound radiation. One has to be a little cautious about such observations, however, since such destructive interference mainly occurs in the far-field pressure. If the sound field close to the structure is of concern then it is the near-field pressure which is important and this is dominated by the local velocity of the structure, which, on average, will depend on the sum of squared amplitudes of all the structural modes. The coupling between the far-field sound radiation from the structural modes can be accounted for by transforming these modes into a set of velocity distributions that radiate sound independently, which are called the radiation modes.

At very low frequencies it is only the lowest order radiation mode which is a significant radiator of sound, and the amplitude of this lowest-order radiation mode corresponds to the net volume velocity of the vibration. Various distributed sensors have been designed to specifically measure the volume velocity, generally using spatially-weighted strain sensors such as those made of piezoplastic, such as PVDF. If such a transducer was used as an actuator, it would generate a uniform force over the structure, provided the structure were constrained not to move at the edges, and the combination of such an actuator and a volume velocity sensor would appear to be an ideal pair of transducers for the active control of low frequency sound radiation, since only a single channel feedback controller would then be required. Unfortunately, in practice, such strain actuators generate in-plane motion on a structure such as a panel, as well as the out-of-plane motion which generates sound. This in-plane motion is detected by a distributed strain sensor and can provide the dominant mechanism of coupling at higher frequencies so that the amplitude of the response between actuator and sensor is high and additional phase shift is introduced. Both of these effects make it difficult to design an efficient feedback controller using such a pair of transducers.

Another disadvantage of strain sensors is that they are unable to detect whole-body motion of the structure, which may be making an important contribution to its sound radiation. An alternative method of estimating the volume velocity, which will detect whole-body vibration, is to electrically integrate the summed output of an array of inertial accelerometers. Practical active control systems using such an array of discrete sensors with a PVDF uniform-force actuator can, however, lack the authority to control realistic vibration levels in a number of applications. Piezoceramic actuators, such as PZT, are able to couple into many structures with much greater authority than piezoplastics such as PVDF, but are too brittle to use as continuously distributed actuators. An array of such PZT transducers could, however, be driven in such a way as to approximate a given continuously-defined excitation.

If one had implemented an array of PZT actuators and an array of inertial velocity sensors, the interesting question then arises as to whether adding the sensor outputs together and connecting them to a weighted array of actuators via a single channel controller is the most efficient or robust method of implementing an active control system.

Classical control theory could be used to design a fully-coupled multichannel feedback control system to connect such an array of sensors to such an array of actuators, using LQG or H. formulations for example. The design of such controllers is, however, dependent on having an accurate model of the structure under control, and the dynamics of many real structures varies considerably with operating conditions. The natural frequency of the first structural mode of a fuselage panel on an aircraft, for example, may vary by a factor of four as the aircraft is pressurised during normal flight. It is very difficult to ensure good performance with a conventional centralised controller while making it robust to such uncertainties in the plant response.

An alternative approach is to reduce the phase shift between individual pairs of actuators and sensors by collocating them, and to implement simple constant-gain feedback controllers locally between each pair of transducers. Such a decentralised control strategy can be shown to be unconditionally stable if ideal force actuators and velocity sensors are used, and is found in practice to also be remarkably robust when applied to collocated arrays of piezoceramic actuators and inertial velocity sensors. Initial simulations and experiments suggest that an array of such locally controlled systems, when set up with an appropriate feedback gain, can give substantial reductions in sound radiation. This is achieved by using far more control channels than is strictly necessary from the point of view of the radiation modes, but has the advantage of always suppressing the vibration of a structure, and hence its near-field pressure, as well as its far-field sound radiation. There may also be considerable advantages to making the actuator, sensor and controller as an integrated and mass-produced module, arrays of which can then be incorporated into the manufacture of larger structures.

There is still considerable debate about the best strategy for implementing active vibroacoustic systems, in terms of which transducers and which control strategy to use. To some extent this will depend on the individual application and the frequency at which control is required. Two extreme examples would be single-channel systems with fully-distributed actuators and sensors controlling only volume velocity, and fully-coupled multi-channel systems with large numbers of discrete actuators and sensors minimising the amplitudes of the radiation modes. An interesting intermediate solution may be the use of multiple local controllers with large numbers of discrete actuators and sensors, whose overall effect is to reduce sound radiation by controlling the velocity at a number of points on the structure.

7. References

1. Fuller C.F., Elliott S.J. and Nelson P.A. (1996) *Active Control of Vibration*, Academic Press.
2. Elliott S.J. (2001) *Signal Processing for Active Control*, Academic Press.
3. Wallace C.E. (1972) Radiation resistance of a rectangular panel. *Journal of the Acoustical Society of America*, **51**, 946-952.
4. Lee C.K. and Moon F.C. (1990) Modal sensors/actuators. *American Society of Mechanical Engineers Journal of Applied Mechanics*, **57**, 434-441.
5. Borgiotti G.V. (1990) The power radiated by a vibrating body in an acoustic fluid and its determination from boundary measurements. *Journal of the Acoustical Society of America*, **88**, 1884-1893.
6. Photiadis D.M. (1990) The relationship of singular value decomposition to wave-vector filtering in sound radiation problems. *Journal of the Acoustical Society of America*, **88**, 1152-1159.
7. Cunefare K.A. (1991) The minimum multimodal radiation efficiency of baffled finite beams. *Journal of the Acoustical Society of America*, **90**, 2521-2529.

8. Baumann W.T., Saunders W.R. and Robertshaw H.H. (1991) Active suppression of acoustic radiation from impulsively excited structures. *Journal of the Acoustical Society of America*, **88**, 3202-3208.
9. Elliott S.J. and Johnson M.E. (1993) Radiation modes and the active control of sound power. *Journal of the Acoustical Society of America*, **94**, 2194-2204.
10. Rex J. and Elliott S.J. (1992) The QWSIS – a new sensor for structural radiation control. *Proceedings of the International Conference on Motion and Vibration Control (MOVIC)*, Yokohama, 339-343.
11. Guigou C. and Berry A. (1993) Design strategy for PVDF sensors in the active control of simply supported plates. *Internal Report*, GAUS, Dept. of Mechanical Engineering, Sherbrooke University.
12. Johnson M.E. and Elliott S.J. (1993) Volume velocity sensors for active control. *Proceedings of the Institute of Acoustics*, **15**(3), 411-420.
13. Johnson M.E. and Elliott S.J. (1995) Active control of sound radiation using volume velocity cancellation. *Journal of the Acoustical Society of America*, **98**, 2174-2186.
14. Hansen C.H. and Snyder S.D. (1997) *Active Control of Noise and Vibration*, E & FN Spon.
15. Clark R.L. and Fuller C.R. (1992) Experiments on the active control of structurally radiated sound using multiple piezoceramic actuators. *Journal of the Acoustical Society of America*, **91**(6), 3313-3320.
16. Fuller C.R. (1991) Active control of sound transmission/radiation from elastic plates by vibration inputs: I. analysis. *Journal of Sound and Vibration*, **136**(1), 1-15.
17. Balas M.J. (1979) Direct velocity feedback control of large space structures. *Journal Guidance Control*, **2**, 232-253.
18. Burke S.E., Hubbard J.E. and Meyer J.E. (1993) Distributed transducers and collocation. *Journal of Mechanical System Signal Processing*, **7**(4), 349-361.
19. Clark R.L. and Fuller C.R. (1992) Modal sensing of efficient acoustic radiators with polyvinylidene floride distributed sensors in active structural acoustic control approaches. *Journal of the Acoustical Society of America*, **91**(6), 3321-3329.
20. Clark R.L., Burdisso R.A. and Fuller C.R. (1993) Design approaches for shaping PVDF sensors in active structural acoustic control. *Journal Intelligent Materials Systems \Structures*, **4**, 3541-365.
21. Carey D.M. and Stulen F.B. (1993) Experiments with a two-dimensional multi-modal sensor. *Proceedings of the International Conference on Recent Advances in Active Control of Sound and Vibration*, Blacksburg, Virginia, USA, 41-52.
22. Gu Y., Clark R.L., Fuller C.R. and Zander A.C. (1994) Experiments on active control of plate vibration using piezoelectric actuators and *polyvinylidene floride (PVDF) modal sensors, ASME* Journal of Vibration and Acoustics **116**, 303-308.
23. Charette F., Guigou C. and Berry A. (1995) Development of volume velocity sensors for plates using PVDF film, *Proceedings of ACTIVE95*, Newport Beach, California, USA, 241-252.
24. Clark R.L., Saunders W.R. and Gibbs G.P. (1998) *Adaptive Structures*, John Wiley & Sons, Inc., New York.
25. Lee C.K., (1990) Theory of laminated piezoelectric plates for the design of distributed sensor/actuators. Part I: governing equations and reciprocal relationships, *Journal of the Acoustical Society of America*, **87**(3), 1144-1158.

26. Gardonio P., Lee Y.-S., Elliott S.J. and Debost S. (2001) A panel with matched polyvinylidene fluoride volume velocity sensor and uniform force actuator for the active control of sound transmission. *Proceedings of the Institution of Mechanical Engineers*, **215,** Part G, 187-206.

27. Gardonio P., Lee Y.-S., Elliott S.J. and Debost S. (2001) A panel with matched polyvinylidene fluoride volume velocity sensor and uniform force actuator for the active control of sound transmission. *Journal of the Acoustical Society of America*, **110**(6), 3025-3031.

28. Lee Y.S., Gardonio P. and Elliott S.J. (2001) Distributed four-layer PVDF sensor and actuator arrangement for the control of beam motion. *Proceedings of SPIE's 8th Annual International Symposium on Smart Structures and Material*, 284-294.

29. Cole D.G. and Clark R.L. (1994) Adaptive compensation of piezoelectric sensori-actuators, *Journal of Intelligent Material Systems and Structures*, **5,** 665-672.

30. Gibbs G.P., Clark R.L., Cox D.E. and Vipperman J.S (2000) Radiation modal expansion: application to active structural acoustic control. *Journal of the Acoustical Society of America*, **107**(1), 332-339.

31. Henrioulle K and Sas P. (2001) A PVDF sensor/actuator pair for the active control of sound transmission. *Proceedings of Internoise 2001, The Hague, The Netherlands*, 284-294.

32. Lee Y-S Gardonio P. and Elliott S.J. (2002) Volume velocity vibration control of a smart panel using an arrangement of a quadratically shaped PVDF actuator and multiple accelerometers. *Proceedings of the Institute of Acoustics Spring Conference 2002, Salford, UK*

33. Francois A., De Man P. and Preumont A. (2001) Piezoelectic array sensing of volume displacement: a hardware demonstration. *of Sound and Vibration*, **244**(3), 395-405.

34. Meirovitch L. (1990) *Dynamics and Control of Structures*. John Wiley and Sons.

35. Preumont A. (1997) *Vibration Control of Active Structures*. Klawer Academic Publishers.

36. Sun J.Q. (1996) "Some observations on physical duality and collocation of structural control sensors and actuators". *Journal of Sound and Vibration*, **194,** 765-770.

37. Elliott S.J., Gardonio P., Sors T.C. and Brennan M.J. (2001) Active vibro-acoustic control with multiple feedback loops, *Journal of the Acoustical Society of America*, **111**(2), 908-915.

38. Joshi S.M. (1989) *Control of Large Flexible Space Structures*. Springer Verlag.

39. Bianchi E., Gardonio P. and Elliott S.J. (2002) Smart panel with decentralised units for the control of sound transmission. Part III Control system implementation, *Proceedings of ACTIVE2002*, 499-510.

ACTUATORS AND SENSORS IN STRUCTURAL DYNAMICS

W. GAWRONSKI
Jet Propulsion Laboratory
California Institute of Technology
Pasadena, CA 91109
U.S.A.

1. Introduction

Flexible structure dynamics depend on the location and gains of actuator and sensors. This fact is often underestimated, although it is an important factor in the planning of structural dynamic tests, and in designing structural controllers. In this work we describe the impact of actuators and sensors gains and locations on structural properties, which includes structural controllability and observability, and structural and modal norms. Using these properties we show how to detect a damage of structural members, how to place actuators and sensors for structural testing and control, and how to tune actuators or sensors to excite a selected mode or a set of selected modes. The reader can find background material in my book [5]. The analysis is conducted in modal state-space coordinates, which are described at the beginning of this paper.

2. Modal State-Space Representation

In this section both the standard and generalized state-space representations of a structure are discussed. Both representations are presented in modal coordinates.

2.1. STANDARD STATE-SPACE REPRESENTATION

Models of a linear time-invariant system are described in a standard form called state space representation, which is of the following form

$$\dot{x} = Ax + Bu$$
$$y = Cx \tag{1}$$

In the above equations the N-dimensional vector x is called the state vector, the s-dimensional vector u is the system input, and the r-dimensional vector y is the system output. The A, B, and C matrices are real constant matrices of appropriate dimensions (A is $N{\times}N$, B is $N{\times}s$, and C is $r{\times}N$).

A structural model, however, is typically represented by the well-known second-order model. Let n_d be a number of degrees of freedom of the system, let r be a number of

83

A. Preumont (ed.), Responsive Systems for Active Vibration Control, 83–132.

outputs, and let s be a number of inputs. A flexible structure in nodal coordinates is represented by the following equation:

$$M\ddot{q} + D\dot{q} + Kq = B_o u$$
$$y = C_{oq}q + C_{ov}\dot{q} \tag{2}$$

In this equation q is the $n_d \times 1$ displacement vector; u is the $s \times 1$ input vector, y is the output vector, $r \times 1$; M is the mass matrix, $n_d \times n_d$, D is the damping matrix, $n_d \times n_d$, and K is the stiffness matrix, $n_d \times n_d$. The input matrix B_o is $n_d \times s$, the displacement output matrix C_{oq} is $r \times n_d$, and the velocity output matrix C_{ov} is $r \times n_d$. The mass matrix is positive definite, and the stiffness and damping matrices are positive semidefinite. On details of the derivation of these types of equations see [14], and [7].

The same equation can be represented in modal coordinates. Define the matrix of mode shapes (or modal matrix) Φ of dimensions $n_d \times n$, which consists of n natural modes of a structure

$$\Phi = \begin{bmatrix} \phi_1 & \phi_2 & \cdots & \phi_n \end{bmatrix} \tag{3}$$

where ϕ_i, $i = 1,...,n$, is the ith modal vector. The modal matrix diagonalizes mass and stiffness matrices M and K, namely

$$M_m = \Phi^T M \Phi, \quad K_m = \Phi^T K \Phi \tag{4}$$

The matrices M_m and K_m are diagonal. The matrix M_m is called modal mass matrix, and K_m is modal stiffness matrix.

The same transformation can be applied to the damping matrix

$$D_m = \Phi^T D \Phi \tag{5}$$

where D_m is the modal damping matrix. This matrix is not always obtained diagonal. A damping matrix that can be diagonalized by the above transformation is called a matrix of proportional damping. For example, a linear combination of the stiffness and mass matrices, $D = \alpha_1 K + \alpha_2 M$, produces a proportional damping matrix, see [2], [7] (α_1 and α_2 are non-negative scalars).

The second-order structural model (2) can be also expressed in modal coordinates by introducing a new variable, q_m, called a modal displacement, such that

$$q = \Phi q_m \tag{6}$$

After some manipulations (see [5]) Eq.(2) is in the form

$$\ddot{q}_m + 2Z\Omega\dot{q}_m + \Omega^2 q_m = B_m u$$
$$y = C_{mq} q_m + C_{mv} \dot{q}_m \qquad (7)$$

In the above equation $\Omega = diag(\omega_1, \omega_2, \ldots, \omega_n)$ is the matrix of natural frequencies, Z is the modal damping matrix, $Z = diag(\zeta_1, \zeta_2, \ldots, \zeta_n)$, where ζ_i is the damping of the ith mode. The modal input matrix B_m is obtained as

$$B_m = M_m^{-1} \Phi^T B_o \qquad (8)$$

and in Eq.(7) we used the following modal displacement and rate matrices

$$C_{mq} = C_{oq} \Phi, \quad C_{mv} = C_{ov} \Phi \qquad (9)$$

Note that equation (7) is a set of uncoupled equations, since the matrices Ω and Z are diagonal. Thus, this set of equations can be re-written as follows

$$\ddot{q}_{mi} + 2\zeta_i \omega_i \dot{q}_{mi} + \omega_i^2 q_{mi} = b_{mi} u$$
$$y_i = c_{mqi} q_{mi} + c_{mvi} \dot{q}_{mi} \qquad (10)$$

$i = 1, \ldots, n$, where $y = \sum_{i=1}^{n} y_i$, b_{mi} is the ith row of B_m, and c_{mqi}, c_{mvi} are the ith columns of C_{mq} and C_{mv}, respectively. In the above equations y_i is the system output due to the ith mode dynamics. Note that the structural response y is a sum of modal responses y_i, which is a key property used to derive structural properties in modal coordinates.

Based on Eq.(10), the modal state-space representation (A_m, B_m, C_m) of a structure can be obtained, see [5]. It is characterized by the block-diagonal state matrix, A_m

$$A_m = \mathrm{diag}(A_{mi}) = \begin{bmatrix} \bullet & \bullet & 0 & 0 & \cdots & \cdots & 0 & 0 \\ \bullet & \bullet & 0 & 0 & \cdots & \cdots & 0 & 0 \\ 0 & 0 & \bullet & \bullet & \cdots & \cdots & 0 & 0 \\ 0 & 0 & \bullet & \bullet & \cdots & \cdots & 0 & 0 \\ \cdots & \cdots & \cdots & \cdots & \cdots & \cdots & \cdots & \cdots \\ \cdots & \cdots & \cdots & \cdots & \cdots & \cdots & \cdots & \cdots \\ 0 & 0 & 0 & 0 & \cdots & \cdots & \bullet & \bullet \\ 0 & 0 & 0 & 0 & \cdots & \cdots & \bullet & \bullet \end{bmatrix} \qquad (11)$$

$i = 1, 2, \ldots, n$, where A_{mi} are 2×2 blocks (with non-zero elements marked by $\bullet$). The modal input and output matrices are divided accordingly

$$B_m = \begin{bmatrix} B_{m1} \\ B_{m2} \\ \vdots \\ B_{mn} \end{bmatrix}, \qquad C_m = \begin{bmatrix} C_{m1} & C_{m2} & \cdots & C_{mn} \end{bmatrix} \qquad (12)$$

where B_{mi} and C_{mi} are $2 \times s$, and $r \times 2$ blocks, respectively.

The state x of the modal representation consists of n independent components, x_i, i.e., $x^T = \{x_1^T, \ x_2^T, \ \ldots \ x_n^T\}$. The ith component represents the ith mode, and consists of two states $x_i^T = \{x_{i1} \ \ x_{i2}\}$. The ith component, or mode, has its own state-space representation (A_{mi}, B_{mi}, C_{mi}), and is independently obtained from the following state equations

$$\dot{x}_i = A_{mi} x_i + B_{mi} u$$
$$y_i = C_{mi} x_i \qquad (13)$$

and $y = \sum_{i=1}^{n} y_i$.

Two modal representations can be distinguished. The first one, called modal model 1, has the component defined as follows

$$x_i = \begin{Bmatrix} \omega_i q_{mi} \\ \dot{q}_{mi} \end{Bmatrix}, \qquad (14)$$

while the second one (modal model 2) as follows

$$x_i = \begin{Bmatrix} \omega_i q_{mi} \\ \zeta_i \omega_i q_{mi} + \dot{q}_{mi} \end{Bmatrix} \qquad (15)$$

where q_{mi} and $\dot{q}_{mi}$ are the ith modal displacement and modal velocity.

The following state-space representation (A_{mi}, B_{mi}, C_{mi}) correspond to the above models

• modal model 1

$$A_{mi} = \begin{bmatrix} 0 & \omega_i \\ -\omega_i & -2\zeta_i \omega_i \end{bmatrix}, \qquad B_{mi} = \begin{bmatrix} 0 \\ b_{mi} \end{bmatrix}, \qquad C_{mi} = \begin{bmatrix} \dfrac{c_{mqi}}{\omega_i} & c_{mvi} \end{bmatrix} \qquad (16)$$

• modal model 2

$$A_{mi} = \begin{bmatrix} -\zeta_i\omega_i & \omega_i \\ -\omega_i & -\zeta_i\omega_i \end{bmatrix}, \quad B_{mi} = \begin{bmatrix} 0 \\ b_{mi} \end{bmatrix}, \quad C_{mi} = \begin{bmatrix} \dfrac{c_{mqi}}{\omega_i} - c_{mvi}\zeta_i & c_{mvi} \end{bmatrix} \quad (17)$$

2.2. GENERALIZED STATE-SPACE REPRESENTATION

Here we consider a more complex model, which is used to describe a system under testing or control. It is called a generalized model, and consists of two *kinds* of inputs, denoted u and w, and of two *kinds* of outputs, denoted y and z, see Fig.1.

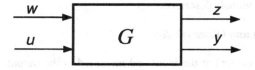

Figure 1. The generalized model consists of two inputs (u – actuator and w – disturbance) and two outputs (y – sensor, and z – performance)

The inputs to the generalized model consist of two vector signals:
- The actuator input, denoted u, which consists of all inputs handled by the controller, or applied as test inputs.
- The disturbance vector, w, which consists of noises and disturbances that are not part of the controller action or are not a part of the test input.

The outputs of the generalized model consist of two vector signals:
- The sensor vector, y: it is the controller feedback signal or the measured test signal.
- The performance vector, z: the signal that represents closed-loop performance or test performance.

The input and output selection in the generalized model are listed in Table 1.

TABLE 1. Inputs and outputs of a generalized model of a structure under testing or control

		Test	Control
Input	u	Test inputs (actuator signals)	Control inputs (produced by controller)
	w	Disturbances, all uncontrolled inputs	Disturbances and commands
Output	y	Measured variables (sensor signals)	Feedback signals accessed by controller
	z	Sensor noise, non-measurable performance outputs	Sensor noise, non-measurable performance outputs

Let A be the state matrix of the system, B_u and B_w are the input matrices of u and w, respectively, and C_y and C_z are the output matrices of y and z, respectively; then the state space representation of the generalized model is as follows

$$\dot{x} = Ax + B_u u + B_w w$$
$$y = C_y x, \qquad\qquad (18)$$
$$z = C_z z$$

This model is transformed into the modal state-space representation the same way as the standard model previously described.

3. Controllability and Observability

System dynamics, excited at the input and measured at the output, are described by the state variables. However, the input may not be able to excite all states; consequently, it cannot fully control the system. Also, not all states may be represented at the output i.e., state dynamics may not be fully observed. Based on these two observations, we call a system controllable if the input excites all states and we call it observable if all the states are represented in the output

3.1. GRAMMIANS

We use grammians to evaluate the system controllability and observability properties. They are defined as follows (see, for example, [9], and [15])

$$W_c = \int_0^\infty e^{A\tau} BB^T e^{A^T\tau} d\tau, \quad W_o = \int_0^\infty e^{A^T\tau} C^T C e^{A\tau} d\tau \qquad (19)$$

and are obtained from the Lyapunov equations:

$$AW_c + W_c A^T + BB^T = 0, \quad A^T W_o + W_o A + C^T C = 0 \qquad (20)$$

For stable A, the obtained grammians W_c and W_o are positive definite.

The eigenvalues of the grammians change during the coordinate transformation. However, the eigenvalues of the grammian product are invariant. These invariants are denoted γ_i,

$$\gamma_i = \sqrt{\lambda_i(W_c W_o)}, \qquad\qquad (21)$$

where $i = 1, ..., N$, and are called the Hankel singular values of the system.

3.2. BALANCED REPRESENTATION – WHERE ACTUATORS AND SENSORS ARE EQUALLY IMPORTANT

A system is balanced when its controllability and observability grammians are equal and diagonal (see [12]) i.e., when

$$W_c = W_o = \Gamma \tag{22}$$

where $\Gamma = \text{diag}(\gamma_1, ..., \gamma_N)$, and $\gamma_i \geq 0$, $i = 1, ..., N$. The diagonal entries γ_i are called Hankel singular values of the system (earlier introduced as eigenvalues of product of the controllability and observability grammians). For a matrix that transforms a system into a balanced representation see [5].

The diagonality means that each state has an independent measure of controllability and observability. The equality means that each state is equally controllable and observable. The diagonality of grammians allows to evaluate each state (or mode) separately, and to determine how important they are for testing and for control purposes. Indeed, if a state is weakly controllable and, at the same time, weakly observable, it can be neglected without impacting the accuracy of analysis, dynamic testing, or control design procedures. On the other hand, if a state is strongly controllable and strongly observable, it must be retained in the system model in order to preserve accuracy of analysis, test, or control system design.

3.3. INPUT AND OUTPUT GAINS AS AN ALTERNATIVE MEASURE OF CONTROLLABILITY AND OBSERVABILITY

Consider the input and output matrices in modal coordinates, as in Eq.(12). Their two-norms, $\|B_m\|_2$ and $\|C_m\|_2$, are called the input and output gains of the structure

$$\|B_m\|_2 = \left(tr(B_m B_m^T)\right)^{\frac{1}{2}}, \quad \|C_m\|_2 = \left(tr(C_{mq}\Omega^{-2}C_{mq}^T + C_{mv}C_{mv}^T)\right)^{\frac{1}{2}} \tag{23}$$

They contain information on structural controllability and observability. Additionally, each mode has its own gain, namely $\|B_{mi}\|_2$, which is the input gain of the ith mode, and $\|C_{mi}\|_2$, which is the output gain of the ith mode, where B_{mi} and C_{mi} are given in modal coordinates, as in (16) and (17); thus, by definition

$$\|B_{mi}\|_2 = \left(tr(B_{mi}B_{mi}^T)\right)^{\frac{1}{2}} = \left(tr(b_{mi}b_{mi}^T)\right)^{\frac{1}{2}} \tag{24}$$

$$\|C_{mi}\|_2 = \left(tr(C_{mi}C_{mi}^T)\right)^{\frac{1}{2}} = \left(tr(\frac{C_{mqi}c_{mqi}^T}{\omega_i^2} + c_{mvi}c_{mvi}^T)\right)^{\frac{1}{2}} \tag{25}$$

It is easy to show that the gains of a structure are the root-mean-square sum of the modal gains

$$\|B_m\|_2 = \sqrt{\sum_{i=1}^n \|B_{mi}\|_2^2} \quad \text{and} \quad \|C_m\|_2 = \sqrt{\sum_{i=1}^n \|C_{mi}\|_2^2} \tag{26}$$

3.4. CONTROLLABILITY AND OBSERVABILITY OF A STRUCTURAL MODEL

In the following approximate relationships are used and denoted with the approximate equality sign "$\cong$". They are applied in the following sense: two variables, x and y, are approximately equal ($x \cong y$) if $x = y + \varepsilon$ and $\|\varepsilon\| \square \|y\|$.

Assuming small damping, the grammians in modal coordinates are diagonally dominant. This is expressed in the following property:

$$W_c \cong \text{diag}(w_{ci}I_2), \quad W_o \cong \text{diag}(w_{oi}I_2) \tag{27}$$

$i = 1,...,n$, where $w_{ci} > 0$ and $w_{oi} > 0$ are the modal controllability and observability coefficients. Using this property the approximate Hankel singular values are obtained as a geometric mean of the modal controllability and observability coefficients

$$\gamma_i \cong \sqrt{w_{cii}w_{oii}} \tag{28}$$

The profiles of grammians and system matrix A in modal coordinates are drawn in Fig. 2.

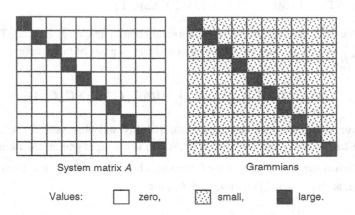

System matrix A — Grammians

Values: ☐ zero, ▨ small, ■ large.

Figure 2. Profiles of the system matrix A (diagonal) and the grammians (diagonally dominant) in the modal coordinates.

Next, we express the grammians of each mode in a closed form. Let B_{mi} and C_{mi} be the $2 \times s$ and $r \times 2$ blocks of B_m and C_m, then the diagonal entries of the controllability and observability grammians are as follows (for the derivation see [5]):

$$w_{ci} \cong \frac{\|B_{mi}\|_2^2}{4\zeta_i\omega_i}, \qquad w_{oi} \cong \frac{\|C_{mi}\|_2^2}{4\zeta_i\omega_i}, \qquad (29)$$

Note from (29) that for a balanced structure the modal input and output gains are approximately equal

$$\|B_{mi}\| \cong \|C_{mi}\| \qquad (30)$$

The approximate Hankel singular values are obtained from

$$\gamma_i \cong \frac{\|B_{mi}\|_2 \|C_{mi}\|_2}{4\zeta_i\omega_i} \qquad (31)$$

Example 1. Comparing the exact and approximate Hankel singular values of the International Space Station structure. The input is a force at node marked by a white dot at the top of Fig. 3, and the output is a rate at this node. The results in Fig. 4 show good coincidence between the exact and approximate Hankel singular values.

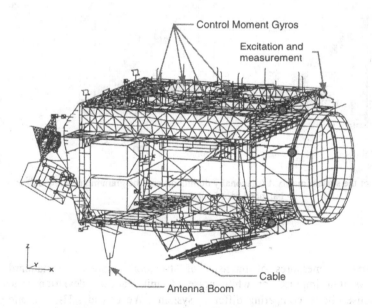

Figure 3. The finite element model of the International Space Station structure.

The closeness of the balanced and modal representations can be also observed in the closeness of the system matrix A in both representations. It was shown already that the matrix A in modal coordinates is diagonal (with 2×2 block on the diagonal). It can also be shown, see [5], that the system matrix A in balanced coordinates is diagonally dominant with 2×2 blocks on the diagonal, while B, and C are divided into $2 \times s$ and $r \times 2$ blocks. The profiles of the grammians and a system matrix A are drawn in Fig. 5.

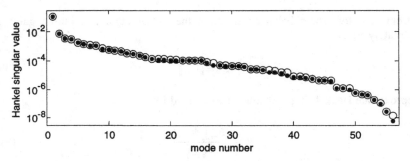

Figure 4. Exact ($\searrow$) and approximate ($\bullet$) Hankel singular values for the International Space Station structure are almost identical.

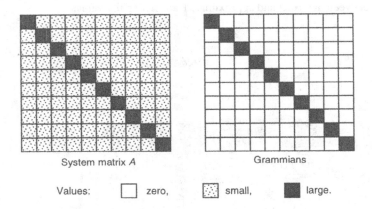

Figure 5. Profiles of the system matrix A (diagonally dominant) and the grammians (diagonal) in balanced coordinates.

4. Norms

System norms serve as measures of intensity of structural response to standard excitations, such as unit impulse, or white noise of unit standard deviation. The standardized response allows comparing different systems. We consider H_2, H_∞, and Hankel norms. It is shown that for flexible structures the H_2 norm has an additive property: it is a root-mean-square (rms) sum of the norms of individual modes. The H_∞ and Hankel norms are also determined from the corresponding modal norms, by selecting the largest one. On the other hand, all three norms of a mode with multiple inputs (or outputs) can be decomposed into the rms sum of norms of a mode with a single input (or output). These two properties allow for the development of unique and efficient model reduction methods and actuator/sensor placement procedures.

4.1. DEFINITIONS OF SYSTEM NORMS

Let (A, B, C) be a state-space representation of a linear system and let $G(\omega) = C(j\omega I - A)^{-1} B$ be its transfer function. The H_2 norm of the system is defined as

$$\|G\|_2^2 = \frac{1}{2\pi} \int_{-\infty}^{\infty} \text{tr}(G^*(\omega)G(\omega))d\omega \qquad (32)$$

It can be interpreted as root-mean-square response of the system, performed over all the elements of the matrix transfer function and over all frequencies. A convenient way to determine the H_2 norm is to use the following formulas

$$\|G\|_2 = \sqrt{\text{tr}(C^T C W_c)}, \quad \text{or,} \quad \|G\|_2 = \sqrt{\text{tr}(BB^T W_o)} \qquad (33)$$

where W_c and W_o are the controllability and observability grammians.

The H_∞ norm is defined as

$$\|G\|_\infty = \max_\omega \sigma_{\max}(G(\omega)) \qquad (34)$$

where $\sigma_{\max}(G(\omega))$ is the largest singular value of $G(\omega)$. The H_∞ norm of a single-input–single-output system is the peak of the transfer function magnitude. It can be computed as a maximal value of ρ, such that the solution S of the following algebraic Riccati equation is positive definite, see [17], p.238

$$A^T S + SA + \rho^{-2} SBB^T S + C^T C = 0 \qquad (35)$$

It is an iterative procedure where one starts with a large value of ρ and reduces it until negative eigenvalues of S appear.

The Hankel norm of a system is a measure of the effect of its past input on its future output [1, p. 103]. It is defined as

$$\|G\|_h = \sup_u \frac{\|y(t)\|_2}{\|u(t)\|_2}, \quad \text{where} \quad \begin{cases} u(t) = 0 & \text{for } t > 0 \\ y(t) = 0 & \text{for } t < 0 \end{cases} \qquad (36)$$

It is determined from the controllability and observability grammians as follows

$$\|G\|_h = \sqrt{\lambda_{\max}(W_c W_o)} \qquad (37)$$

where $\lambda_{\max}(.)$ denotes the largest eigenvalue. Thus, the Hankel norm of the system is therefore largest Hankel singular value of the system, $\|G\|_h = \gamma_{\max}$.

4.2. NORMS OF A SINGLE MODE

For structures in the modal representation each mode is independent, thus the norms of a single mode are independent as well. Consider the ith natural mode and its state-space representation (A_{mi}, B_{mi}, C_{mi}), see(13). For this representation one obtains the following closed-form expression for the H_2 norm, see [5]:

$$\|G_i\|_2 \cong \frac{\|B_{mi}\|_2 \|C_{mi}\|_2}{2\sqrt{\zeta_i \omega_i}} \tag{38}$$

The H_∞ norm of a natural mode can be approximately expressed in the closed-form as follows:

$$\|G_i\|_\infty \cong \frac{\|B_{mi}\|_2 \|C_{mi}\|_2}{2\zeta_i \omega_i} \tag{39}$$

In order to prove this, note that the largest amplitude of the mode is approximately at the ith natural frequency, thus

$$\|G_i\|_\infty \cong \sigma_{\max}(G_i(\omega_i)) = \frac{\sigma_{\max}(C_{mi} B_{mi})}{2\zeta_i \omega_i} = \frac{\|B_{mi}\|_2 \|C_{mi}\|_2}{2\zeta_i \omega_i}$$

The Hankel norm is approximately obtained from the following formula:

$$\|G_i\|_h = \gamma_i \cong \frac{\|B_{mi}\|_2 \|C_{mi}\|_2}{4\zeta_i \omega_i} \tag{40}$$

Comparing the above equations one obtains the approximate relationships between H_2, H_∞, and Hankel norms of the ith mode

$$\|G_i\|_\infty \cong 2\|G_i\|_h \cong \sqrt{\zeta_i \omega_i} \|G_i\|_2 \tag{41}$$

The above relationship is illustrated in Fig.6.

Example 2. The determination of the H_2 norm for a simple system, as in Fig.7, is illustrated. For this system $m_1 = 11$, $m_2 = 5$, $m_3 = 10$, $k_1 = 10$, $k_2 = 50$, $k_3 = 55$, and $k_4 = 10$. The damping matrix is proportional to the stiffness matrix $D = 0.01K$. The single input u is applied to the three masses, such that $f_1 = u$, $f_2 = 2u$, $f_3 = -5u$, and the output is a linear combination of the mass displacements $y = 2q_1 - 2q_2 + 3q_3$, where q_i is the displacement of the ith mass and f_i is the force applied to that mass.

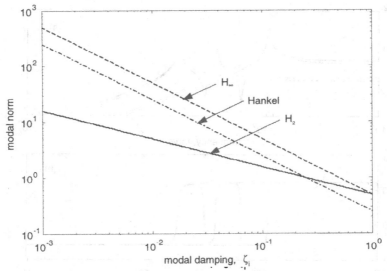

Figure 6. Modal norms versus modal damping

The transfer function of the system and of each mode is shown in Fig.8. It is observed that each mode is dominant in the neighborhood of the mode natural frequency, thus the system transfer function coincides with the mode transfer function near this frequency. The shaded area shown in Fig.9a is the H_2 norm of the second mode. Note that this area is shown in the logarithmic scale for visualization purposes and that most of the actual area is included in the neighborhood of the peak.

The H_2 norms of the modes determined from equation (38) are: $\|G_1\|_2 = 1.9399$, $\|G_2\|_2 = 0.3152$, $\|G_3\|_2 = 0.4405$.

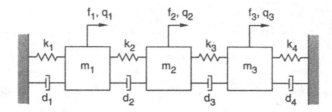

Figure 7. A simple structure.

Example 3. The determination of the H_∞ norm of a simple structure and its modes, as in Example 2, is illustrated. The H_∞ norm of the second mode is shown in Fig.9a as the height of the second resonance peak. The H_∞ norm of the system is shown in Fig.9b as the height of the highest (first in this case) resonance peak. The H_∞ norms of the modes, determined from (39) are: $\|G_1\|_\infty \cong 18.9229$, $\|G_2\|_\infty \cong 1.7454$, $\|G_3\|_\infty \cong 1.2176$.

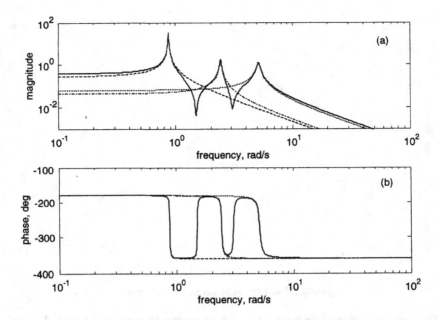

Figure 8. The transfer function of the system and of each mode.

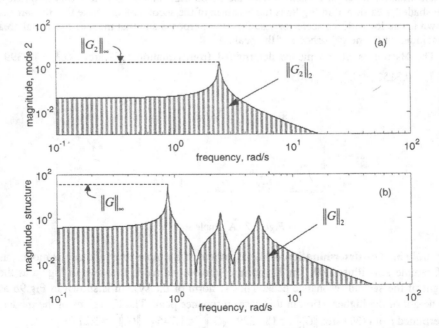

Figure 9. H_2 and H_∞ norms (a) of the second mode and (b) of the structure (single-input-single output case).

Example 4. Determination of the H∞ norm of a single mode from the Ricccati equation (35). This norm is equal to the smallest positive parameter ρ for which the solution S of this equation is positive definite

Due to almost-independence of modes, the solution S of the Riccati equation is diagonally dominant, $S \cong diag(s_1, s_2, ..., s_n)$, where, by inspection, one can find s_i as a solution of the following equation

$$s_i(A_{mi} + A_{mi}^T) + s_i^2 \rho_i^{-2} B_{mi} B_{mi}^T + C_{mi}^T C_{mi} \cong 0,$$

where $i = 1, 2, ..., n$, and A_{mi} is given by Eq.(17), B_{mi} is two-row block of B_m corresponding to block A_{mi} of A_m, and C_{mi} is two-column block of C_m corresponding to block A_{mi} of A_m. For the balanced mode the Lyapunov equations (20) are

$$\gamma_i(A_{mi} + A_{mi}^T) + B_{mi} B_{mi}^T \cong 0, \qquad \gamma_i(A_{mi} + A_{mi}^T) + C_{mi}^T C_{mi} \cong 0$$

Introducing them to the previous equation we obtain

$$s_i(A_{mi} + A_{mi}^T) - s_i^2 \rho_i^{-2} \gamma_i(A_{mi} + A_{mi}^T) - \gamma_i(A_{mi} + A_{mi}^T) \cong 0$$

or, for a stable system

$$s_i^2 - \frac{\rho_i^2}{\gamma_i} s_i + \rho_i^2 \cong 0$$

with two solutions $s_i^{(1)}$ and $s_i^{(2)}$

$$s_i^{(1)} = \frac{\rho_i^2(1 - \beta_i)}{2\gamma_i}, \quad s_i^{(2)} = \frac{\rho_i^2(1 + \beta_i)}{2\gamma_i}, \quad \text{and} \quad \beta_i = \sqrt{1 - \frac{4\gamma_i^2}{\rho_i^2}}$$

For $\rho_i = 2\gamma_i$ one obtains $s_i^{(1)} = s_i^{(2)} = 2\gamma_i = \rho_i$. Moreover, $\rho_i = 2\gamma_i$ is the smallest ρ_i for which a positive solution s_i exists. It is indicated in Fig.10 by plots of $s_i^{(1)}$ (solid line) and $s_i^{(2)}$ (dashed line) versus ρ_i, and for $\gamma_i = 0.25, 0.5, 1, 2, 3$, and 4; the circle "O" denotes locations for which $\|G_i\|_\infty = \rho_{i \max} = 2\gamma_i$.

In order to obtain positive definite S, all s_i must be positive. Thus, the largest ρ_i from the set $\{\rho_1, \rho_2, \cdots \rho_n\}$ is the smallest one for which S is positive definite, which can be easily verified in Fig.10. Thus,

$$\|G\|_\infty = \max_i \rho_i \cong 2\gamma_{\max}.$$

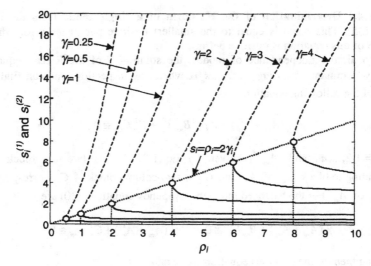

Figure 10. Solutions $s_i^{(1)}$ (solid lines) and $s_i^{(2)}$ (dashed lines); note that $\rho_i = 2\gamma_i$ at locations marked "O".

Example 5. H_∞, H_2 and Hankel norms of a truss. Consider a truss presented in Fig.11. Vertical control forces are applied at nodes 9 and 10, and the output rates are measured in the horizontal direction at nodes 4 and 5. For this structure, the H_2 and H_∞ norms of each mode are given in Fig.12a.

From (41) it follows that the ratio of the H_∞ and H_2 norms is

$$\frac{\|G_i\|_2}{\|G_i\|_\infty} \cong \sqrt{\zeta_i \omega_i} = 0.707\sqrt{\Delta\omega_i}$$

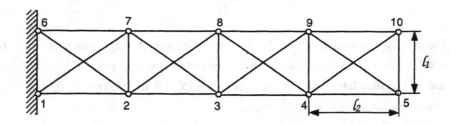

Figure 11. A 2D truss structure.

hence, the relationship between the H_∞ and H_2 norms depends on the width of the resonance. For a wide resonant peak the H_2 norm of the ith mode is larger than the corresponding H_∞ norm. For a narrow resonant peak the H_∞ norm of the ith mode is larger than the corresponding H_2 norm. This is visible in Fig.12a, where neither norm is dominant.

The exact Hankel singular values and the approximate values, obtained from (37) and (40), respectively, are shown in Fig. 12b. A good coincidence between the exact and approximate values is observed.

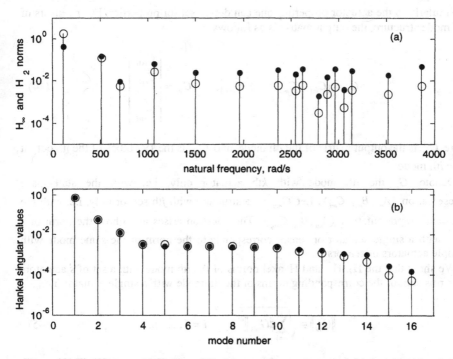

Figure 12. The 2D truss: (a) H_2 (✲) and H_∞ (•) approximate norms, and (b) the exact (✲) and the approximate (•) Hankel singular values.

4.3. NORMS OF A MODE WITH MULTIPLE ACTUATORS AND SENSORS

Above we considered norms of modes with multiple actuators and sensors. Here we continue to consider the same problem, but we decompose the norms into more elementary norms. Consider a flexible structure with s actuators and n modes, so that the modal input matrix B consists of n block-rows of dimension $2 \times s$

$$B_m = \begin{bmatrix} B_{m1} \\ B_{m2} \\ \vdots \\ B_{mn} \end{bmatrix} \tag{42}$$

and the ith block-row B_{mi} of B_m that corresponds to the ith mode has the form

$$B_{mi} = \begin{bmatrix} B_{mi1} & B_{mi2} & \cdots & B_{mis} \end{bmatrix} \tag{43}$$

where B_{mik} corresponds to the kth actuator at the ith mode.

Similarly to the actuator properties one can derive sensor properties. For r sensors of an n mode structure, the output matrix is as follows:

$$C_m = \begin{bmatrix} C_{m1} & C_{m2} & \cdots & C_{mn} \end{bmatrix}, \qquad C_{mi} = \begin{bmatrix} C_{m1i} \\ C_{m2i} \\ \vdots \\ C_{mri} \end{bmatrix} \tag{44}$$

where C_{mi} is the output matrix of the ith mode, and C_{mji} is the 1×2 block of the jth output at the ith mode.

Denote G_{mik} the ith mode with kth actuator only, i.e. with the state-space representation $(A_{mi}, B_{mik}, C_{mi})$. Let G_{mij} be a structure with jth sensor only, i.e. with the state-space representation $(A_{mi}, B_{mi}, C_{mij})$. The question arises as to how the norm of a mode with a single actuator or sensor corresponds to the norm of the same mode with multiple actuators or sensors.

We show that the H$_2$, H$_\infty$, and Hankel norms of the ith mode with a set of s actuators is the rms sum of the corresponding norms of the ith mode with a single actuator, i.e.,

$$\|G_{mi}\| \cong \sqrt{\sum_{k=1}^{s} \|G_{mik}\|^2}, \qquad i = 1, \ldots, n. \tag{45}$$

where $\|.\|$ denotes either H$_2$, H$_\infty$, or Hankel norm. In order to show it, note that the norm of the ith mode with the kth actuator and the norm of the ith mode with all actuators are

$$\|G_{mik}\| \cong \frac{\|B_{mik}\|_2 \|C_{mi}\|_2}{\alpha_i}, \qquad \|G_{mi}\| \cong \frac{\|B_{mi}\|_2 \|C_{mi}\|_2}{\alpha_i}$$

where $\alpha_i = 2\sqrt{\zeta_i \omega_i}$ for the H$_2$ norm, $\alpha_i = 2\zeta_i \omega_i$ for the H$_\infty$ norm, and $\alpha_i = 4\zeta_i \omega_i$ for the Hankel norm. But, from the definition of the norm and from (43) it follows that

$$\|B_{mi}\|_2 = \sqrt{\sum_{k=1}^{s} \|B_{mik}\|_2^2}.$$

Introducing the above equation to the previous one, one obtains (45).

Similarly to the actuator properties one can derive sensor properties. For r sensors of n modes the output matrix is as in Eq.(44). For this output matrix the H$_2$, H$_\infty$, and Hankel norms of the ith mode of a structure with a set of r sensors is the rms sum of the corresponding norms of the mode with each single actuator from this set, i.e.,

$$\|G_{mi}\| \cong \sqrt{\sum_{j=1}^{r}\|G_{mij}\|^2} \qquad (46)$$

$i = 1, \ldots, n$. These properties are illustrated in Fig.13a,b.

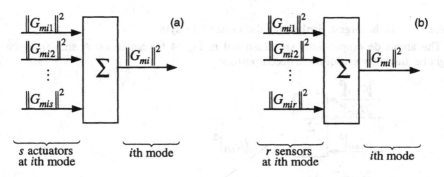

Figure 13. Decomposition of the H_2, H_∞, and Hankel norms of a mode into (a) actuator norms, and (b) sensor norms

4.4. NORMS OF AN ENTIRE STRUCTURE

The H_2, H_∞, and Hankel norms of the entire structure are expressed in terms of the norms of its modes. Let $G(\omega) = C_m(j\omega I - A_m)^{-1}B_m$ be the transfer function of a structure, and let (A_m, B_m, C_m) be its modal state-space representation. The system H_2 norm is, approximately, the rms sum of the modal norms

$$\|G\|_2 \cong \sqrt{\sum_{i=1}^{n}\|G_{mi}\|_2^2} \qquad (47)$$

where n is the number of modes.

In order to show it, note that the controllability grammian W_c in modal coordinates is diagonally dominant, thus the H_2 norm can be expanded as follows:

$$\|G\|_2^2 = \text{tr}(C_m^T C_m W_c) \cong \sum_{i=1}^{n}\text{tr}(C_{mi}^T C_{mi} W_{ci}) = \sum_{i=1}^{n}\|G_{mi}\|_2^2.$$

Consider H_∞ norm of a structure in modal coordinates. Due to the independence of the modes, the system H_∞ norm is the largest of the mode norms, i.e.,

$$\|G\|_\infty \cong \max_i \|G_{mi}\|_\infty, \quad i = 1, \ldots, n. \qquad (48)$$

This property says that for a single-input-single-output system the largest modal peak response determines the worst-case response.

The Hankel norm of the structure is the largest norm of its modes, i.e.,

$$\|G\|_h \cong \max_i \|G_{mi}\|_h = \gamma_{max} = 0.5\|G\|_\infty \qquad (49)$$

where γ_{max} is the largest Hankel singular value of the system.

The above decompositions are illustrated in Fig.14 for actuators. A similar figure might be drawn for sensor norm decomposition.

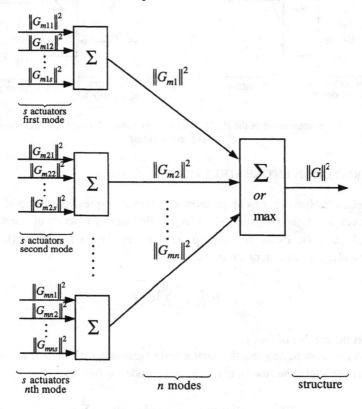

Figure 14. Combined decomposition of norms of a structure (H₂, H∞, and Hankel) into modal norms and actuator norms

Example 6. Obtaining norms of the entire structure from modal norms. Using modal norms from Examples 2 and 3 find the H₂ and H∞ norms of the structure.

The H₂ norm of the entire structure represented by the shaded area in Fig.8b, is approximately a sum of areas of each mode. We find from Eq.(47) that the approximate value of the system norm is the rms sum of modal norms, that is

$\|G\|_2 \cong \sqrt{2.0141^2 + 0.3152^2 + 0.4405^2} = 2.0141$. This value is equal to the exact value of the system norm.

The H_∞ norm is the largest of the modal norms, see Eq.(48); using the results of Example 3 we found that the system norm is equal to the largest (first) modal norm, i.e., that $\|G\|_\infty = \|G_1\|_\infty \cong 18.9619$; it is equal to the exact value of the system norm computed independently.

Example 7. Using norms to detect structural damage. We illustrate the application of H_2 modal and sensor norms to determine damage locations. In particular, we localize damaged elements, and assess the modes particularly impacted by the damage.

Denote norm of the jth sensor of a healthy structure by $\|G_{shj}\|_2$, and norm of the jth sensor of a damaged structure by $\|G_{sdj}\|_2$. The jth sensor index of the structural damage is defined as a weighted difference between the jth sensor norms of a healthy and damaged structure, i.e.,

$$\sigma_{sj} = \frac{\left| \|G_{shj}\|_2^2 - \|G_{sdj}\|_2^2 \right|}{\|G_{shj}\|_2^2}$$

The sensor index reflects the impact of the structural damage on the jth sensor.

Similarly, denote the norm of the ith mode of a healthy structure by $\|G_{mhi}\|_2$, and the norm of the ith mode of a damaged structure by $\|G_{mdi}\|_2$. The ith mode index of the structural damage is defined as a weighted difference between the ith mode norm of a healthy and damaged structure, i.e.,

$$\sigma_{mi} = \frac{\left| \|G_{mhi}\|_2^2 - \|G_{mdi}\|_2^2 \right|}{\|G_{mhi}\|_2^2}$$

The ith mode index reflects impact of the structural damage on the ith mode.

A structure with fixed ends as in Fig.15 is analyzed. The cross section area of the steel beams is 1 cm². Two damage cases are considered. First, as a 20% reduction of the stiffness of the beam No 5, and the second case as a 20% reduction of the stiffness of the beam No17. The structure was more densely divided near the damage locations to better reflect the stress concentration. Nineteen strain-gage sensors are placed at the beams 1 to 19. A vertical force at node P excites the structure.

For the first case the sensor and the modal indices are shown in Fig.16a,b. The sensor indices in Fig.16a indicate that the sensor No 5, located at the damaged beam, suffered the most changes. The modal indices in Fig.16b show that the first mode was heavily affected by the damage.

The sensor and modal indices for the second case are shown in Fig.17a,b. Fig.17a shows the largest sensor index at location No 17, of the damaged beam. The modal indices in Fig.17b show that the tenth and the second modes were the most affected by the damage.

104

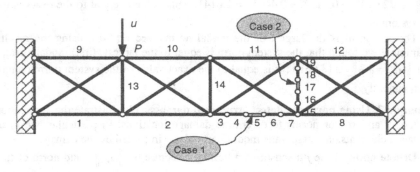

Figure 15. The beam structure: healthy elements are black, damaged elements are white, and numbers refer to the sensors.

4.5. NORMS OF A GENERALIZED STRUCTURE

Consider a generalized structure as in Fig.1, with inputs w and u, and outputs z and y. Let G_{wz} be the transfer matrix from w to z, let G_{wy} be the transfer matrix from w to y, let G_{uz} be the transfer matrix from u to z, and let G_{uy} be the transfer matrix from u to y. Let G_{wzi}, G_{uyi}, G_{wyi}, and G_{uzi} be the transfer functions of the ith mode. The following multiplicative properties of modal norms hold:

$$\left\|G_{wzi}\right\|\left\|G_{uyi}\right\| \cong \left\|G_{wyi}\right\|\left\|G_{uzi}\right\| \tag{50}$$

In order to prove it, denote by B_{mw} and B_{mu} the modal input matrices of w and u, respectively; let C_{mz} and C_{my} be the modal output matrices of z and y, respectively; and let B_{mwi}, B_{mui}, C_{mzi}, and C_{myi} be their ith blocks related to the ith mode. The H_∞ norms are approximately determined from (39) as

$$\left\|G_{wzi}\right\| \cong \frac{\left\|B_{mwi}\right\|_2 \left\|C_{mzi}\right\|_2}{\alpha_i}, \qquad \left\|G_{uyi}\right\| \cong \frac{\left\|B_{mui}\right\|_2 \left\|C_{myi}\right\|_2}{\alpha_i},$$

$$\left\|G_{wyi}\right\| \cong \frac{\left\|B_{mwi}\right\|_2 \left\|C_{myi}\right\|_2}{\alpha_i}, \qquad \left\|G_{uzi}\right\| \cong \frac{\left\|B_{mui}\right\|_2 \left\|C_{mzi}\right\|_2}{\alpha_i}.$$

where $\alpha_i = 2\sqrt{\zeta_i \omega_i}$ for the H_2 norm, $\alpha_i = 2\zeta_i \omega_i$ for the H_∞ norm, and $\alpha_i = 4\zeta_i \omega_i$ for the Hankel norm. Introducing the above equations to (50), the approximate equality is proven by inspection.

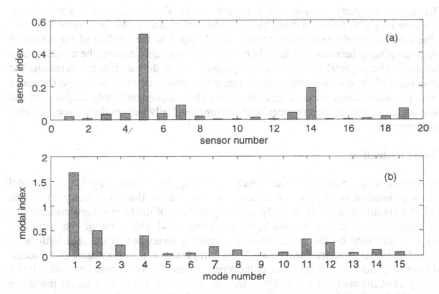

Figure 16. Sensor and modal indices for damage case 1: sensor index for the damaged element No 5 is high, modal index shows that the first mode is predominantly impacted.

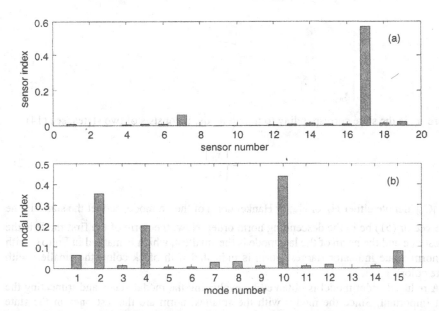

Figure 17. Sensor and modal indices for damage case 2: sensor index for the damaged element No 17 is high, modal index shows that the modes No 10, 2, and 4 are predominantly impacted.

The above property shows that for each mode the product of norms of the performance loop (i.e. from the disturbance to the performance) and the control loop (i.e. from the actuators to the sensors) is approximately equal to the product of the norms of the cross-couplings: between the disturbance and sensors, and between the actuators and performance. The physical meaning of this property lies in the fact that by increasing the actuator–sensor link, one increases automatically the cross-link: the actuator-performance link and the disturbance-sensors link. It shows that sensors not only respond to the actuator input but also to disturbances, and actuators not only impact the sensors but also the performance.

5. Model Reduction

Model reduction is a part of dynamic analysis, testing, and control. Typically, a model with a large number of degrees of freedom (developed for the static analysis) causes numerical difficulties in dynamic analysis, to say nothing of the high computational cost. In the system identification approach, on the other hand, the order of the identified system is determined by the reduction of the initially oversized model that includes a noise model. Finally, in structural control, the complexity and performance of a model-based controller depends on the order of the structural model. In all cases the reduction is a crucial part of the analysis and design. Thus, the reduced-order system solves the above problems if it acquires the essential properties of the full-order model.

In this section we consider structural model in modal coordinates, namely modal models 1 and 2, as in Eq.(16) and (17). The states of the model are as follows

$$x = \left\{ \begin{array}{c} x_1 \\ x_2 \\ \vdots \\ x_n \end{array} \right\} \quad \longleftarrow \quad \text{norm indicator of the components} \qquad (51)$$

where x_i is the state corresponding to the ith mode. It consists of two states, see (14)

$$x_i = \left\{ \begin{array}{c} x_{i1} \\ x_{i2} \end{array} \right\} \qquad (52)$$

Let $\|G_i\|$ denote either H_2, or H_∞, or Hankel norm of the ith mode, and let the states in the state vector (51) be in the descending norm order. Now, the norm of the first mode is the largest one and the norm of the last mode is the smallest, which is marked in Eq.(51) with the norm value indicator (largest norm is indicated with black color, the smallest with white color).

A reduced-order model is obtained by evaluating the modal states and truncating the least important. Since the modes with the smallest norm are the last ones in the state vector, a reduced-order model is obtained here by truncating the last states in the modal vector. How many of them? It depends on the system requirements, and is up to design engineer to check the reduced model accuracy. Let (A_m, B_m, C_m) be the modal representation corresponding to the modal state vector x as in (51). Let x be partitioned as follows:

$$x = \left\{ \begin{matrix} x_r \\ x_t \end{matrix} \right\} \tag{53}$$

where x_r is the vector of the retained states, and x_t is a vector of truncated states. If there are $k < n$ retained modes, x_r is a vector of $2k$ states, and x_t is a vector of $2(n-k)$ states. Let the state triple (A_m, B_m, C_m) be partitioned accordingly,

$$A_m = \begin{bmatrix} A_{mr} & 0 \\ 0 & A_{mt} \end{bmatrix}, \qquad B_m = \begin{bmatrix} B_{mr} \\ B_{mt} \end{bmatrix}, \qquad C_m = \begin{bmatrix} C_{mr} & C_{mt} \end{bmatrix} \tag{54}$$

The reduced model is obtained by deleting the last $2(n-k)$ rows of A_m and B_m, and the last $2(n-k)$ columns of A_m and C_m.

Modal reduction by a truncation of a stable model always produces a stable reduced model, since the poles of the reduced model have not been changed.

Example 8. Reduction of the truss model. Consider a 2D truss as in Example 5. Its model is reduced in modal coordinates using the H_∞ norm. The approximate norms of the modes are shown in Fig. 18. From this figure the system norm (the largest of the mode norms) is $\|G\|_\infty = 1.6185$. Based on the modal norms it was decided that in the reduced-order model we reject all modes of the H_∞ norm less than 0.00003. The area where the H_∞ norm is less than 0.00003 lies below the dashed line in Fig.18, and the modes that have the H_∞ norm in this area are deleted. Consequently, the reduced model consists of three modes (No. 1,2, and 4). The transfer function of the full and reduced models (from the second input to the second output) is shown in Fig. 19a and the corresponding impulse response is shown in Fig. 19b. Both figures indicate a small reduction error, which is obtained as $(\|G\|_\infty - \|G_r\|_\infty)/\|G\|_\infty \cong \|G_4\|_\infty / \|G\|_\infty = 0.0040$.

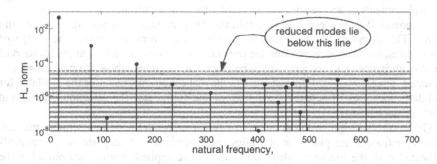

Figure 18. H_∞ norms of the 2D truss modes.

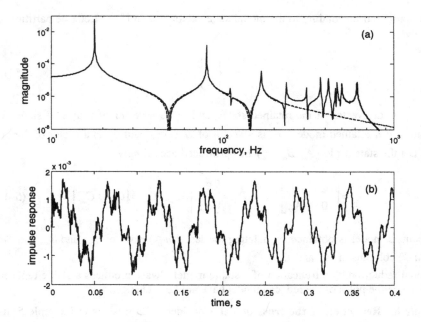

Figure 19. Magnitude of the transfer function (a), and impulse responses (b) of the full (solid line) and reduced (dashed line) truss models show that mainly the high-frequency modes were deleted, and overlapped impulse responses.

6. Actuator and Sensor Placement

A typical actuator and sensor location problem for structural testing can be described as a structural test plan. The plan is based on the available information on the structure itself, on disturbances acting on the structure, and on the expected structural performance. The preliminary information on structural properties is typically obtained from the finite-element model. The disturbance information includes disturbance location and disturbance spectral contents. The structure performance is commonly evaluated through the displacements or accelerations of selected structural locations.

In general it is not possible to duplicate the dynamics of a real structure during testing. This happens not only due to physical restrictions or limited knowledge of disturbances, but also because often the test actuators cannot be located at the location of actual disturbances and sensors cannot be placed at locations where the performance is evaluated. Thus, to conduct the test close to the real conditions, one uses the available (or candidate) locations of actuators and sensors and formulates the selection criteria and selection mechanisms.

The control design problem of a structure can be defined in a similar manner. Namely, actuators are placed at the allowable locations and they are not necessarily collocated with the locations where disturbances are applied; sensors are placed at the sensor allowable locations, which are generally outside the locations where performance is evaluated.

For simple test articles, an experienced test engineer can determine the appropriate sensor or actuator locations in an ad hoc manner. However, for first-time testing of large and complex structures the placement of sensors and actuators is neither an obvious nor a simple task. In practice, heuristic means are combined with engineering judgment and simplified analysis to determine actuator and sensor locations. In most cases the locations vary during tests (in a trial and error approach) to obtain acceptable data to identify target modes. The actuator and sensor placement problem was investigated by many researchers, as reported in a review article [16].

For a small number of sensors or actuators a typical solution to the location problem is found through a search procedure. For large numbers of locations the search for the number of possible combinations is overwhelming, time consuming and does not necessarily give the optimal solution. The approach proposed here consists of the determination of the norm of each sensor (or actuator) for selected modes and then grade them according to their participation in the system norm. This is a computationally fast (non-search) procedure with a clear physical interpretation.

6.1. PROBLEM STATEMENT

Given a larger set of sensors and actuators, the placement problem consists of determining the locations of a smaller subset of sensors or actuators such that the H_2, H_∞, or Hankel norms of the subset are as close as possible to the norm of the original set. In this chapter this placement problem is solved in the modal coordinates.

Let $\mathcal{R}$ and $\mathcal{S}$ be the sets of the candidate sensor and actuator locations, chosen in advance as allowable locations of actuators of population S and as allowable locations of sensors of population R. The placement of s actuators from S actuator candidate locations, and the placement of r sensors from R sensor candidate locations is considered. Of course, the number of candidate locations is larger than the number of final locations, i.e., $R > r$ and $S > s$.

6.2. PLACEMENT INDICES AND MATRICES

Actuator and sensor placement problems are solved independently, however both procedures are similar. Denote the H_2, H_∞, or Hankel norms of the ith mode with the jth actuator only or the ith mode with the kth sensor only

$$\left\| G_{ij} \right\| = \frac{\left\| B_{mij} \right\|_2 \left\| C_{mi} \right\|_2}{\alpha_i}, \qquad \left\| G_{ik} \right\| = \frac{\left\| B_{mi} \right\|_2 \left\| C_{mki} \right\|_2}{\alpha_i} \tag{55}$$

where $\alpha_i = 2\sqrt{\zeta_i \omega_i}$ for the H_2 norm, $\alpha_i = 2\zeta_i \omega_i$ for the H_∞ norm, and $\alpha_i = 4\zeta_i \omega_i$ for the Hankel norm. Denoted by G is the transfer function of the system with all S candidate actuators. The placement index σ_{ki} that evaluates the kth actuator at the ith mode in terms of either H_2, H_∞, or Hankel norm is defined with respect to all the modes and all admissible actuators

$$\sigma_{ki} = w_{ki} \frac{\|G_{ki}\|}{\|G\|} \tag{56}$$

where $\|.\|$ denotes H_2, H_∞, or Hankel norm, respectively, and $k = 1, ..., S$, $i = 1, ..., n$, and where $w_{ki} \geq 0$ is the weight assigned to the kth actuator and the ith mode n is a number of modes and G_{ki} is the transfer function of the ith mode and of the kth actuator. The weight reflects the importance of the mode and the actuator in applications. We evaluate actuator importance using a placement matrix in the following form

$$\Sigma = \begin{bmatrix} \sigma_{11} & \sigma_{12} & \cdots & \sigma_{1k} & \cdots & \sigma_{1S} \\ \sigma_{21} & \sigma_{22} & \cdots & \sigma_{2k} & \cdots & \sigma_{2S} \\ \cdots & \cdots & \cdots & \cdots & \cdots & \cdots \\ \sigma_{i1} & \sigma_{i2} & \cdots & \sigma_{ik} & \cdots & \sigma_{iS} \\ \cdots & \cdots & \cdots & \cdots & \cdots & \cdots \\ \sigma_{n1} & \sigma_{n2} & \cdots & \sigma_{nk} & \cdots & \sigma_{nS} \end{bmatrix} \leftarrow i\text{th mode} \tag{57}$$

$$\uparrow$$
$$k\text{th actuator}$$

The kth column of the above matrix consists of indexes of the kth actuator for every mode and the ith row is a set of the indexes of the ith mode for all actuators.

Similarly to actuators, the sensor placement index σ_{ki} evaluates the kth sensor at the ith mode

$$\sigma_{ki} = w_{ki} \frac{\|G_{ki}\|}{\|G\|} \tag{58}$$

$k = 1, ..., R$, $i = 1, ..., n$, where $w_{ki} \geq 0$ is the weight assigned to the kth sensor and the ith mode, n is a number of modes, and G_{ki} is the transfer function of the ith mode and kth sensor. The sensor placement matrix is defined as follows

$$\Sigma = \begin{bmatrix} \sigma_{11} & \sigma_{12} & \cdots & \sigma_{1k} & \cdots & \sigma_{1R} \\ \sigma_{21} & \sigma_{22} & \cdots & \sigma_{2k} & \cdots & \sigma_{2R} \\ \cdots & \cdots & \cdots & \cdots & \cdots & \cdots \\ \sigma_{i1} & \sigma_{i2} & \cdots & \sigma_{ik} & \cdots & \sigma_{iR} \\ \cdots & \cdots & \cdots & \cdots & \cdots & \cdots \\ \sigma_{n1} & \sigma_{n2} & \cdots & \sigma_{nk} & \cdots & \sigma_{nR} \end{bmatrix} \leftarrow i\text{th mode} \tag{59}$$

$$\uparrow$$
$$k\text{th sensor}$$

where the kth column consists of indexes of the kth sensor for every mode and the ith row is a set of the indexes of the ith mode for all sensors.

The placement matrix gives an insight into the placement properties of each actuator since the placement index of the kth actuator is determined as the rms sum of the kth column of Σ. The vector of the actuator placement indices is defined as $\sigma_a = \begin{bmatrix} \sigma_{a1} & \sigma_{a2} & \cdots & \sigma_{aS} \end{bmatrix}^T$ and its kth entry is the placement index of the kth actuator. In the case of the H_2 norm, it is the rms sum of the kth actuator indexes over all modes

$$\sigma_{ak} = \sqrt{\sum_{i=1}^{n} \sigma_{ik}^2}, \qquad k = 1, \dots, S \qquad (60)$$

In the case of the H_∞ and Hankel norms, it is the largest index over all modes

$$\sigma_{ak} = \max_i(\sigma_{ik}), \qquad i = 1, \dots, n, \quad k = 1, \dots, S \qquad (61)$$

Similarly, the vector of the sensor placement indices is defined as $\sigma_s = \begin{bmatrix} \sigma_{s1} & \sigma_{s2} & \cdots & \sigma_{sR} \end{bmatrix}^T$ and its kth entry is the placement index of the kth sensor. In the case of the H_2 norm, it is the rms sum of the kth sensor indexes over all modes

$$\sigma_{sk} = \sqrt{\sum_{i=1}^{n} \sigma_{ik}^2}, \qquad k = 1, \dots, R \qquad (62)$$

In the case of the H_∞ and Hankel norms, it is the largest index over all modes

$$\sigma_{sk} = \max_i(\sigma_{ik}), \qquad i = 1, \dots, n, \quad k = 1, \dots, R \qquad (63)$$

The vector of the mode indices is defined as follows: $\sigma_m = \begin{bmatrix} \sigma_{m1} & \sigma_{m2} & \cdots & \sigma_{mn} \end{bmatrix}^T$, and its ith entry is the index of the ith mode. This entry is an rms sum of the ith mode indices over all actuators

$$\sigma_{mi} = \sqrt{\sum_{k=1}^{S} \sigma_{ik}^2}, \qquad i = 1, \dots, n \qquad (64)$$

or an rms sum of the ith mode indices over all sensors

$$\sigma_{mi} = \sqrt{\sum_{k=1}^{R} \sigma_{ik}^2}, \qquad i = 1, \dots, n \qquad (65)$$

- The actuator placement index, σ_{ak}, is a nonnegative contribution of the kth actuator at all modes to the H_2, H_∞, or Hankel norms of the structure.

112

- The sensor placement index, σ_{sk}, is a nonnegative contribution of the kth sensor at all modes to the H_2, H_∞, or Hankel norms of the structure.
- The mode index, σ_{mi}, is a nonnegative contribution of the ith mode for all actuators (or all sensors) to the H_2, H_∞, or Hankel norms of the structure.

The determination of the H_∞ actuator and modal indices for the pinned beam is illustrated in Fig.20. Six actuators are located on the beam and four modes are considered. The second mode index is the rms sum of indices of all actuators for this mode and the third actuator index is the largest of these actuator indices over four modes. From the above properties it follows that the index σ_{ak} (σ_{sk}) characterizes the importance of the kth actuator (sensor), thus it serves as the actuator (sensor) placement index. Namely, the actuators (sensors) with small index σ_{ak} (σ_{sk}) can be removed as the least significant ones. Note also that the mode index σ_{mi} can also be used as a reduction index. Indeed, it characterizes the significance of the ith mode for the given locations of sensors and actuators. The norms of the least significant modes (those with the small index σ_{mi}) should either be enhanced by the reconfiguration of the actuators or sensors, or be eliminated.

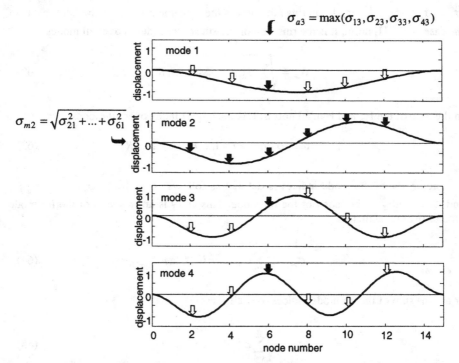

Figure 20. The determination of the H_∞ actuator and modal indices of a pinned beam ($\Downarrow$ – actuator location; and $\blacktriangledown$ – actuators used for the calculation of the indices); the second mode index is the rms sum of indices of all six actuators for this mode, while the third actuator index is the largest of this actuator indices over all four modes.

Example 9. Placing sensors and actuators on a beam to detect up to four modes.
Consider a beam as shown in Fig.21 divided into $n = 30$ elements, with a vertical force
at nodes 1, 2, ... up to 29. By using the placement technique presented above and the H_∞
norm, one shall find the best places for displacement sensors in the y-direction to sense
the first, second, third, and fourth mode; and to sense simultaneously the first two modes,
the first three modes, and the first four modes. Do it for actuator locations at nodes 1, 2,
... up to 29.

Each node of a beam has 3 degrees of freedom: horizontal displacement x, vertical
displacement y, and rotation in the figure plane θ. Denote a unit vector
$e_i = [0,0,\ldots,1,\ldots,0]$ that has all zeros except 1 at the ith location, then the displacement
output matrix for sensors located at ith node is $C_{qi}^T = e_{3i-1}$. The input matrix for the
actuator at jth node is $B_{oj} = e_{3j-1}^T$.

The H_∞ norm $\|G_{ki}\|_\infty$ for the kth mode (k=1,2,3,4) and ith sensor location is obtained
from (55) using B_o and C_{qi} as above. From these norms the sensor placement indices
for kth mode and for jth actuator are obtained from (58), using weight such that
$\max_i(\sigma_{\infty kij}) = 1$.

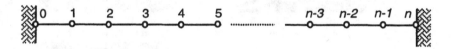

Figure 21. Clamped beam divided into n elements.

The plots of $\sigma_{\infty kij}$ are shown in Fig.22a,b,c,d. The plot of the sensor indices for the
first mode is shown in Fig.22a. It shows the maximum at node 15, indicating that the
sensors shall be placed at these nodes. The plot of the sensor placement indices for the
second mode is shown in Fig.22b. It shows two maximal values, at nodes 9 and 21,
indicating these two locations as the best for sensing the second mode. The plot of the
sensor placement indices for the third mode is shown in Fig.22c, showing two maximal
values at nodes 6 and 24, indicating that these two locations are the best for sensing the
third mode. Finally, the plot of the sensor placement indices for the fourth mode is shown
in Fig.22d. It shows 4 maximal values at nodes 5, and 25, indicating that these four
locations are the best for sensing the fourth mode.

Next, the indices for the first two modes are determined using Eq.(61), namely
$\sigma_{\infty 12i} = \max(\sigma_{\infty 1i}, \sigma_{\infty 2i})$. The plot of this index is shown in Fig.23b with the index
reaching its maximum at three locations: 9, 15, and 21. These locations are the best for
sensing the first and the second modes. Clearly, location 15 serves for first mode sensing,
while locations 9 and 21 serve for second mode sensing.

Next, the indices for the first three modes are determined, using Eq.(61), i.e.,
$\sigma_{\infty 123i} = \max(\sigma_{\infty 1i}, \sigma_{\infty 2i}, \sigma_{\infty 3i})$. The plot of this index is shown in Fig.23c with the index
reaching its maximum at five locations: 6, 9, 15, 21 and 24. These locations are the best
for sensing the first, second, and third modes. Obviously, location 15 serve for first mode

114

sensing, locations 9 and 21 serve for second mode sensing, and locations 6 and 24 serve for third mode sensing.

Finally, the indices for the first four modes are determined using Eq.(61), $\sigma_{\infty 1234i} = \max(\sigma_{\infty 1i}, \sigma_{\infty 2i}, \sigma_{\infty 3i}, \sigma_{\infty 4i})$. The plot of this index is shown in Fig.23d, with the index reaching its maximum at seven locations: 5, 6, 9, 15, 21, 24, and 25. These locations are the best for sensing the first, second, third, and fourth modes. Location 15 serves for first mode sensing, locations 9 and 21 serve for second mode sensing, locations 6 and 24 serve for third mode sensing, and locations 5 and 25 serve for fourth mode sensing.

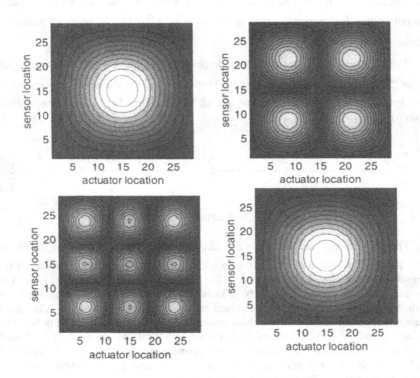

Figure22. H_∞ indices as a function of sensor locations: (a) for the first beam mode, (b) for the second beam mode, (c) for the third beam mode, and (d) for the fourth beam mode.

So far in this example we used the H_∞ norms and indices. It is interesting to compare the sensor placement using the H_2 norms and indices. First, the H_2 norm $\|G_{ki}\|_2$ for the kth mode (k=1,2,3,4) and ith sensor location is obtained from(55) using B_{oj} and C_{qi} as above. The H_2 indices are shown in Fig.24. They are quite similar to the H_∞ indices from Fig.22. Next, the H_2 indices for the first two modes are determined using Eq.(62), namely

$\sigma_{2,12i} = \sqrt{\sigma_{2,1i}^2 + \sigma_{2,2i}^2}$. The plot of this index is shown in Fig.25b, where the index reaches its maximum at locations 10 and 20. Next, the indices for the first three modes are determined, using Eq.(62), i.e. $\sigma_{2,123i} = \sqrt{\sigma_{2,1i}^2 + \sigma_{2,2i}^2 + \sigma_{2,3i}^2}$. The plot of this index is shown in Fig.25c, where the index reaches its maximum at locations 7, 15 and 23. Finally, the indices for the first four modes are determined using Eq.(62), i.e. $\sigma_{2,1234i} = \sqrt{\sigma_{2,1i}^2 + \sigma_{2,2i}^2 + \sigma_{2,3i}^2 + \sigma_{2,4i}^2}$. The plot of this index is shown in Fig.25d, where the index reaches its maximum at locations 6 and 24.

A comparison of the H_∞ and H_2 indices in Fig.23 and Fig.25 shows that the H_2 index variation is more dramatic with the change of sensor location, while the H_∞ index variations are flatter (as the result of selection of maximal values). Thus the first index is a more sensitive measure of the sensor (or actuator) location. Due to the flattening action of the H_∞ norm, the H_∞ indices show slightly different sensor locations.

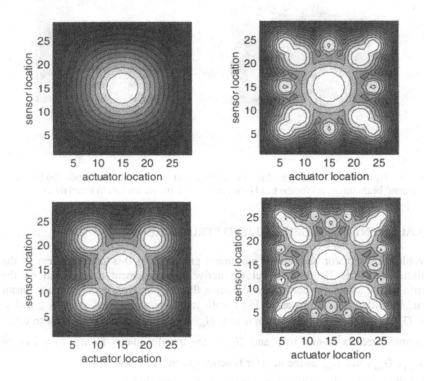

Figure 23. H_∞ sensor placement indices as a function of sensor locations: (a) for the first mode, (b) for the first two modes, (c) for the first three modes, and (d) for the first four modes.

116

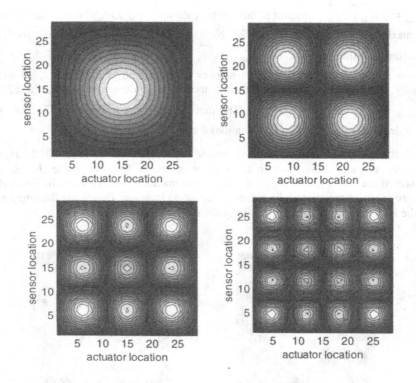

Figure 24. H$_2$ indices as a function of sensor locations: (a) for the first beam mode, (b) for the second beam mode, (c) for the third beam mode, and (d) for the fourth beam mode.

6.3. PLACEMENT FOR A GENERALIZED STRUCTURE

The problem of actuator and sensor placement presented in this section refers to the generalized structure. For this model we derive the placement rules based on the properties of the structural norms and illustrate their application with the truss sensor location. Consider a structure as in Fig.1 with inputs w and u, and outputs z and y. Denote G_{wz} as the transfer matrix from w to z, G_{wy} as the transfer matrix from w to y, G_{uz} as the transfer matrix from u to z, and G_{uy} as the transfer matrix from u to y. Denote G_{wzi}, G_{uyi}, G_{wyi}, and G_{uzi} as the transfer functions of the ith mode.

The following multiplicative properties of modal norms hold:

$$\left\|G_{wzi}\right\|\left\|G_{uyi}\right\| \cong \left\|G_{wyi}\right\|\left\|G_{uzi}\right\|$$ (66)

for $i = 1, \ldots, n$, where $\|\cdot\|$ denotes either H$_2$, H$_\infty$, or Hankel norms.

Define the actuator transfer function as $G_{ui} = \begin{bmatrix} G_{uyi} & G_{uzi} \end{bmatrix}$. Introducing (45) and the modal norms (55) to (66) one obtains the following property:

$$\left\| G_{ui} \right\|^2 \cong \alpha_{wi}^2 \sum_{k=1}^{S} \left\| G_{u_k yi} \right\|^2 \qquad (67)$$

where $G_{u_k yi}$ is the transfer function of the ith mode from the kth actuator to the output y and α_{wi} is the disturbance weight of the ith mode, defined as

$$\alpha_{wi} = \sqrt{1 + \frac{\left\| G_{wzi} \right\|^2}{\left\| G_{wyi} \right\|^2}} \qquad (68)$$

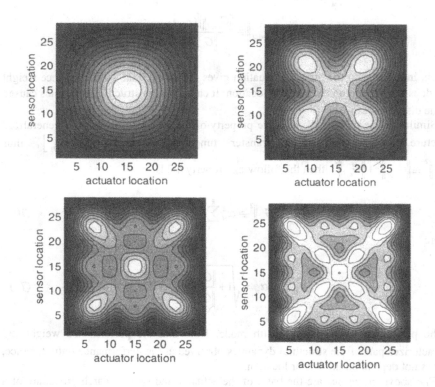

Figure 25. H$_2$ sensor placement indices as a function of sensor locations: (a) for the first mode, (b) for the first two modes, (c) for the first three modes, and (d) for the first four modes.

To show it, note that

$$\left\|G_{ui}\right\|^2 \cong \left\|G_{uzi}\right\|^2 + \left\|G_{uyi}\right\|^2 \tag{69}$$

Note also that from Eq.(45) one obtains

$$\left\|G_{uzi}\right\|^2 \cong \sum_{k=1}^{S}\left\|G_{u_k zi}\right\|^2, \quad \text{and} \quad \left\|G_{uyi}\right\|^2 \cong \sum_{k=1}^{S}\left\|G_{u_k yi}\right\|^2$$

where $G_{u_k zi}$ is the transfer function of the ith mode from the kth actuator to the performance z. Introducing the above equations to Eq.(69) we obtained the following relationship: $\left\|G_{ui}\right\|^2 \cong \sum_{k=1}^{S}\left(\left\|G_{u_k zi}\right\|^2 + \left\|G_{u_k yi}\right\|^2\right)$. Next, Eq.(66) gives

$$\left\|G_{u_k zi}\right\| \cong \frac{\left\|G_{u_k yi}\right\|\left\|G_{wzi}\right\|}{\left\|G_{wyi}\right\|}$$

which, introduced to the previous equation gives (67). Note that the disturbance weight α_{wi} does not depend on the actuator location: it characterizes structural dynamics caused by the disturbances w.

Similarly, one obtains the additive property of the sensor locations of a generalized structure. Define the sensor transfer function as $G_{yi} = \begin{bmatrix} G_{wyi} & G_{uyi} \end{bmatrix}$, thus $\left\|G_{yi}\right\|^2 \cong \left\|G_{wyi}\right\|^2 + \left\|G_{uyi}\right\|^2$, then the following property holds:

$$\left\|G_{yi}\right\|^2 \cong \alpha_{zi}^2 \sum_{k=1}^{R}\left\|G_{uy_k i}\right\|^2 \tag{70}$$

where

$$\alpha_{zi} = \sqrt{1 + \frac{\left\|G_{wzi}\right\|^2}{\left\|G_{uzi}\right\|^2}} \tag{71}$$

is the performance weight of the ith mode. Note that the performance weight α_{zi} characterizes part of the structural dynamics observed at the performance output; hence, it does not depend on the sensor location.

The above properties are the basis of the actuator and sensor search procedure of a generalized structure. The actuator index evaluates the actuator usefulness in test, and is defined as follows:

$$\sigma_{ki} = \frac{\alpha_{ui}\left\|G_{u_k yi}\right\|}{\left\|G_u\right\|} \tag{72}$$

where $\left\|G_u\right\|^2 = \left\|G_{uy}\right\|^2 + \left\|G_{uz}\right\|^2$, while the sensor index is

$$\sigma_{ki} = \frac{\alpha_{yi}\left\|G_{uy_ki}\right\|}{\left\|G_y\right\|} \qquad (73)$$

where $\left\|G_y\right\|^2 = \left\|G_{uy}\right\|^2 + \left\|G_{wy}\right\|^2$.

The indices are the building blocks of the actuator placement matrix Σ

$$\Sigma = \begin{bmatrix} \sigma_{11} & \sigma_{12} & \cdots & \sigma_{1k} & \cdots & \sigma_{1S} \\ \sigma_{21} & \sigma_{22} & \cdots & \sigma_{2k} & \cdots & \sigma_{2S} \\ \cdots & \cdots & \cdots & \cdots & \cdots & \cdots \\ \sigma_{i1} & \sigma_{i2} & \cdots & \sigma_{ik} & \cdots & \sigma_{iS} \\ \cdots & \cdots & \cdots & \cdots & \cdots & \cdots \\ \sigma_{n1} & \sigma_{n2} & \cdots & \sigma_{nk} & \cdots & \sigma_{nS} \end{bmatrix} \leftarrow i\text{th mode} \qquad (74)$$

$$\uparrow$$
$$k\text{th actuator}$$

and a similar matrix for sensor placement.

The placement index of the kth actuator (sensor) is determined from the kth column of Σ. In the case of the H_2 norm, it is the rms sum of the kth actuator indexes over all modes,

$$\sigma_k = \sqrt{\sum_{i=1}^{n}\sigma_{ik}^2} \qquad (75)$$

$k = 1,...,S$ or R, and in the case of the H_∞ and Hankel norms it is the largest index over all modes

$$\sigma_k = \max_i(\sigma_{ik}) \qquad (76)$$

$i = 1, ..., n,\ k = 1, ..., S$ or R.

This property shows that the index for the set of sensors/actuators is determined from the indexes of each individual sensor or actuator. This decomposition allows for the evaluation of an individual sensor/actuator through its participation in the performance of the whole set of sensors/actuators.

Example 10. Sensor placement for a generalized structure. Consider the 3D truss as in Fig.26. The disturbance w is applied at node 7 in the horizontal direction. The performance z is measured as rates of all nodes. The input u is applied at node 26 in the vertical direction, and the candidate sensor locations are at the nodes 5, 6, 7, 12, 13, 14, 19, 20, 21, 26, 27, and 28 in all three directions (total of 36 locations). Using the first 50 modes, the task is to select a low number of sensors that would measure, as close as possible, the disturbance-to-performance dynamics.

First, the H_∞ norms of each mode of G_{wz}, G_{wy}, G_{uz}, and G_{uy} are determined and presented in Fig.27a,b. Next, the validity of Eq.(66) is checked. Indeed, it holds since the plots of $g_1(k) = \|G_{wzk}\|_\infty \|G_{uyk}\|_\infty$ and of $g_2(k) = \|G_{wyk}\|_\infty \|G_{uzk}\|_\infty$ overlap in Fig.28.

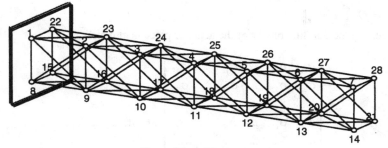

Figure 26. A 3D truss structure.

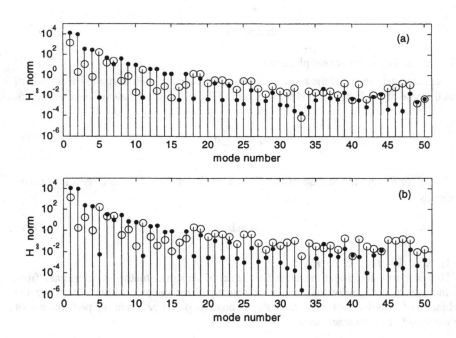

Figure 27. The H_∞ norms of the 3D truss modes: (a) G_{wz} (•) and G_{uy} (ᴎ); and (b) G_{wy} (•) and G_{uz} (ᴎ)

Next, the sensor weights α_{yi} are determined for each mode, and shown in Fig.29. The placement indices σ_k for each sensor are determined from (76), and their plot is shown in Fig.30. Note that some sensors with a high value of σ_k are highly correlated. After removing the highly correlated locations, the two locations $k = 29$ and $k = 30$ remained. They correspond to node 14 in the y- and z-directions.

6.4. SIMULTANEOUS PLACEMENT OF ACTUATORS AND SENSORS

Simultaneous selection of sensor and actuator locations is an issue of certain importance, since fixing the locations of sensors while placing actuators (or vice versa) limits the improvement of system performance. The presented placement algorithm is developed for structures using either H_2, or H_∞, or Hankel modal norms. The algorithm consists of determination of either H_2, or H_∞, or Hankel norms for a single mode, single actuator, and single sensor. Based on these norms the sensor and actuator placement matrices are generated for each considered mode to evaluate sensor and actuator combinations, and to determine the simultaneous actuator and sensor locations that maximize each modal norm.

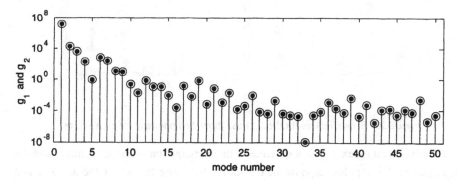

Figure 28. Overlapped plots of $g_1(k)$ (•) and $g_2(k)$ (↘) show that Eq.(66) holds.

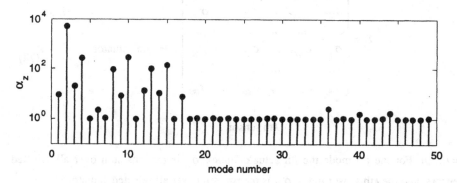

Figure 29. Modal weights for the 3D truss to accommodate disturbances in the generalized model.

In this section the symbol $\|\cdot\|$ will denote either H_2, or H_∞, or Hankel norm. For the set R of the candidate actuator locations, one shall select a subset r of actuators, and concurrently for the set S of the candidate sensor locations, one shall select a subset s of sensors. The criterion is the maximization of the system norm.

Recall that the norm $\|G_{ijk}\|$ characterizes the ith mode equipped with the jth actuator and the kth sensor. Previously we defined the placement index for actuators and for sensors separately, see Eqs.(56) and (58). Here, for each mode, we define the actuator and sensor placement index

$$\sigma_{ijk} = \frac{\|G_{ijk}\|}{\|G_{mi}\|}, \qquad i = 1,...,n \tag{77}$$

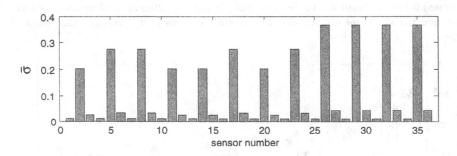

Figure 30. Sensor indices for the 3D truss show the importance of each sensor.

The placement index σ_{ijk} is a measure of the participation of the jth actuator and the kth sensor in the impulse response of the ith mode. Using this index the actuator and sensor placement matrix of the ith mode is generated

$$\Sigma_i = \begin{bmatrix} \sigma_{i11} & \sigma_{i12} & \cdots & \sigma_{i1k} & \cdots & \sigma_{i1S} \\ \sigma_{i21} & \sigma_{i22} & \cdots & \sigma_{i2k} & \cdots & \sigma_{i2S} \\ \cdots & \cdots & \cdots & \cdots & \cdots & \cdots \\ \sigma_{ij1} & \sigma_{ij2} & \cdots & \sigma_{ijk} & \cdots & \sigma_{ijS} \\ \cdots & \cdots & \cdots & \cdots & \cdots & \cdots \\ \sigma_{iR1} & \sigma_{iR2} & \cdots & \sigma_{iRk} & \cdots & \sigma_{iRS} \end{bmatrix} \leftarrow j\text{th actuator} \tag{78}$$

$$\uparrow$$

$$k\text{th sensor}$$

$i=1,...,n$. For the ith mode the jth actuator index σ_{aij} is the rms sum over all selected sensors, and the kth sensor index σ_{sik} is the rms sum over all selected actuators

$$\sigma_{aij} = \sqrt{\sum_{k=1}^{s} \sigma_{ijk}^2}, \qquad \sigma_{sik} = \sqrt{\sum_{j=1}^{r} \sigma_{ijk}^2} \qquad (79)$$

These indices, however, cannot be readily evaluated since for evaluation of the actuator index one needs to know the sensor locations (which have not been yet selected), and vice versa. This difficulty can be overcome by using the property similar to Eq.(66). Namely, for the placement indices we obtain

$$\sigma_{ijk}\sigma_{ilm} \cong \sigma_{ijm}\sigma_{ilk} \qquad (80)$$

This property can be proven by the substitution of the norms as either in Eq.(38), (39) or (40) into the above equation.

It follows from this property that by choosing the two largest indices for the ith mode, say σ_{ijk} and σ_{ilm} (such that $\sigma_{ijk} > \sigma_{ilm}$), the corresponding indices σ_{ijm} and σ_{ilk} are also large. In order to show it, note that $\sigma_{ilm} \le \sigma_{ijm} \le \sigma_{ijk}$ holds and $\sigma_{ilm} \le \sigma_{ilk} \le \sigma_{ijk}$ also holds as a result of (80) and the fact that $\sigma_{ijm} \le \sigma_{ijk}$ and $\sigma_{ilk} \le \sigma_{ijk}$. In consequence, by selecting individual actuator and sensor locations with the largest indices one automatically maximize the indices (79) of the sets of actuators and sensors.

The determination of locations of large indices is illustrated with the following example. Let σ_{124}, σ_{158}, and σ_{163} be the largest indices selected for the first mode. They correspond to 2, 5, and 6 actuator locations and 3, 4, and 8 sensor locations. They are marked in dark color in Fig.31. According to (80) the indices σ_{123}, σ_{128}, σ_{153}, σ_{154}, σ_{164}, and σ_{168} are also large. They are marked in white color with white spots in Fig.31. Now we see that the rms summation for actuators is over all selected sensors (3, 4, and 8), the rms summation for sensors is for over all selected actuators (2, 5, and 6), and that both summations maximize the actuator and sensor indices.

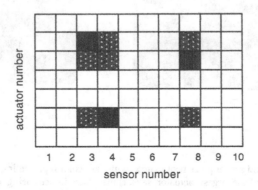

Figure 31. An example of the actuator and sensor placement matrix for the first mode; the largest indices are dark, and the large indices are dark with white spots.

Example 11. Simultaneous placement of actuator and sensor for a clamped beam as in Fig.21. The candidate actuator locations are the vertical forces at nodes 1 to 29 and the candidate sensor locations are the vertical rate sensors located at nodes 1 to 29. Using the H_∞ norm and considering the first four modes, we shall determine at most 4 actuator and 4 sensor locations (one for each mode).

In this example $n=4$ and $R=S=29$. Using Eqs.(79) the placement matrices for the first four modes were determined and plotted in Figs.31a-d. Before the placement procedure is applied the accuracy of Eq.(80) is checked. For this purpose the second mode is chosen, i.e., $i=2$, and the following actuator and sensor locations are selected: $j=k=3$, $l=m=q$, and $q=1,\ldots,29$. For these parameters Eq.(80) is as follows

$$\sigma_{233}\sigma_{2qq} \cong \sigma_{23q}\sigma_{2q3}$$

$q=1,\ldots,29$. The plots of the left- and right-hand side of the above equations are shown in Fig.33, showing good coincidence.

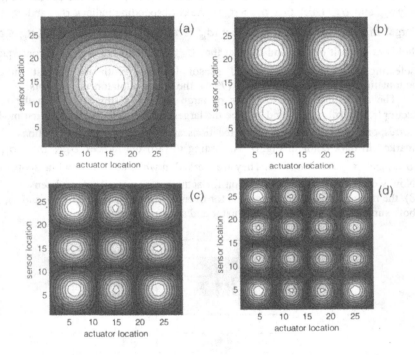

Figure 32. Actuator and sensor placement matrix, with dark color representing the high value of its entries, and locations of the largest (actuator, sensor) placement indices being: (a) (15,15) for mode 1, (b) (9,9), (9,21), (21,21), (21,9) for mode 2, (c) (6,6), (6,24), (24,6), (24,24) for mode 3, and (d) (5,5), (5,25), (25,5), (25,25) for mode 4.

The maximal values of the actuator and sensor index in the placement matrix determine the preferred location of the actuator and sensor for each mode. Note that for each mode four locations – two sensor locations and two actuator locations – have the same maximal value. Moreover, they are symmetrical with respect to the beam center, see Fig.32. We selected four collocated sensors and actuators at the left-hand side of the beam center, one for each mode. Namely, for mode 1 – node 15, for mode 2 – node 9, for mode 3 – node 6, and for mode 4 – node 5.

7. Modal Actuators and Sensors

In some structural tests it is desirable to isolate (i.e., excite and measure) a single mode. Such technique considerably simplifies the determination of modal parameters, see Ref.[13]. This was first achieved by using the force appropriation method, called also the Asher method, see [11], or phase separation method, see [3]. In this method a spatial distribution and the amplitudes of a harmonic input force are chosen to excite a single structural mode. Modal actuators or sensors were presented also in [4], [10], and [8] with application to structural acoustic problems. This section follows Ref.[6].

In this section we present a technique to determine the actuator or sensor locations and their gains to excite and sense a target mode or a set of targeted modes. The technique is based on the relationship between the modal and nodal coordinates of the actuator or sensor locations. Being distinct from the force appropriation method it does not require the input force to be a harmonic one. Rather, it determines the actuator locations and actuator gains, while the input force time history is irrelevant (modal actuator or sensor acts as a filter). The locations and gains, for example, can be implemented as a width-shaped piezoelectric film. Finally, the method allows for excitation and observation of not only a single structural mode but also of a set of selected modes, each one with the assigned amplitude.

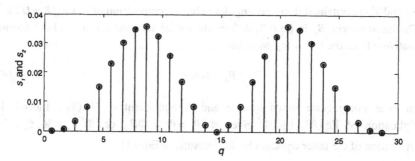

Figure 33. The verification of Eq.(80): O denotes $s_1 = \sigma_{233}\sigma_{2qq}$, • denotes $s_2 = \sigma_{23q}\sigma_{2q3}$, and

$$s_1 = s_2.$$

A structural model in this section is described by the second order modal model, see Eq.(7). In modal coordinates the equations of motion of each mode are decoupled, as in Eq.(10). Thus if the modal input gain is zero, the mode is not excited; if the modal output gain is zero, the mode is not observed. This simple physical principle is the base for the more specific description of the problem in the following sections.

7.1. MODAL ACTUATORS

The task in this section is to determine the locations and gains of the actuators such that n_m modes of the system are excited with approximately the same amplitude, where $1 \leq n_m \leq n$, and n is the total number of considered modes. This task is solved using the modal equations (7) or (10). Note that if the ith row, b_{mi}, of the modal input matrix, B_m, is zero the ith mode is not excited. Thus, by assigning the entries of b_{mi} either 1 or 0 one makes the ith mode either excited or not. For example, if one wants to excite the first mode only, B_m is a one-column matrix of a form $B_m = \begin{bmatrix} 1 & 0 & \cdots & 0 \end{bmatrix}^T$. On the other hand, if one wants to excite all modes independently and equally, one assigns a unit matrix, $B_m = I$.

Given (or assigned) the modal matrix B_m, the nodal matrix B_o is derived from Eq.(8). Equation (8) can be re-written as follows

$$B_m = R B_o \tag{81}$$

where $R = M_m^{-1} \Phi^T$ and matrix R is of dimensions $n \times n_d$. Recall that the number of assigned modes is $n_m \leq n$. If the assigned modes are controllable, i.e. the rank of R is n_m, the least-square solution of Eq.(81) is

$$B_o = R^+ B_m \tag{82}$$

In the above equation R^+ denotes the pseudoinverse of R, i.e., $R^+ = V \Sigma^{-1} U^T$ where U, Σ, and V are obtained from the singular value decomposition of R, i.e., $R = U \Sigma V^T$.

The input matrix B_o in Eq.(82) defines the modal actuator and it can be determined alternatively from the following equation

$$B_o = M \Phi B_m \tag{83}$$

which does not require pseudoinverse and is equivalent to Eq.(81). Indeed, left-multiplication of Eq.(83) by Φ^T gives $\Phi^T B_o = \Phi^T M \Phi B_m$ or $\Phi^T B_o = M_m B_m$. Left-multiplication of the latter equation by M_m^{-1} results in Eq.(81).

Example 12. Modal actuator for a beam, single mode case. Consider a clamped beam as in Fig.21, divided into $n = 15$ elements. The vertical displacement sensors are located at nodes from 2 to 15 and the single output is a sum of the sensor readings. Actuator locations shall be determined such that the second mode with modal gain of 0.01 is excited while the remaining modes are not excited. The first nine modes are considered.

The assigned modal matrix is in this case $B_m^T = \begin{bmatrix} 0 & 0.01 & 0 & 0 & 0 & 0 & 0 & 0 & 0 \end{bmatrix}^T$. From (82), for this modal input matrix, a nodal input matrix B_o is determined. It contains gains for the vertical forces at the nodes from 2 to 15. The gain distribution of the actuators is shown in Fig.34a. Note that this distribution is proportional to the displacements of the second mode shape. This distribution can be implemented as an actuator with a gain proportional to its width. Thus the shape of an actuator that excites the second mode is shown in Fig.34b.

For the above input and output the magnitude of the transfer function are presented in Fig.35. The plot shows clearly that only the second mode is excited.

If one wants to excite the ith mode with certain amplitude, say a_i, the H_∞ norm can be used as a measure of the amplitude of the ith mode. In case of a single-input-single-output system, the H_∞ norm of the ith mode is equal to the height of the ith resonance peak. In case of multiple inputs (or outputs) the H_∞ norm of the ith mode is approximately equal to the root-mean-square sum of the ith resonance peaks corresponding to each input (or output). It is approximately determined as follows

$$\|G_i\|_\infty \cong \frac{\|B_{mi}\|_2 \|C_{mi}\|_2}{2\zeta_i\omega_i} = \frac{\|b_{mi}\|_2 \|c_{mi}\|_2}{2\zeta_i\omega_i} \tag{84}$$

see (39).

Assume a unity input gain for the current mode, i.e. $\|b_{mi}\|_2 = 1$, so that the current amplitude a_{oi} is

$$a_{oi} = \frac{\|c_{mi}\|_2}{2\zeta_i\omega_i} \tag{85}$$

In order to obtain amplitude a_i one has to multiply a_{oi} by the weight w_i, such that $a_i = w_i a_{oi}$. Introducing (85) to the latter equation one obtains the weight

$$w_i = \frac{2a_i\zeta_i\omega_i}{\|c_{mi}\|_2} \tag{86}$$

Define the weight matrix $W = diag(w_1, w_2, \ldots, w_n)$, then the matrix that sets the required output modal amplitudes is

$$B_{mw} = WB_m \tag{87}$$

Example 13. Modal actuator for a beam, multiple mode case. All nine modes of the same beam need to be excited with the amplitude of 0.01 by a single actuator. For this case, the matrix B_m is as follows $B_m^T = 0.01 \times [1 \ 1 \ 1 \ 1 \ 1 \ 1 \ 1 \ 1 \ 1]^T$ and the weighting matrix is obtained from Eq.(86). The resulting gains of the nodal input matrix B_o (shown in Fig.36a) do not follow any particular mode shape. The width of an actuator that corresponds to the input matrix B_o and excites all nine modes is shown in Fig.36b.

The plot of the transfer function of the single-input system with the input matrix B_o is shown in Fig.37. The plot shows that all nine modes are indeed excited, with approximately the same amplitude of 0.01 cm.

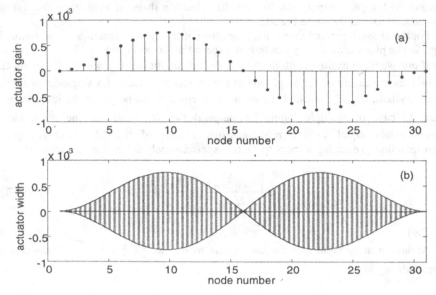

Figure 34. Actuator gains (a) and the corresponding actuator width (b) that excite the second mode.

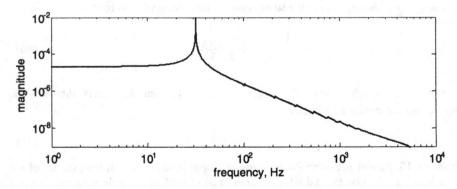

Figure 35. Magnitude of a transfer function with the second-mode modal actuator: only the second mode is excited.

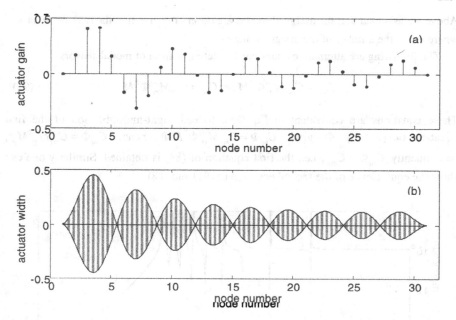

Figure 36. Actuator gains (a), and corresponding actuator width that excites all nine modes.

7.2. MODAL SENSORS

The modal sensor determination is similar to the determination of modal actuators. The governing equations are derived from Eq.(9)

$$C_{mq} = C_{oq}\Phi, \quad C_{mv} = C_{ov}\Phi \tag{88}$$

If one wants to observe a single mode only (say ith mode), one assumes the modal output matrix is in the form $C_{mq} = \begin{bmatrix} 0 & \ldots & 0 & 1 & 0 & \ldots & 0 \end{bmatrix}$, where 1 stands at the ith position. If one wants to observe n_m modes one assumes the modal output matrix is in the form $C_{mq} = \begin{bmatrix} c_{q1}, c_{q2}, \ldots, c_{qn} \end{bmatrix}$, where $c_{qi} = 1$ for selected modes, otherwise $c_{qi} = 0$. The corresponding output matrix is obtained from (88)

$$C_{oq} = C_{mq}\Phi^+ \tag{89}$$

where Φ^+ is the pseudoinverse of Φ. Similarly, one obtains the rate sensor matrix C_{ov} for the assigned modal rate sensor matrix C_{mv}

$$C_{ov} = C_{mv}\Phi^+ \tag{90}$$

130

Above we assumed that the assigned modes are observable, i.e. that the rank of Φ is n_m, where n_m is the number of the assigned modes.

The following are alternative equations for determination of modal sensors

$$C_{oq} = C_{mq}M_m^{-1}\Phi^T M, \quad C_{ov} = C_{mv}M_m^{-1}\Phi^T M \tag{91}$$

These equations are equivalent to Eqs.(88). Indeed, right-multiplication of the first equation of (91) by Φ gives $C_{oq}\Phi = C_{mq}M_m^{-1}\Phi^T M\Phi$, hence $C_{oq}\Phi = C_{mq}M_m^{-1}M_m$, consequently $C_{oq}\Phi = C_{mq}$, i.e., the first equation of (88) is obtained. Similarly one can show the equivalence of the second equation of (91) and (88).

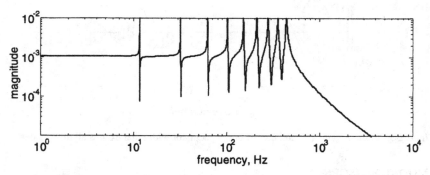

Figure 37. Magnitude of a transfer function for the nine-mode modal actuator shows the nine resonances of the excited modes.

Multiple modes with assigned modal amplitudes a_i are obtained using the sensor weights. The weighted sensors are obtained from equation similar to Eq.(86). Namely, the ith weight is determined from the following equation

$$\|c_{mi}\|_2 = \frac{2\zeta_i\omega_i a_i}{\|b_{mi}\|_2} \tag{92}$$

Example 14. Modal sensors for a beam and for all modes. The beam from Fig.21 divided into $n = 15$ elements, with three vertical force actuators located at nodes 2, 7, and 12 is considered. The task is to find the displacement output matrix C_{oq} such that the first nine modes have equal contribution to the measured output with amplitude 0.01.

The matrix C_{mq} that excites the first nine modes is the unit matrix of dimension 9 of amplitude $a_i=0.01$, i.e., $C_{mq} = 0.01 \times W \times I_9$. The gains that make the mode amplitudes approximately equal are determined from Eq.(92). The output matrix C_{oq} is determined from Eq.(89). For this matrix the magnitudes of the transfer functions of the 9 outputs in Fig.38 show that all nine of them have a resonance peak of 0.01.

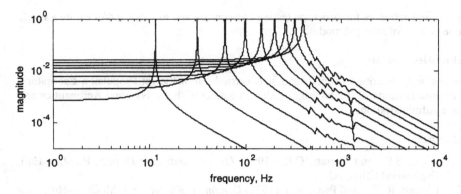

Figure 38. Magnitude of the transfer function with the nine single-mode sensors.

Example 15. Modal sensors for a beam, for all but one mode. The beam with actuators as in Example 14 is considered. Find the nodal rate sensor matrix C_{ov} such that all nine modes but mode 2 contribute equally to the measured output with the amplitude of 0.01.

The matrix C_{mv} that gives in the equal resonant amplitudes of 0.01 is as follows: $C_{mv} = 0.01 \times W \times [1 \quad 0 \quad 1 \quad 1 \quad 1 \quad 1 \quad 1 \quad 1 \quad 1]$, where the weight W is determined from (92) and the output matrix C_{ov} is obtained from (90). For this matrix the magnitude of the transfer function is shown in Fig.39, dashed line. This magnitude is compared with the magnitude of the transfer function for the output that contains all the 9 modes (solid line). It is easy to notice that the second resonance peak is missing in the plot.

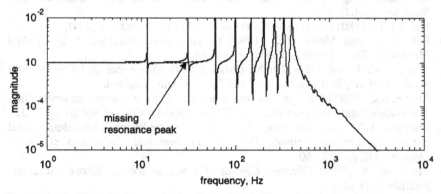

Figure 39. Magnitude of the transfer function with the nine-mode sensor (solid line), and with the eight-mode sensor where the second mode is missing (dashed line).

8. Conclusions

In this paper we discussed the importance of locations and gains of sensors and actuators in structural dynamics. They impact the controllability and observability of a structure and its norms. Using modal state space representation the norms can be used to locate sensors and actuators for structural testing and control, and to sense and/or excite a single mode or a set of selected modes.

Acknowledgements

The research described in this paper was carried out at the Jet Propulsion Laboratory, California Institute of Technology, under a contract with the National Aeronautics and Space Administration.

References

1. Boyd, S.P., and Barratt, C.H. (1991) *Linear Controller Design*, Prentice Hall, Englewood Cliffs, NJ.
2. Clough, R. W., and Penzien, J. (1975) *Dynamics of Structures*, McGraw-Hill, New York.
3. Cooper, J.E. and Wright, J.R. (1997) To fit or to tune? That is the question, *Proc. 15th International Modal Analysis Conf.*, Orlando, FL, pp.1353-1359.
4. Friswell, M.I. (2001) On the design of modal actuators and sensors, *Journal of Sound and Vibration*, vol.241, No.3, pp.361-372.
5. Gawronski, W. (1998) *Dynamics and Control of Structures. A Modal Approach.* Springer, New York.
6. Gawronski, W. (2000) Modal actuators and sensors, *Journal of Sound and Vibration*, vol. 229, no.4, pp.1013-1022.
7. Hatch, M.R. (2000) *Vibration Simulation Using Matlab and ANSYS*, Chapman and Hall/CRC, Boca Raton.
8. Hsu, C.-Y., Lin, C.-C. and Gaul, L., (1998) Vibration and sound radiation controls of beams using layered modal sensors and actuators, *Smart Materials and Structures*, vol.7, pp.446-455.
9. Kailath, T. (1980) *Linear Systems*, Prentice Hall, Englewood Cliffs, NJ.
10. Lee, C.-K. and Moon, F.C. (1990) Modal sensors/actuators, *J. of Applied Mechanics, Trans. ASME*, vol. 57, pp.434-441.
11. Maia, N.M.M., and Silva, J.M.M. (Eds.), (1997) *Theoretical and Experimental Modal Analysis*, Research Studies Press Ltd., Taunton, England.
12. Moore, B.C. (1981) Principal component analysis in linear systems, controllability, observability and model reduction, *IEEE Trans. Autom. Control*, vol.26, pp.17-32.
13. Phillips, A.W., and Allemang, R.J. (1996) Single degree-of-freedom modal parameter estimation methods. *Proc. 14th International Modal Analysis Conf.*, Deaborn, MI, pp.253-260.
14. Preumont, A. (2002) *Vibration Control of Active Structures*, Kluwer Academic Publishers, Dordrecht.
15. Skogestad, S., and Postlethwaite, I. (1996) *Multivariable Feedback Control*, Wiley, Chichester, England.
16. Van de Wal, M., and de Jager, B. (2001) A review of methods for input/output selection, *Automatica*, vol.37, pp.487-510.
17. Zhou, K., and Doyle, J.C. (1998) *Essentials of Robust Control*, Prentice Hall, Upper Saddle River, NJ.

ROBUST CONTROL OF VIBRATING STRUCTURES

K. B. LIM
NASA Langley Research Center
Mail Stop 161, Hampton, Virginia, USA, 23681
k.b.lim@larc.nasa.gov

Abstract. This article gives a broad heuristic overview of some concepts in the application of robust control to flexible structures. It includes the modeling, analysis, and design of controllers and the sort of results that can be expected in a realistic setting for a flexible structure control problem, beyond single input/single output. A goal of this article is to convince, by heuristic means, those who are profoundly skeptical concerning robust control, at the least the intimate connection that exists between classical and multivariable robust control theory. Indeed, the classical robustness concepts in a single loop generalizes to multivariable robust control theory for multiple loops. An underlying theme of this article is the fact that important practical limitations originate from limitations in the accuracy of mathematical models representing flexible structures, not because of computational or numerical limitations or even a lack of sound ideas in mathematical modeling, but rather because of the lack of reliable detailed knowledge necessary in practice to accurately quantify the dynamical process in a given physical system. To this end, recent results are highlighted in the construction of uncertainty bounds directly from measured data.

1. Background

The ubiquity of classical control laws seems inevitable noting the fact that basically only the gain and phase response of a given dynamical system are needed to devise effective yet simple control laws for single loop systems. Furthermore, these responses can usually be determined empirically for a given dynamical system, often without requiring a detailed understanding of the governing internal dynamics of the system or a derivation based on first principles. Consequently, a large body of knowledge exists today for analysis and design of classical controllers for a broad scope of applications (see, for example, [1]). However, two significant limitations became evident concerning multivariable systems: the inconvenience and ad-hoc nature in using a loop at a time analysis and multiple gain tuning procedures, and a rigorous quantification of the sensitivity and robustness under simultaneous uncertainties in multiple loops.

The advent of state space and optimization theories combined with a rapid development of computing power in the '60s and '70s, enabled detailed characterizations and predictions of the dynamics of multivariable systems using "internal" state variables, and the development of optimal control systems. Since then, after extensive application of modern control, it became clear that although elaborate mathematical models can be readily constructed from first principles and subsequently optimal control laws designed and analyzed, the predicted optimal performance is not usu-

A. Preumont (ed.), Responsive Systems for Active Vibration Control, 133–179.
© 2002 *Kluwer Academic Publishers. Printed in the Netherlands.*

ally realizable in practice, and the gap between predicted and realized performance increases with increasing dynamical complexity in the system. This mismatch in the response, namely the robustness problem, is well appreciated in classical control by means of graphical analysis in the frequency domain. However, classical control robustness ideas are more suited for nonparametric errors in a single loop system, and cannot easily account for simultaneous multiple parametric and nonparametric errors in multivariable systems. This motivated the development of robust control theory for multivariable systems in the '80s. A popular view of robust control theory today is to describe all system uncertainties including unknown random exogenous disturbances as a set of mathematical models, and then to design a robust controller that will optimize a worst case performance over this set of mathematical models. This naturally led to a worst case design philosophy and the intense interest in optimizing H_∞ norms over the past two decades.

Robustness issues due to spillover instabilities of higher frequency structural modes, have been reported early in [2], and in a broader context for large flexible space structures in [3]. Indeed, the control of large flexible space structures had been an area of much interest beginning in the late '70s, evidenced for example by a series of symposiums [4]. Subsequently, NASA and the Department of Defense had been keen in the development of key technologies in the area of Controls-Structures Interaction (CSI) [5]. The flexible structure example used to illustrate robust control ideas in this article, is described in a broader context in [6].

2. An overview of robust control theory

In this section, some key concepts underlying much of linear time-invariant robust control is reviewed briefly, for those readers who are familiar with classical control concepts but not with robust control theory. Since the scope of the issues selectively reviewed is extensive due to its generality and richness, it is hoped that the brief outline is sufficiently simple to expose the key ideas but not overly simplified to be misleading.

2.1. PERFORMANCE SPECIFICATION USING WEIGHTED NORMS

To illustrate the need to quantify signals and systems, consider the control of an aircraft as shown in Figure 1. To specify a performance measure related to passenger comfort, one could specify a small vertical acceleration due to gust. To specify a performance measure related to the guidance of an aircraft along a flight path, one could specify a small tracking error about a nominal path. Of course the question then is, what is "small" acceleration for the gust response and path error for the guidance ? The "smallness" of the signal, y, can be specified in the sense of a signal norm $\|y\|$ and which norm is appropriate depends on its intended physical application. For example, if a peak acceleration is to be improved for passenger comfort, then the L_∞ norm (loosely speaking a measure of a signal peak) should be used. On the other hand, for mean tracking error the L_2 norm (loosely speaking an RMS measure of a signal) is obviously the right choice.

An important concern in controller design is what to do if the aircraft dynamics is not accurately known. This robustness concern is addressed by first introducing

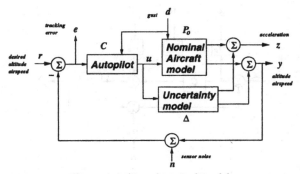

Figure 1. Aircraft control model.

and then quantifying an uncertainty model as shown in Figure 1. The "size" of model error, Δ, in a system should reflect its capacity to amplify a physical input signal, (of course the input/output signals themselves must be quantified by some signal norm) i.e., should define an "induced" norm. To quantify multivariable systems, we frequently make use of induced norms, and in particular for induced 2-norms, which is identical to

$$\sup \frac{\|y\|_2}{\|u\|_2} = \sup_{\omega \in R} \bar{\sigma}[G(j\omega)] := \|G\|_\infty \qquad (1)$$

The H_∞ norm is also a useful way to define a bounded set of a system uncertainty, say Δ, of transfer function matrices. This of course quantifies the erroneous input signal amplification due to the system uncertainty (see Figure 1). It turns out that for robust control purposes the H_∞ norm can be "generalized" to μ to conveniently quantify performance robustness under general uncertainty structure, as described later.

Performance specification example. Figure 2 show how one could specify a tracking performance over some bandwidth. This is typical of loop shaping concept in classical control. The tracking error is $e = Sr$, where $S := (1 + PC)^{-1}$ is the sensitivity

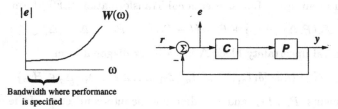

Figure 2. Tracking performance specification example.

function, and we want the tracking error to satisfy the inequality

$$|e(\omega)| = |S(\omega)r(\omega)| \leq W(\omega), \quad \forall \omega, \ |r(\omega)| \leq 1$$
$$\Leftrightarrow \sup_\omega |W^{-1}(\omega)S(\omega)| = \|W^{-1}(\omega)S(\omega)\|_\infty \leq 1$$

136

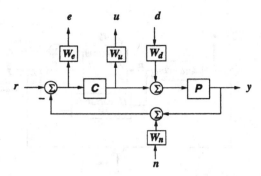

Figure 3. Performance specification based on four weighted transfer functions.

So, tracking performance can be specified by a frequency weighted H_∞-norm inequality. As a slightly more general performance specification example, consider specifying requirements for sensitivity and complementary sensitivity, see Figure 3. The folowing four weighted transfer functions are frequently used in control design traded-off studies:

$$\left\{\begin{matrix} e \\ u \end{matrix}\right\} = -\left[\begin{matrix} (W_e W_d P)S & (W_e W_n)S \\ (W_u W_d)T & (\frac{W_u W_n}{P})T \end{matrix}\right]\left\{\begin{matrix} d \\ n \end{matrix}\right\} \tag{2}$$

where $S := (1 + L)^{-1}$, $T := (1 + L)^{-1}L$, $L := PC$ the loop gain, and W_e, W_u, W_d, W_n, denote weights for tracking error, control penalty, disturbance, and noise, respectively. The loop gain used in classical control to shape S and T can be extended to multivariable systems via H_∞ loop-shaping (see for example [7, 8]), a generalization using singular values to handle vector signals and frequency weighted transfer matrices.

2.2. REPRESENTING PLANT UNCERTAINTIES

To quantify the effects of uncertainties on a control system, we first need to model and quantify uncertainties. The basic idea is to represent uncertainties in a system by a set of models. For general linear multivariable feedback systems, a set of model is described in an upper Linear Fractional Transformation (LFT) form

$$\mathcal{F}_u(P, \Delta) := P_{22} + P_{21}\Delta(I - P_{11}\Delta)^{-1}P_{12}, \quad \Delta : \|\Delta\|_\infty \le 1 \tag{3}$$

and the general uncertainty block Δ has a block diagonal form,

$$\Delta = \{\mathrm{diag}[\delta_1 I_{n_1}, \ldots, \delta_r I_{n_r}, \Delta_1, \ldots, \Delta_F] : \delta_i, \ \Delta_j \in H_\infty\}$$

The components, P_{11}, P_{12}, and P_{21}, describe the interconnection of the uncertainty block to the nominal plant, P_{22}, together forming the augmented plant, P. Usually a "best" single model available is used for the component, P_{22} ($= \mathcal{F}_u(P, 0)$). So, in general we define input-output mappings over a set of plants as[1]

$$y = \mathcal{F}_u(P, \Delta)u, \quad \Delta \in \mathcal{D}_W \tag{4}$$

---

[1] $\mathcal{D}_W$ denote a block diagonal set of uncertainties with sizes denoted by W

Unknown but bounded gain uncertainty example. Suppose

$$y = kP_o(s)u, \quad \text{where} \quad \underline{k} \leq k \leq \bar{k} \tag{5}$$

This set of uncertain gains can be written as

$$k = k_o(1 + W\Delta), \quad |\Delta| \leq 1 \tag{6}$$

where $k_o := (\bar{k} + \underline{k})/2$ denotes nominal or average gain and $W := (\bar{k} - \underline{k})/2k_o$ denotes its uncertainty radius. This can be represented as a lower LFT as shown in Figure 4. Hence, we can write the following input-output relations for the above set of plants

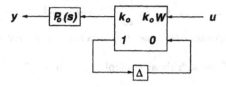

Figure 4. A gain uncertainty represented as an upper LFT.

with uncertain gains

$$y = \mathcal{F}_l(\underbrace{\begin{bmatrix} k_o & k_o W \\ 1 & 0 \end{bmatrix}}_{k}, \Delta) \, P_o(s)u, \quad |\Delta| \leq 1 \tag{7}$$

Notice that the uncertainty radius W is included in the augmented plant, so that Δ is unity bounded.

Mass-spring-dashpot example. Slightly more general,[2] consider

$$\ddot{x} + \frac{c}{m}\dot{x} + \frac{k}{m}x = \frac{f}{m} \tag{8}$$

with the following unknown but bounded uncertainties

$$m = m_o(1 + W_m\delta_m), \quad m_o := \underbrace{\frac{1}{2}(\bar{m} + \underline{m})}_{nominal}, \quad W_m := \underbrace{\frac{1}{2}(\frac{\bar{m} - \underline{m}}{m_o})}_{radius}, \quad |\delta_m| \leq 1$$

$$c = c_o(1 + W_c\delta_c), \quad c_o := \frac{1}{2}(\bar{c} + \underline{c}), \quad W_c := \frac{1}{2}(\frac{\bar{c} - \underline{c}}{c_o}), \quad |\delta_c| \leq 1$$

$$k = k_o(1 + W_k\delta_k), \quad k_o := \frac{1}{2}(\bar{k} + \underline{k}), \quad W_k := \frac{1}{2}(\frac{\bar{k} - \underline{k}}{k_o}), \quad |\delta_k| \leq 1$$

Similar to the gain uncertainties illustrated earlier, the uncertainties in the damping, stiffness, and inverse mass parameters can be represented as the following LFT's

$$c = \mathcal{F}_l\left(\underbrace{\begin{bmatrix} c_o & c_o W_c \\ 1 & 0 \end{bmatrix}}_{C_1}, \delta_c\right), \quad k = \mathcal{F}_l\left(\underbrace{\begin{bmatrix} k_o & k_o W_k \\ 1 & 0 \end{bmatrix}}_{K_1}, \delta_k\right), \quad \frac{1}{m} = \mathcal{F}_l\left(\underbrace{\begin{bmatrix} \frac{1}{m_o} & -\frac{W_m}{m_o} \\ 1 & -W_m \end{bmatrix}}_{M_1}, \delta_m\right) \tag{9}$$

---

[2] From page 171- [9].

By replacing these uncertain parameters by their respective LFT's, we obtain the dynamic interconnections as illustrated in Figure 5. One can then "pull out" the Δ's

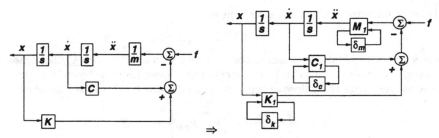

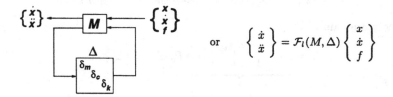

Figure 5.   Dynamic interconnections for a mass-spring-dashpot system.

to obtain a single LFT for a "robust control design model"

$$\text{or} \qquad \left\{ \begin{matrix} \dot{x} \\ \ddot{x} \end{matrix} \right\} = \mathcal{F}_l(M, \Delta) \left\{ \begin{matrix} x \\ \dot{x} \\ f \end{matrix} \right\}$$

So, any interconnection of LFT's is simply a larger LFT, which is a very convenient feature in modeling. In addition, non-parametric uncertainties can usually be representated as LFTs. With some idea of uncertainty models, we examine robustness in more detail.

2.3. NOTION OF ROBUSTNESS

Suppose a plant P belongs to a set of models

$$P \in \mathcal{P}_\Delta := \{\mathcal{F}_l(M, \Delta), \|\Delta\| \le 1\} \tag{10}$$

and consider a characteristic, Ω, of a feedback system. A controller C is said to be "robust" with respect to Ω over $\mathcal{P}_\Delta$ if Ω holds for the controller/plant system (C, P) for every plant $P \in \mathcal{P}_\Delta$. We begin by reviewing stability margin concepts used in classical control and then outline a generalization of this classical control ideas to multivariable systems as described in more detail in [10].

Gain and Phase Margins.   Consider the robustness of internal stability for a single loop system. The feedback loop is said to be internally stable if the closed loop transfer function has no poles in right hand plane (RHP), or equivalently, the loop gain, $PC(s) \ne -1$, *critical point*, for all s in RHP. Using Cauchy's principle, we can construct a graphical check for internal stability: a closed loop is internal stable if its Nyquist plot of $PC(s)$ encircles -1 point exactly the number of unstable poles in $PC(s)$, which is the well known Nyquist stability criteria.

More importantly for robustness analysis, we can use Nyquist stability criteria to define *stability margins* by indicating *how close* we are to crossing the critical point (i.e. instability). This motivates the definitions for *gain margin* of $PC(s)$, which is the largest gain k such that the feedback loop remains stable, and the *phase margin* of $PC(s)$, which is the largest phase angle ϕ such that feedback loop remains stable. Figure 6 ilustrates these well known concepts. Notice that gain and phase

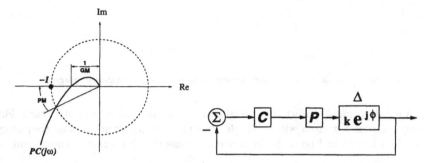

Figure 6. Nyquist plot of $PC(s)$ illustrating robustness with respect to independent gain and phase errors in loop.

margins measure distances of the Nyquist plot to the critical point in *certain specific directions*. A consequence of this directional dependency is that, even if a system possess both large gain and phase margins, a small *simultaneous* or *directionally independent* perturbation could destabilize it, i.e. it may still not be robust. Hence, good gain and phase margins are only necessary conditions for stability robustness. This motivates a *disk* type of stability margin about loop gain which *will not* be directionally dependent.

Directionally independent loop gain stability margin. Consider the minimum distance of loop gain from critical point as a stability margin,

$$\inf_{\omega} |1 + L(j\omega)| = \left(\sup_{\omega} \frac{1}{|1 + L(j\omega)|} \right)^{-1} = \|S\|_{\infty}^{-1}. \tag{11}$$

Notice that this minimum distance is the inverse H_{∞} norm of the sensitivity transfer function, so that a large sensitivity norm implies that the system is near an instability. A perturbed system will remain stable if its loop gain, $\tilde{L}(j\omega)$, satisfies this minimum distance margin

$$|\tilde{L}(j\omega) - L(j\omega)| < \|S\|_{\infty}^{-1}, \quad \forall \omega \tag{12}$$

Notice from Figure 7 that this stability margin is conservative because, (i) a constant stability margin is applied over all frequencies, and (ii) it is based only on the loop gain structure.

Frequency & structure dependent stability margin. For these reasons, we reformulate the previous directionally independent stability margin by incorporating (i)

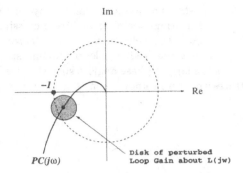

Figure 7. A minimum distance directionally independent stability margin.

a frequency dependent uncertainty weight, and, (ii) an uncertainty structure. For example, consider a system with *multiplicative* uncertainty structure and frequency weight W as shown in Figure 8. From a robustness standpoint, it is convenient to

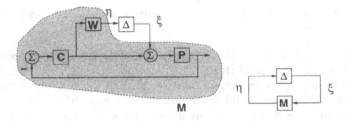

Figure 8. System with frequency weighted and multiplicative uncertainty structure.

group the non-delta terms, i.e., the *nominal model*, as M so that we can look at any system with a frequency weighted and structured uncertainties as having an $\Delta - M$ interconnection, where in this example, $M := WT$, $T := \frac{PC}{1+PC}$, the *complementary sensitivity*.

Note that for robustness analysis, it is necessary to assume that controller, C, stabilizes the nominal plant, M. Furthermore, we restrict the uncertainties, Δ, $|\Delta| < 1$, to a set of stable perturbations so that perturbed plant $(1 + W\Delta)P$ will have same number of unstable poles as nominal plant, P, as required by Nyquist stability criteria. We also select a stable uncertainty radius filter, W, by choice.

Under the above assumptions, we apply Nyquist stability condition to this new loop transfer function, $M\Delta$. The system will be stable for a particular multiplicative uncertainty, Δ, if the new loop gain $M\Delta$ does not encircle -1. Therefore, for robust stability (RS), $M\Delta$ should not cross -1 for all $|\Delta| \leq 1$, i.e.

$$RS \iff |1 + M\Delta| \neq 0, \quad \forall |\Delta| \leq 1, \quad \forall \omega$$
$$\cdots$$
$$\iff |M| < 1, \quad \forall \omega \qquad \equiv \|M\|_\infty < 1$$

So, the robust stability of a system with multiplivative uncertainty is defined by an inequality on a weighted H_∞ norm of its complementary sensitivity. More importantly, one can derive the stability robustness condition of any uncertainty structure in the same way, once we define $\Delta - M$ for the system.[3]

Robust Performance.   A controller is said to satisfy robust performance if the controller guarantees internal stability and performance to a specified degree, for a specified set of plants. For example, consider a system with multiplicative uncertainty, $|\Delta| \leq 1$, as shown in Figure 9 where r and e denotes the input command and

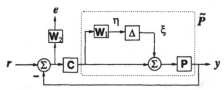

Figure 9.   Tracking performance under multiplicative uncertainty.

tracking error, respectively, while W_1 and W_2 denotes the weights for uncertainty and tracking error, respectively. Assume that a controller C stabilizes a nominal plant, P, so that we can define a nominal performance (NP) for tracking, say

$$NP \quad \Leftrightarrow \quad \text{at } \Delta = 0, \sup \frac{\|e\|_2}{\|r\|_2} < 1 \quad \equiv \quad \|W_1 S\|_\infty < 1 \qquad (13)$$

where $S := (1+PC)^{-1}$ is nominal sensitivity. The robust performance (RP) question then is, what happens to tracking performance if $\Delta \neq 0$?

It turns out that the following conditions can be derived for RP (see page 56-7 of [10])

$$
\begin{aligned}
RP \quad &:= \quad RS \ \& \ \|W_1 \tilde{S}\|_\infty < 1, \ \forall \Delta : |\Delta| \leq 1 \\
&\Leftrightarrow \quad RS \ \& \ |W_1| < |1 + \tilde{L}|, \ \forall \tilde{L} = (1 + W_2 \Delta) L : |\Delta| \leq 1, \forall \omega \\
&\cdots \\
&\Leftrightarrow \quad \||W_1 S| + |W_2 T|\|_\infty < 1
\end{aligned}
$$

where $\tilde{S} := (1 + \tilde{P}C)^{-1}$ and $\tilde{L}$ denote the perturbed sensitivity and loop gain respectively. Notice that RP implies NP and RS, but not vice-versa, and it is hard to test RP condition by checking over all Δ's. The condition dependence over all Δ is satisfied by using worst case Δ. A classical control interpretation of this H_∞ inequality RP condition is illustrated in Figure 10. The nominal performance condition, $\|W_1 S\|_\infty < 1 \equiv |W_1| < |1 + L|, \forall \omega$, means that the nominal loop gain lies outside $|W_1(j\omega)|$ disk for all frequencies. The robust stability condition, $\|W_2 T\|_\infty < 1 \equiv |W_2 L| < |1 + L|, \forall \omega$, means that the critical point lies outside $|W_2(j\omega)L(j\omega)|$ disk for all frequencies. The robust performance condition, $|W_1| + |W_2 L| < |1 + L|, \forall \omega$, means that the disks are disjoint for all frequencies.

It is important to note that for this example the RP test is easy, a simple check on H_∞ norm. However, for general RP problems *there is no similar simple test.* Subsequently, the so called Structured Singular Value or μ was introduced by Doyle

---

[3] A table of robust stability tests is given in page 55 of [10].

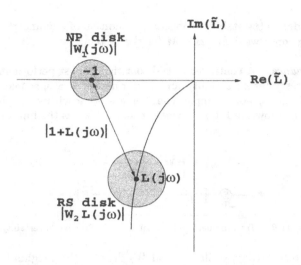

Figure 10. Nyquist interpretation of the H_∞ norm conditions for SISO example.

in 1982 [11], to methodically formulate and test for RP (and RS) - which is equivalent to checking over all Δ's, a nontrivial task.

Robustness under multiple uncertainties. Consider for example, the RS conditions for a system having multiplicative and feedback uncertainties as shown in Figure 11 where

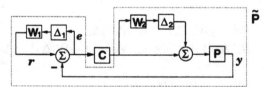

Figure 11. A system with multiple uncertainties.

$$\tilde{L} := \tilde{P}C, \quad \tilde{P} := P\left(\frac{1 + W_2\Delta_2}{1 - W_2\Delta_1}\right), \quad \forall\Delta_i : |\Delta_i| \le 1, i = 1, 2. \quad (14)$$

The following RS condition derived using the Nyquist criteria assumes that the nominal closed loop system is stable and all perturbed plants have same number of unstable poles as nominal plant.

$$RS \quad \Leftrightarrow \quad \text{avoid crossing -1 by some } \tilde{L}, \forall\Delta_i : |\Delta_i| \le 1, i = 1, 2, \forall\omega$$
$$\Leftrightarrow \quad |1 + L - W_1\Delta_1 + LW_2\Delta_2| > 0, \quad \forall\Delta_i : |\Delta_i| \le 1, i = 1, 2, \forall\omega$$
$$\cdots$$
$$\Leftrightarrow \quad \||W_1S| + |W_2T|\|_\infty < 1$$

Note that the dependence on Δs in the above inequality conditions is removed by considering its *worst case*. In summary, we see that the robust tracking performance problem from r to e with respect to Δ_2

$$RP \quad := \quad RS \ \& \ \|W_1\tilde{S}\|_\infty < 1, \ \forall \Delta_2, |\Delta_2| \le 1 \tag{15}$$

and the corresponding robust stability problem with respect to two uncertainties, Δ_1, and Δ_2

$$RS \quad := \quad |1 + \tilde{L}| > 0, \quad \forall \Delta_i : |\Delta_i| \le 1, i = 1, 2, \ \forall \omega \tag{16}$$

have *identical* conditions $\||W_1S| + |W_2T|\|_\infty < 1$. In fact, this holds for the more general LFT interconnections and this equivalence is used in μ analysis to handle both RP and RS problems in a single framework.

To summarize, we have reviewed that classical robustness and sensitivity concepts in the forms of gain and phase margins in a single loop generalizes naturally to robust control theory of the H_∞ variety, for multivariable systems. The usefulness of H_∞ norm and its function space is clearly evident in the characterization of worst case response in signals and systems for the purpose of quantifying robustness due to larger than infinitesimal uncertainties. In the next two sections, we give a brief overview of the theory underlying robustness analysis and design.

2.4. μ PERSPECTIVE ON ROBUSTNESS

Any linear interconnection of inputs (including command, noise, disturbance) r, outputs (all signals to attenuate) e, augmented plant P, controller K, block diagonal uncertainty $\Delta \in \mathcal{D}$, can be represented as $\Delta - P - K$ LFT interconnection as illustrated in Figure 12. The uncertainty has the block-diagonal structure

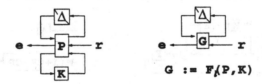

Figure 12.   General Interconnection.

$$\mathcal{D} = \left\{ \mathrm{diag}(\delta_1 I_{n_1}, \ldots, \delta_r I_{n_r}, \Delta_1, \ldots, \Delta_{n_m}) \mid \delta_i \in \mathcal{F}, \Delta_j \in \mathcal{F}^{k_j \times k_j} \right\} \tag{17}$$

where $\mathcal{F}$ is either the field of real or complex numbers depending on its usage. Assuming that K stabilizes P so that the nominal closed loop system, G, is stable, we define the stability and performance requirements in terms of LFT's and H_∞ norms:

$$\left.\begin{array}{lll} NP & \Leftrightarrow & \|G_{22}\|_\infty < 1 \\ RS & \Leftrightarrow & F_u(G, \Delta) \text{ stable}, \forall \Delta \in B_\Delta \\ RP & \Leftrightarrow & RS \ \& \ \|F_u(G, \Delta)\|_\infty < 1, \forall \Delta \in B_\Delta \end{array}\right\} \tag{18}$$

where $B_\Delta := \{\Delta \in \mathcal{D} : \|\Delta\|_\infty \le 1\}$ represents a structured unit ball of uncertainty. The main question for robust stability is, under what conditions does $\Delta - G_{11}$ loop remain stable for $\forall \Delta \in B_\Delta$?[4] Using the MIMO Nyquist stability criteria (see p. 146- [12]):

Robust Stability (Determinant condition, general).

$$\Delta - G_{11} \text{ loop is } RS \; \forall \Delta \in B_\Delta \quad \Leftrightarrow \quad \det(I - G_{11}\Delta) \neq 0, \forall \omega, \forall \Delta \in B_\Delta \qquad (19)$$

This determinant condition is general and applies for any set B_Δ, but *is hard to test.* As a special case, if Δ is structured but consists of complex elements only:

Robust Stability (Spectral radius condition, complex, structured).

$$\Delta - G_{11} \text{ loop is } RS \; \forall \Delta \in B_\Delta, \Delta \text{ complex} \quad \Leftrightarrow \quad \max_{\Delta \in B_\Delta} \rho(G_{11}\Delta) < 1, \forall \omega \qquad (20)$$

For a full complex Δ, we can define an unstructured complex unit ball $B_F := \{\Delta \in C^{m \times n} : \|\Delta\|_\infty \leq 1\}$ and show that [13]:

Robust Stability (H_∞ norm condition, complex, unstructured).

$$\Delta - G_{11} \text{ loop is } RS \; \forall \Delta \in B_F \quad \Leftrightarrow \quad \|G_{11}\|_\infty < 1 \qquad (21)$$

Notice that it is easy to construct RS test for any unstructured complex uncertainties, namely, construct G_{11} and compute its H_∞ norm.

Example 1: Robust Stability for complex, unstructured uncertainty. Consider RS of a system with unstructured, complex, multiplicative uncertainty, i.e., $\Delta \in B_F$, at input

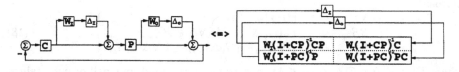

$$\begin{array}{c} RS \\ \Delta \in B_F \text{ complex} \end{array} \quad \Leftrightarrow \quad \|W(I + CP)^{-1}CP\|_\infty < 1 \qquad (22)$$

This is all nice and easy but what if there are *two* unstructured uncertainty blocks ?

Example 2: Robust Stability for complex, structured uncertainty. With two unstructured, complex, multiplicative uncertainty at the (plant) input $\Delta_I \in B_F$, and output $\Delta_o \in B_F$, where $\Delta := \text{diag}(\Delta_I, \Delta_o) \in B_\Delta$,

[figure]

---

[4] Since G is assumed stable, only the stability of $\Delta - G_{11}$ loop is in question.

we have to use the spectral radius condition for RS since Δ *is structured although individual components are unstructured*, so that

$$\begin{matrix} RS \\ \Delta \in B_\Delta, \text{ complex} \end{matrix} \quad \Leftrightarrow \quad \max_{\Delta \in B_\Delta} \rho\left(\begin{bmatrix} W_I(I + CP)^{-1}CP & W_I(I + CP)^{-1}C \\ W_o(I + PC)^{-1}P & W_o(I + PC)^{-1}PC \end{bmatrix} \Delta \right) < 1, \forall \omega \tag{23}$$

Obviously, this is difficult to test, so we need to examine further.

Structured singular value. As a means to arrive at a usable necessary and sufficient condition for robustness with respect to structured uncertainties, Doyle (1982) [11] introduced the "structured" singular value idea known as μ

$$\mu(M) := \frac{1}{\min_{\Delta \in \mathcal{D}} \{\bar\sigma(\Delta) \mid \det(I - M\Delta) = 0\}} \tag{24}$$

and if no $\Delta \in \mathcal{D}$ exists such that $\det(I - M\Delta) = 0$, then $\mu(M) \equiv 0$. Notice that $\mu(M)$ is a non-negative real-valued function and satisfies

$$\mu(M) < \frac{1}{\gamma} \quad \Leftrightarrow \quad \forall \Delta \in \mathcal{D}, \ \bar\sigma(\Delta) \le \gamma \mid \det(I - M\Delta) \ne 0 \tag{25}$$

so that $\frac{1}{\mu(M)}$ corresponds to the smallest "destabilizing" $\Delta \in \mathcal{D}$ for the $\Delta - M$ loop. To remind us that $\mu(M)$ depends also on the *uncertainty structure*, we sometimes write $\mu_\Delta(M)$. Notice also that $\mu_\Delta(M)$ also generalizes the singular value, namely, for the special case of complex, unstructured uncertainty, $\mu(M) = \bar\sigma(M)$, hence, the name *structured* singular value. However, the computation of μ from the definition is in general difficult since it requires a search over $\mathcal{D}$. Before we describe how μ, or rather its bound, is computed in practice, we briefly consider how μ is used to characterize robustness. The following results refer to the system in Figure 12:

Robust Stability (μ conditions). Consider the following arguments:

Suppose $\quad \mu(G_{11}) < 1$ at frequency ω

$\Rightarrow$ smallest $\bar\sigma(\Delta), \Delta \in \mathcal{D} \mid \det(I - G_{11}\Delta) = 0$ is > 1

$\Rightarrow \det(I - G_{11}\Delta) \ne 0, \forall \Delta \in B_\Delta$

$\Rightarrow RS$ satisfied at ω.

Suppose $\quad \mu(G_{11}) \ge 1$ at frequency ω

$\Rightarrow$ smallest $\bar\sigma(\Delta), \Delta \in \mathcal{D} \mid \det(I - G_{11}\Delta) = 0$ is ≤ 1

$\Rightarrow \exists \Delta \in B_\Delta \mid \det(I - G_{11}\Delta) = 0$

$\Rightarrow RS$ violated at ω.

Since the above argument holds over any given frequency,

$$RS \ \forall \Delta \in B_\Delta \quad \Leftrightarrow \quad \mu(G_{11}) < 1, \ \forall \omega \tag{26}$$

where $G := F_l(P, K)$.

Robust Performance (μ conditions). The robust performance condition can also be written as a μ condition. Consider RP of system $G := F_l(P, K)$ with an LFT uncertainty $\Delta_1 \in B_\Delta$, i.e.,

$$\begin{matrix} RP \\ \Delta_1 \in B_{\Delta_1} \end{matrix} \quad \Leftrightarrow \quad \begin{matrix} RS \\ \Delta_1 \in B_{\Delta_1} \end{matrix} \quad \& \quad \max_{\Delta_1 \in B_{\Delta_1}} \bar{\sigma}(F_u(G, \Delta_1)) < 1, \forall \omega \qquad (27)$$

Introduce a complex unstructured block uncertainty Δ_2 ("performance block"), so that $\bar{\sigma}(F_u(G, \Delta_1)) = \mu_{\Delta_2}(F_u(G, \Delta_1))$ and recall *RS* of $\Delta_1 \in B_{\Delta_1}$ written as a μ condition, i.e., $\mu_{\Delta_1}(G_{11}) < 1$, $\forall \omega$. Using the above two facts

$$\begin{matrix} RP \\ \Delta_1 \in B_{\Delta_1} \end{matrix} \quad \Leftrightarrow \quad \underbrace{\mu_{\Delta_1}(G_{11}) < 1, \ \forall \omega \quad \& \quad \max_{\Delta_1 \in B_{\Delta_1}} \mu_{\Delta_2}(F_u(G, \Delta_1)) < 1, \forall \omega}_{\mu_\Delta(G) < 1, \forall \omega} \qquad (28)$$

where $\Delta_1 \in \mathcal{D}$, $\Delta_2 \in \mathcal{C}^{n_e \times n_r}$. The latter equivalence holds by "Main Loop Theorem". [5] Notice that the μ condition for RP is similar to the μ condition for RS except with the closed fictitious "performance block". These conditions as summarized below are extremely succinct when compared to the conditions given in equations 18.

Summary of μ conditions for NP, RS, and RP.

$$\left. \begin{matrix} NP & \Leftrightarrow & \bar{\sigma}(G_{22}) = \mu_{\Delta_2}(G_{22}) < 1, \ \forall \omega \\ RS & \Leftrightarrow & \mu_{\Delta_1}(G_{11}) < 1, \ \forall \omega \\ RP & \Leftrightarrow & \mu_\Delta(G) < 1, \ \forall \omega \end{matrix} \right\} \qquad (29)$$

We can use μ to check for robustness and the meaning is clear for RS in the event that $\mu_{\Delta_1}(G_{11}) \not< 1$. However, for RP what does it mean if $\sup_\omega \mu_\Delta(G) = \gamma$? From definition, one can show that

$$\mu_\Delta(G) \quad \Leftrightarrow \quad \begin{matrix} RS \\ \Delta_1 \in B_{1/\gamma} \end{matrix} \quad \text{and} \quad \|F_u(G, \Delta_1)\|_\infty < \gamma$$

This means that a large γ implies poor RS *and* performance, and vice-versa. Moreover, since the worst case performance of $\|F_u(G, \Delta_1)\|_\infty < \gamma$ is guaranteed only over a scaled uncertainty size of $1/\gamma$, the worst case performance over a unit ball uncertainty will not in general be γ. So, if one is interested in the worst case performance over a *given* uncertainty size, we need to compute the *skewed-μ*.

Worst case performance. The worst case performance is defined as

$$\max_{\Delta_1 \in \mathcal{D}_1} \|F_u(G, \Delta_1)\|_\infty$$

It turns out that the above worst case H_∞ norm can be computed indirectly by computing a series of scaled μ, i.e., "skewed" by a factor $\mu^s(\omega)$ such that

$$\mu\left(\begin{bmatrix} I_1 & \\ & \frac{1}{\mu^s} I_2 \end{bmatrix} G\right) = 1 \qquad (30)$$

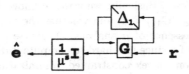

Figure 13. Worst case performance calculation as a scaled μ problem.

As illustrated in Figure 13, this is a scaled μ problem on Δ_2, specifically, a linear search over real positive scalar, μ^s. If the search converges, we are guaranteed

$$\sup_r \frac{\|\hat{e}\|_2}{\|r\|_2} = \frac{1}{\mu^s}\|F_u(G, \Delta_1)\|_\infty < 1, \ \forall \Delta_1 \in B_1 \tag{31}$$

so μ^s gives the worst case performance over $\Delta_1 \in B_1$ i.e. $\|F_u(G, \Delta_1)\|_\infty < \mu^s$.

2.5. ROBUST CONTROLLER DESIGN BY μ-SYNTHESIS

The first step in a robust controller design and analysis is to select weighting transfer functions for all input and output signals to reflect desired performance characteristics, such as attenuation factors, bandwidth, and uncertainty structure and size, etc, such that $\mu < 1$ means desired performance is satisfied. This means we are given the augmented plant P with all its uncertainty and performance weights, and its interconnections to a generally block diagonal uncertainty Δ_1, i.e., a robust control design model as illustrated in Figure 14. The set of plants to be controlled is given

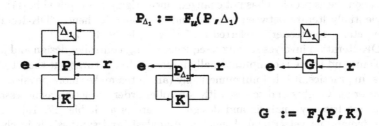

Figure 14. Robust control design model.

by $\mathcal{P} := \{P_{\Delta_1} : \Delta_1 \in B_{\Delta_1}\}$, and the RP design goal is to find a stabilizing controller, K, such that

$$\|F_l(P_{\Delta_1}, K)\|_\infty < 1, \ \forall \Delta_1 \in B_{\Delta_1} \quad \Leftrightarrow \quad \mu_\Delta(G) < 1, \quad \forall \omega \tag{32}$$

where $\Delta := \text{diag}(\Delta_1, \Delta_2)$, $\Delta_1 \in \mathcal{D}$, and $\Delta_2 \in \mathcal{C}^{n_e \times n_r}$. More practically, the μ-synthesis goal is to minimize peak value of $\mu_\Delta(G)$ of the appended closed loop transfer function G over all stabilizing controller K, i.e.

$$\min_K \max_\omega \mu_\Delta(F_l(P, K)(j\omega)).$$

---

[5] For more details and proof of this important result, refer to page 197- of [9].

μ analysis and computation. Basically, we want to know if the closed loop system is robustly stable, i.e., $\mu_{\Delta_1}(G_{11}) < 1$, and whether the performance is robust, i.e., $\mu_\Delta(G) < 1$. This is the essence of μ analysis. Of course as noted earlier, a major issue is the computation of μ since it is hard to compute it directly, a task equivalent in general to solving a non-convex constrained minimization problem over Δ. The conventional approach is to circumvent this computational burden by calculating bounds for μ which are easier to compute. In a nutshell, explicit expressions for the lower an upper bounds of μ for a general uncertainty structure are used. These expressions corresponds to the most constrained (diagonal) and the least constrained (fully populated) uncertainty structure. The idea then is to tighten these bounds by scaling and transformations that do not affect $\mu_\Delta(G)$ itself but affect its bounds, $\rho(G)$ and $\bar\sigma(G)$, resulting in the following

$$\max_{U \in \mathcal{U}} \rho(GU) \leq \mu_\Delta(G) \leq \min_{D \in \mathcal{D}} \bar\sigma(DGD^{-1}) \tag{33}$$

These bounds or its variations [11, 14, 15] are used extensively in μ analysis and synthesis.

μ-synthesis by DK Iteration. For computational feasibility, we replace the direct minimization of μ_Δ peak by minimizing its least upper bound and so, the μ-synthesis problem becomes an optimization problem with respect to K and D

$$\min_K \min_{D \in \mathcal{D}} \|DF_l(P, K)D^{-1}\|_\infty$$

where D is a frequency dependent, minimum-phase transfer matrix. The basic idea to solve the μ synthesis problem has not changed since originally proposed in [11], which is to sequentially iterate between D scale and controller K, hence "DK-Iteration". For more details, the reader is referred to, for example, [9, 12, 16].

The DK-Iteration involves solving a sequence of H_∞ controller design and μ analysis problems and hence is computationally intensive, but it turns out that it usually converges. In practice, it is helpful numerically (and often makes more physical sense) to use low order weights and scales, with controller order reduction as necessary. In addition, user friendly analysis and design software is available [17, 16]. Perhaps the more demanding yet crucial element in μ-analysis and synthesis is in choosing physically sensible weight matrices to capture competing multiobjective performance goals and model uncertainties and disturbances. For this task, classical loop-shaping ideas and knowledge of the dynamics of the physical system under consideration is helpful. Moreover, if input and output measurement data is available, one can utilize it to estimate uncertainty weight matrices for an assumed interconnection structure.

3. Uncertainty bounds from data

We have seen that robust control theory is very elegent and computationally reliable but depends heavy on the uncertainty structure and weight, which is implicit in the augmented nominal plant and its LFT interconnectons. However, even after almost two decades of well established theory, there is a persistent gap across robust control theory and technology. This means that its usefulness to practical systems is limited as is painfully evident based on studies ranging from simulation studies of simple toy models to applications to more realistic systems. The underlying reasons are that *a priori* knowledge of a physical application is imperfect and that there are obvious limitations to measurements. Interestingly the exploitation of a given uncertainty model, with explicit assumptions on the system structure, is the key advantage in robust control as well as its main predicament in applications. With a given uncertainty model, robustness analysis can address "what if" questions such as worst case response which can be used as a means to trade-off system robustness with performance. Furthermore, with a given uncertainty model, robust control synthesis can build-in robustness with respect to this uncertainty model. In the end however, the *physical* significance of these analysis and synthesis results depends on the relevance of the given uncertainty model with respect to the physical system in question.

In the following sections, we outline a novel approach available recently [18, 19] for the determination of uncertainty weight matrices from data for robust control applications and then demonstrate it on a vibration control example. The goal is to improve the usefulness of robust control methods by constructing more reliable uncertainty model bounds based on both *a priori* information and test data. The latter is usually not used due to the lack of a sensible methodology.

3.1. FRAMEWORK

Given a set of plants, a set of exogenous signals, and a test input signal, the basic question in model validation is whether there exists a plant and a signal in these given sets, whose predicted output response matches measured outputs. In order to formulate a more complete theory, a set of assumptions are required limiting the class of signals and systems that are applicable. For a more complete discussion, the interested reader can refer to [19]. Figure 15 illustrates the finite discrete-time measured signals assumed available for uncertainty bound estimation. The physical

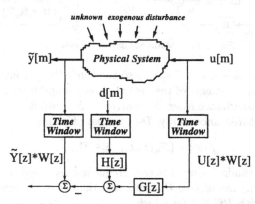

Figure 15. Schematic of signals collected from a physical system.

system is described by a causal, asymptotically stable, n-th order linear, time-invariant, discrete time state space model. In addition, the system is assumed to be in steady state during measurements. With these assumptions, the frequency domain response is

$$Y[z] = G[z]U[z] + H[z]D[z] \tag{34}$$

where $Y[z] := \sum_{k=-\infty}^{\infty} y[k]z^{-k}$ is the z-transform of an everlasting signal, $\{y\}$, and $G[z]$ and $H[z]$ denote the discrete frequency responses of system model and noise filter, respectively. With time-limited data, $\{y\}^N$ and $\{u\}^N$, we cannot compute the z-transforms, $Y[z]$, $U[z]$ so we rewrite in terms of DFT's $Y_N[z_r]$, $U_N[z_r]$ of samples $\{y\}^N$ and $\{u\}^N$ at discrete frequency points, $z_r = e^{i\frac{2\pi}{N}r}$, $r = 0, \ldots, N-1$,

$$Y_N[z_r] = G[z_r]U_N[z_r] + \frac{1}{\sqrt{N}}H[z_r]D[z_r] + R_N[z_r] \tag{35}$$

The residual, $R_N[z_r]$ is due to time-limited data [20], and in model validation experiments, we try to satisfy N-periodic assumption on measured input and output signals so that $R_N[z_r] = 0, \forall r$, by implementing a proper test input sequence. The effect of unknown exogenous disturbances and noise, $\frac{1}{\sqrt{N}}H[z_r]D[z_r]$, is accounted for by prescribing allowances in the periodogram.

Figure 16 shows the basic LFT framework and its canonical form for model validation. Notice that for a closed loop system model validation where r and K are

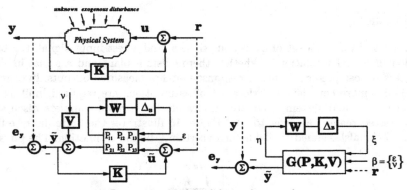

Figure 16. Model validation framework.

assumed known, $e_u := \tilde{u} - u = 0$ if $e_y = 0$ so that only the output error need to be considered. Given measurements of the output, y, command input, r, an augmented nominal plant and disturbance filter, P, controller, K, output noise filter, V, and a set of bounded structured uncertainty, $\mathcal{D}_W$, the set of plants

$$\mathcal{P}_W := \{\mathcal{F}_u(P, \Delta), \Delta \in \mathcal{D}_W\} \tag{36}$$

is said to be a model validating set if at each frequency, $\omega_k = \frac{2\pi k}{\tau N}, k = 0, \ldots, N-1$, it contains an uncertainty model $\Delta \in \mathcal{D}_W$ such that there exists exogenous disturbance signals, $\beta := (\epsilon, \nu)$ with $\|\beta\| \leq 1$ for which

$$y = \mathcal{F}_u(G(P, K, V), \Delta) \left\{ \begin{array}{c} \epsilon \\ \nu \\ r \end{array} \right\} \tag{37}$$

and

$$\mathcal{D} := \{\Delta \in C^{m \times n} : \Delta = \text{diag}(\delta_1 I_{m_1}, \ldots, \delta_r I_{m_r}, \Delta_{r+1}, \ldots, \Delta_\tau), \delta_i \in F_i, \Delta_k \in C^{m_k \times n_k}\}$$
$$W := \text{diag}(w_1 I_1, \ldots, w_\tau I_\tau)$$
$$\mathcal{D}_W := \{\Delta \in \mathcal{D} : \Delta = W\Delta_B\}$$
$$\Delta_B := \{\Delta \in \mathcal{D} : \bar{\sigma}(\Delta) \leq 1\}$$

The set of signals y, r, η, ν, denote the DFT's of their respective time-limited samples while systems, P, K, V, G, W, denote transfer function matrices evaluated at the same uniformly spaced frequencies up to Nyquist frequency, and $G(P, K, V)$ denotes the augmented closed loop transfer function matrix containing both noise and disturbance filters.

This question was first posed in [21] where it is shown that the validation test can be formulated analogous to a skewed-μ or alternately as a quadratic optimization problem. For the "most commonly applied cases: additive and input or output multiplicative" [22] it has been shown (and in time domain [23]) that the validation test takes the form of a convex feasibility problem when the fictitious signal η does not depend on ξ. More recently, [24] has shown that by choosing a plant uncertainty model based on coprime factorizations of the nominal plant and known controller, all closed loop transfer functions can be made affine in the coprime factor uncertainties so that the convexity of the validation test can be preserved through enforcing $G(P, K, V)_{11} = 0$. In dealing with closed loop validation in [24], the noise model enters as an additive output outside the loop and hence only G_{11} is relevant. On the other hand, in the work reported in [23], an open loop model validation problem with additive output noise is considered so that only P_{11} is relevant.

A key problem arises if one wishes to test a more general uncertainty structure for model validation, namely, the loss of convexity in the feasibility problem. This means that from a numerical implementation standpoint, the validation test is effectively a sufficient condition. Furthermore, if the validation test is passed for a particular plant set, there exists an infinity of other sets which will also be validating, irrespective of its convexity. In our view, this non-uniqueness diminishes the significance of a validation result for a particular set.

For these reasons, we take an alternate view of the model validation problem. Instead of trying to validate a specific set with respect to given data by asking "Is $\mathcal{D}_W$ model (in)validating?" we ask instead "Does a model validating set $\mathcal{D}_W$ exist?" The former question is a test on a specific uncertainty set resulting in a positive or negative answer while the latter question is a test to determine whether some uncertainty set with the given LFT structure exists that can satisfy model validation conditions. An affirmative answer to the latter question is of course a necessary condition to the former and it turns out to be much simpler to test. Furthermore, if a finite size model validating set $\mathcal{D}_W$ exist, then it turns out that *all* model validating sets having the same LFT uncertainty structure can be readily parameterized [18].

3.2. MODEL VALIDATION AND PARAMETERIZATION

A new approach breaks up the model validation question into 3 parts:[6]

1. Does a pair of signals (ξ, β) where $\|\beta\| \leq 1$ exists such that $e_y = 0$?
2. Parameterize signal set $S_{\phi,\psi} := \{\xi(\phi,\psi), \beta(\phi,\psi), \eta(\phi,\psi)\}$ such that $e_y = 0$.
3. Does a $\Delta \in \mathcal{D}$ exists such that $\xi = \Delta\eta$ for a set $(\xi, \beta, \eta) \in S_{\phi,\psi}$?

Figure 17 illustrates the question posed in Step 1. Such a pair of signals exists if and

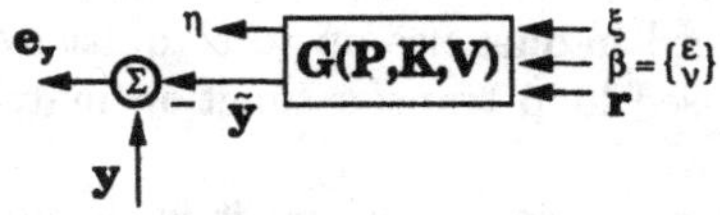

Figure 17. Does (ξ, β) exists, where $\|\beta\| \leq 1$, such that $e_y = 0$?

only if

$$\left. \begin{array}{c} e_y^o \in \mathrm{Im}(M) \\ \|T_2^H(M^+)_\beta e_y^o\| \leq 1 \end{array} \right\} \quad \forall \omega_k \tag{38}$$

where the constant matrices are given by

$$\begin{aligned} e_y^o &:= y - G_{23}r \\ M &:= [G_{21}, G_{22}] \\ \mathrm{Im}(N_M) &= \mathrm{Ker}(M) \\ T_2 &:= \mathrm{Im}((N_M)_\nu)^\perp \end{aligned}$$

Notice that M is a crucial matrix which captures the uncertainty freedoms to be used for model validation. This test involves only a constant matrix check at discrete frequency points of interest in contrast to a (non)convex feasibility problem. If the above necessary condition cannot be satisfied, this means that no matter how large the uncertainty radii W (imbedded in G) are, the model is invalidated because either the a priori uncertainty structure is too restrictive or the noise allowance is too small. Hence to proceed in trying to construct a model validating set, one has to improve the nominal plant model, P, and/or modify the uncertainty structure $\mathcal{D}$ to increase its richness, and/or increase the noise allowance by modifying the noise filter model V.

If the above constant matrix test passes, then one can proceed to Step 2 and parameterize the set of signals

$$S_{\phi,\psi} := \{[\xi(\phi,\psi), \beta(\phi,\psi), \eta(\phi,\psi)] : e_y = 0, \ \phi \in \Phi, \psi \in \Psi\} \tag{39}$$

where all feasible triples (ξ, ϵ, ν) are given by

$$\begin{pmatrix} \xi \\ \beta \end{pmatrix} = \begin{pmatrix} \xi_o \\ \beta_o \end{pmatrix} + \Omega \begin{pmatrix} \phi \\ \psi \end{pmatrix} \tag{40}$$

---

[6] For more details see [18]

where ψ is arbitrary and ϕ satisfies

$$\|\phi\| \le b_o := \sqrt{1 - \|T_2^H (M^+)_\beta e_y^o\|^2} \tag{41}$$

and

$$\Omega := N_M U \begin{bmatrix} \Sigma_1^{-1} & 0 \\ 0 & I_{n_\psi} \end{bmatrix} \tag{42}$$

$$\begin{pmatrix} \xi_o \\ \beta_o \end{pmatrix} := [M^+ - N_M ((N_M)_\beta)^+ (M^+)_\beta] \, e_y^o \tag{43}$$

$$(N_M)_\beta := \begin{bmatrix} T_1 & T_2 \end{bmatrix} \begin{bmatrix} \Sigma_1 & 0 \\ 0 & 0 \end{bmatrix} U^H \tag{44}$$

Figure 18 illustrates the final step of the existence test, which is to check whether a $\Delta \in \mathcal{D}$ exists such that $\xi = \Delta \eta$ for some signal pair $(\xi, \eta(\xi, \beta)) \in \mathcal{S}_{\phi,\psi}$ where $\eta = G_{11}\xi + G_{12}\beta + G_{13}r$. It turns out that this part of the existence test is satisfied if and

Figure 18. Final step of existence test

only if there exists $(\xi, \beta, \eta) \in \mathcal{S}_{\phi,\psi}$ such that (ξ, η) is $\mathcal{D}$-realizable[7] and $(\xi_i, \eta_i(\xi, \beta))$ are collinear for each repeated uncertainty block. For the special class of uncertainty structure with no repeated uncertainties, i.e. $r = 0$ but not requiring $G(P, K, V)_{11} = 0$, with a satisfaction of the necessary conditions for existence in equation 38, it is almost certain that there exists $(\xi, \beta) \in \mathcal{S}_{\phi,\psi}$ such that $\eta_i(\xi, \beta) \ne 0, \forall i$. This means that for this important special class of uncertainty structure, the necessary conditions in equation 38 are actually necessary and almost sufficient for a model validating set $\mathcal{D}_W$ to exist.

If a model validating set $\mathcal{D}_W$ exists, then *all* model validating sets of plants are given by

$$\mathcal{P}_{W\phi\psi} := \{\mathcal{F}_u(P, \Delta), \Delta \in \mathcal{D}_W\} \tag{45}$$

where $\psi \in \mathcal{C}^{n_\psi}$, $\phi \in \mathcal{C}^{n_\phi}$, $\|\phi\| \le b_o$, and $W := \operatorname{diag}(w_1 I_{n_1}, \ldots, w_\tau I_{n_\tau})$ is any matrix satisfying

$$\frac{\|\xi_i(\phi, \psi)\|}{\|\eta_i(\phi, \psi)\|} \le |w_i|, \quad i = 1, \ldots, \tau \tag{46}$$

(ξ, η) is $\mathcal{D}$-realizable and are parameterized by

$$\xi_i = \xi_{o,i} + \Omega_i \left\{ \begin{matrix} \phi \\ \psi \end{matrix} \right\} \tag{47}$$

$$\eta_i = \eta_{o,i} + [G_{11} \quad G_{12}]_i \Omega \left\{ \begin{matrix} \phi \\ \psi \end{matrix} \right\} \tag{48}$$

$\operatorname{dist}^{(\mathcal{F}_i)}(\xi_i, \eta_i) = 0, i = 1, \ldots, r$, and

$$\eta_o := [G_{11} \quad G_{12}] \left\{ \begin{matrix} \xi_o \\ \beta_o \end{matrix} \right\} + G_{13}r. \tag{49}$$

---

[7] i.e., $\xi_i = 0$ or $\eta_i(\xi, \beta) \ne 0, \forall i$

154

Note that if model validating sets exists, they are highly non-unique.

Validation as a feasibility problem. Using the previous parameterization, we re-examine the earlier question, "Is $\mathcal{D}_W$ model (in)validating?" Suppose a model validating set $\mathcal{D}_W$ exists. Since all model validating sets for the chosen uncertainty structure are given by the above parameterization, the original question can be posed as follows:

Given W, at each frequency does a (ϕ, ψ), $\|\phi\| \leq b_o$ exists such that

$$\frac{\|\xi_i(\phi, \psi)\|}{\|\eta_i(\phi, \psi)\|} \leq |w_i|, \quad i = 1, \ldots, \tau \tag{50}$$

where (ξ, η) is $\mathcal{D}$-realizable and $\mathrm{dist}^{(\mathcal{F}_i)}(\xi_i, \eta_i) = 0$, $i = 1, \ldots, r$?

The $\mathcal{D}$-realizability and collinearity conditions can be implicitly satisfied by the following form:

Given W, at each frequency does a (ϕ, ψ), $\|\phi\| \leq b_o$ exists such that

$$\left. \begin{array}{l} \xi_i(\phi, \psi) - \delta_i \eta_i(\phi, \psi) = 0, \quad \delta_i \in \mathcal{F}_i \\ |\delta_i| - |w_i| \leq 0 \end{array} \right\} \quad i = 1, \ldots, r \tag{51}$$

$$\|\xi_j(\phi, \psi)\|^2 - |w_j|^2 \|\eta_j(\phi, \psi)\|^2 \leq 0, \tag{52}$$

where $j = r + 1, \ldots, \tau$?

Although ξ_i and η_i are affine in ϕ and ψ as given in equations 47 and 48, the feasibility conditions are in general nonconvex. Specifically, the collinearity requirement due to repeated (and/or real) uncertainty involves a quadratic equality since δ_i is unknown equation 51 and the norm inequalities for the nonrepeated uncertainties, although quadratic, involves a difference in norms (equation 52). Nevertheless, one is hopeful of having a workable methodology if one observes from equations 51 and 52 that a feasible solution can be generated by first focusing on finding a pair (ϕ, ψ) which satisfies the collinearity condition in equation 51 and then easily selecting sufficiently large weights w_i, $i = 1, \ldots, \tau$, to satisfy equations 51 and 52.

Optimizing for smallest scaled set. A useful set of model validating plants must be *small but not smaller*, i.e., it should not be an unnecessarily large set of plants which will limit performance but it must contain the "true" plant. Given a candidate set $\mathcal{D}_W$ of uncertainty norm radii, W, we seek a smallest scaled factor x for which the scaled set $\mathcal{D}_{xW}$ is model validating. This question can be posed by imbedding the previous feasibility problem in the following optimization:

$$\min_{\phi, \psi, \delta_1, \ldots, \delta_r, x^2} x^2 \tag{53}$$

subject to

$$\left. \begin{array}{l} \xi_i(\phi, \psi) - \delta_i \eta_i(\phi, \psi) = 0, \quad \delta_i \in \mathcal{F}_i \\ |\delta_i|^2 - x^2 |w_i|^2 \leq 0 \end{array} \right\} \quad i = 1, \ldots, r \tag{54}$$

$$\|\xi_i(\phi, \psi)\|^2 - x^2 |w_i|^2 \|\eta_i(\phi, \psi)\|^2 \leq 0, \quad i = r + 1, \ldots, \tau \tag{55}$$

$$\|\phi\|^2 \leq b_o^2 \tag{56}$$

Although the above problem is not convex in general,[8] the conditions involve poly-nominals in $\phi, \psi, \delta_1, \ldots, \delta_r, x^2$ with are at most cubic order. The existence of a feasible point $(\phi, \psi, \delta_1, \ldots, \delta_r, x^2)$ in the above optimization problem is equivalent to the existence of a model validating set $\mathcal{D}_{xW}$. Of course when $x < 1$, the original candidate uncertainty set, $\mathcal{D}_W$ satisfies model validation conditions. Notice that if there are no repeated uncertainty ($r = 0$), any choice of ψ and ϕ where $\|\phi\| \leq b_o$ will likely admit a model validating set $\mathcal{D}_{xW}$, since a larger uncertainty radii, W will admit a feasible point. However, even without repeated uncertainties, the constraints are in general not convex in ϕ and ψ and therefore the numerical optimization task is likely nontrivial.

The important issue of how one would simultaneously choose *a priori* weights for parametric ($w_i, i = 1, \ldots, r$) and nonparametric ($w_i, i = r + 1, \ldots, \tau$) uncertainties to form candidate uncertainty weight is beyond the scope of this work. Whatever is chosen in an application, these candidate uncertainty weights should reflect the relative importance of the uncertainty components by design, based on additional *a priori* information on the particular application.

Smallest unmodeled dynamics subject to parametric uncertainties Consider para-metric uncertainties which for physical reasons are considered to be constants but unknown. One approach to handling these unknown constants is to consider them as independent parameter identification problems in the context of system identifi-cation, *a la* [25, 26, 27, 20, 28]. This of course means avoiding model validation and LFT uncertainties entirely. The alternative viewpoint we take uses these unknown constants as additional variables in minimizing the norms of model validating LFT uncertainties.

If parameters imbedded in the plant change (as in parameter identification prob-lem) such that the crucial model validation design freedom matrix, $M = [G_{21}, G_{22}]$ changes, this can cause cascading changes which are difficult to evaluate in a nu-merical optimization context. Hence, we try to avoid changing the augmented plant directly by viewing the imbedded plant changes or parametric uncertainties as LFT uncertianties which are allowed to vary without changing the augmented plant.

In the case where we have competing unmodeled dynamics and parametric uncer-tainties, we suggest fixing an allowance for parametric uncertainties while minimizing non-parametric uncertainties (of course in addition to a fixed noise allowance). This may be more reasonable physically than the approach whereby a priori candidate uncertainty weights are fixed for both parametric and non-parametric uncertainties and then determining a smallest scaled set. The former approach amounts to a slight modification of the optimization problem for a smallest scaled set in the previous section and choosing the x scale in equation (54) to unity.

Unknown but bounded exogenous signals. To deal with unknown exogenous signals such as process and measurement noise, we consider them as unknown but bounded. No assumptions on the statistics and the independence of the unknown signals are required. In trying to satisfy the model validation conditions, these signals are treated as allowances to be used in finding the minimal norm LFT model uncertainty nec-

---

[8] For discussions on the special case where $[G_{11}, G_{12}] = 0$ leads to a convex problem, see [19].

essary. As such, erronously assuming an overly conservative level of noise will likely lead to optimistic levels (smaller than actual) of model validating LFT uncertainty. On the other hand, assuming a lower level of noise allowance (than actual) will likely lead to pessimistic levels of (larger than true value) model validating LFT uncertainty. Consequently, it is important to specify a reasonably accurate model of the noise and disturbances in the system for model validation and uncertainty bound determination purposes.

For convenience, the combined noise/disturbance allowance in the canonical form is given by

$$\|\beta(\omega)\|_2 \leq 1 \ \forall \omega \quad \Leftrightarrow \quad \|\beta\|_{l_\infty} := \|V_\beta^{-1}\hat{\beta}\|_{l_\infty} \leq 1 \tag{57}$$

where $V_\beta := \mathrm{diag}(V_\epsilon, V_\nu)$ denotes the filter matrix for disturbance and noise allowances and $\hat{\beta}$ denotes the exogenous signal as it enters the loop. Specifically, V_β is designed to reflect anticipated spectra of $\hat{\beta}$ and to normalize the unknown norm bound of β to unity as in equation (57). The assumed block diagonality in V_β implies the independence of ϵ and ν. In some systems, reasonably accurate models for V_ϵ and V_ν may be available. In other systems where typical spectrums of $\hat{\epsilon}(\omega)$ and $\hat{\nu}(\omega)$ are available and if their individual channels are known to be independent, a stable discrete time filter can usually be fitted for each channel and realized as a diagonal filter.

3.3. UNCERTAINTY BOUND IDENTIFICATION TOOLBOX

A convenient to use set of software subroutines called Uncertainty Bound Identification Toolbox (UBID) currently exists [19]. This toolbox consists of necessary subroutines to compute a smallest model validating set from data and a priori uncertainty structure and a priori exogenous disturbance/noise information. More specifically the toolbox performs the following operations:

- Pre-process sample data by dividing a time record into windowed-overlapped-zero appended samples.
- Define noise and disturbance allowance filters from raw noise/disturbance response signals or from its statistics.
- Form augmented plant with uncertainty structure in the standard form:

$$\left\{ \begin{array}{c} \eta \\ y \end{array} \right\} = \left[\begin{array}{ccc} G_{11} & G_{12} & G_{13} \\ G_{21} & G_{22} & G_{23} \end{array} \right] \left\{ \begin{array}{c} \xi \\ \left\{ \begin{array}{c} \epsilon \\ \nu \end{array} \right\} \\ r \end{array} \right\}$$

where $\xi = \Delta \eta$. The user can also load the augmented plant and uncertainty structure from a file. A particularly relevant feature for flexible structure application is the option to form augmented state space plant with eigenvalue uncertainties, as shown in Figure 19.
- Compute smallest model validating set

  - Performs existence test for model validation.
  - Form objective and constraint functions and their analytical gradients
  - Perform sufficiency test for convexity based on norm test of $[G_{11}, G_{12}]$

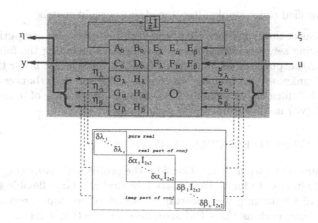

Figure 19. Augmented state space plant with eigenvalue uncertainties in UBID Toolbox.

- Solve optimization problem

  * If convex, set up convex program and solve using LMI Toolbox:

  $$\min_{z} \quad c^T z$$

  subject to LMI constraints

  $$\begin{bmatrix} Q_i z - \|\xi_{o,i}\|^2 & \text{sym} \\ S_i z & I \end{bmatrix} > 0, \quad i = 1, \ldots, \tau$$

  $$\begin{bmatrix} b_o^2 & \text{sym} \\ Lz & I \end{bmatrix} > 0$$

  where $z := [\text{Re}(\psi); \text{Im}(\psi); \text{Re}(\phi); \text{Im}(\phi); x^2] \in \mathcal{R}^{2n_\psi + 2n_\phi + 1}$.

  * If non-convex, solve NLP using an SQP algorithm from Optimization Toolbox: solves for a smallest scale model validating set based on the following nonlinearly constrained optimization formulation:

  $$\min_{\phi \in \mathcal{C}^{n_\phi}, \psi \in \mathcal{C}^{n_\psi}, \delta_1, \ldots, \delta_r, x^2 \in \mathcal{R}_+} \quad x^2$$

  subject to

  $$\left. \begin{array}{l} \xi_i(\phi, \psi) - \delta_i \eta_i(\phi, \psi) = 0, \quad \delta_i \in \mathcal{R}_i \\ |\delta_i|^2 - |w_i|^2 \leq 0 \end{array} \right\} \quad i = 1, \ldots, r$$

  $$\|\xi_i(\phi, \psi)\|^2 - x^2 |w_i|^2 \|\eta_i(\phi, \psi)\|^2 \leq 0, \quad i = r+1, \ldots, \tau$$

  $$\|\phi\|^2 \leq b_o^2$$

Each element in $w_1, \ldots, w_\tau$ of the user specified frequency varying column vector, defines a radius for each individual uncertainty block. The scaling factor, x, appears only with the radii of the nonrepeated complex blocks, $x \cdot w_i, i = r+1, \ldots, \tau$, implying that the scaling minimization is applied only to the nonrepeated complex blocks.

— Fit a specified order stable filter uncertainty weight matrix

The limitations of the algorithms that form UBID Toolbox, for computing a smallest model validating set, arise from various assumptions including the following: (i) N-periodic input/output discrete time signal samples, (ii) discrete linear time-invariant systems, (iii) unknown but bounded exogenous noise and disturbances, (iv) nominal model, (v) LFT uncertainty structure, (vi) relative importance of initial uncertainty weights, and (vii) nonlinear optimizer performance.

3.4. APPLICATION TO FLEXIBLE STRUCTURE

True model and simulation data. Consider the problem of estimating model uncertainty bounds from data for active vibration control of a large flexible structure. For the purpose of illustrating the uncertainty bound estimation approach for a large flexible structure, such as the CEM structure (see Section 4 for more details), let the first 30 structural modes obtained from FEM be a truth model and assume three thrust actuators and displacement encoders be distributed on the structure in mutually perpendicular directions. Hence, define a 60-th order true state space model discretized at sampling interval T_s

$$\left\{ \begin{array}{c} x_{k+1} \\ y_k \end{array} \right\} = \left[\begin{array}{cc} A_{true} & B \\ C & D \end{array} \right] \left\{ \begin{array}{c} x_k \\ u_k \end{array} \right\} \tag{58}$$

Discrete time response measurements are simulated using this truth model. All three input test signals are generated by independently driving a low pass filter of bandwidth 5 Hertz with a white (bandwidth of Nyquist frequency) Gaussian random signal of unit standard deviation. A fictitious additive output noise is similarly chosen to be a white Gaussian random signal with a standard deviation of .01. The test data consists of two periods each about 408 seconds duration. The test signal is only applied during the first 204 seconds followed by free response for another 204 seconds. The second period begins after most of the transient free response decays to the output noise levels. This is designed to mitigate the erronous residual effects of nonperiodicity in the computation of DFT's for use in model validation equations.

Nominal model. To simulate a nominal model, or a good working model in a realistic application, we begin by defining a perturbed system matrix,

$$A_{pert} := A_{true} + \Delta A \tag{59}$$

such that A_{true} and ΔA matrices have the following block diagonal forms

$$A_{true} := \left[\begin{array}{ccc} \Lambda_1 & & 0 \\ & \ddots & \\ 0 & & \Lambda_{30} \end{array} \right] \in \mathcal{R}^{60 \times 60}, \quad \Lambda_i := \left[\begin{array}{cc} \alpha_i & \beta_i \\ -\beta_i & \alpha_i \end{array} \right] \in \mathcal{R}^{2 \times 2}$$

$$\Delta A := \left[\begin{array}{ccc} \delta\Lambda_1 & & 0 \\ & \ddots & \\ 0 & & \delta\Lambda_{30} \end{array} \right] \in \mathcal{R}^{60 \times 60}, \quad \delta\Lambda_i := \left[\begin{array}{cc} \delta\alpha_i & \delta\beta_i \\ -\delta\beta_i & \delta\alpha_i \end{array} \right] \in \mathcal{R}^{2 \times 2}$$

for $i = 1, \ldots, 30$. The eigenvalue perturbations are assumed constants such that A_{pert} is stable, i.e., $|\alpha_i + j\beta_i| < 1$, $|\alpha_i + \delta\alpha_i + j(\beta_i + \delta\beta_i)| < 1$, $i = 1, \ldots, 30$. In this example, all eigenvalues are assumed to occur in complex conjugate pairs although in general eigenvalues on the real axis in the z-plane can also be included as discussed in Section 4.2.2.

The perturbed model, (A_{pert}, B, C, D), is then internally balanced using a similarity transformation, P, so that

$$\begin{Bmatrix} \tilde{x}_{k+1} \\ y_k \end{Bmatrix} = \begin{bmatrix} P^{-1}A_{pert}P & P^{-1}B \\ CP & D \end{bmatrix} \begin{Bmatrix} \tilde{x}_k \\ u_k \end{Bmatrix} \tag{60}$$

where $\tilde{x}_k := Px_k$ denotes the balanced state at time k. The nominal model is then obtained by keeping only the 12 largest Hankel singular values. Table I shows the eigenvalues of the true and nominal plant models and their true errors in the real and imaginary components.[9] Notice that the real and imaginary component eigenvalue

TABLE I. True and nominal eigenvalues of a 30 mode flexible structure.

Mode	Freq(r/s)	Damp	True	Nominal	Error(real)	Error(imag)
1			0.9954 + 0.0462i			
2	.941	.081	0.9951 + 0.0467i	0.9950 + 0.0472i	0.0001	-0.0005
3	.973	.099	0.9940 + 0.0482i	0.9943 + 0.0476i	-0.0003	0.0006
4	4.62	.020	0.9689 + 0.2280i	0.9701 + 0.2264i	-0.0012	0.0016
5			0.9673 + 0.2342i			
6			0.9555 + 0.2754i			
7	9.41	.020	0.8830 + 0.4490i	0.8838 + 0.4457i	-0.0008	0.0033
8			0.7363 + 0.6552i			
9	15.2	.019	0.7109 + 0.6816i	0.7121 + 0.6800i	-0.0012	0.0016
10			0.7021 + 0.6903i			
11	16.2	.020	0.6758 + 0.7150i	0.6759 + 0.7144i	-0.0001	0.0006
12			0.5948 + 0.7718i			
13			0.5179 + 0.8319i			
14			0.5155 + 0.8333i			
15			0.3825 + 0.8989i			
⋮			⋮			
30			-0.7389 + 0.6002i			

errors which are boxed in the Table I corresponds to two groups of dominant resonant modes near .9 and 4.6 rad/sec (see Figure 20). Mode 7 with the largest error of .0033 in its imaginary component corresponds to the resonance near 9.4 rad/sec but is not included as an uncertain parameter because its nominal response is only secondary.

The procedure used above to obtain a reduced order nominal model from the perturbed model is typical of robust control model of flexible structures in practice.

---

[9] Of course in an actual physical application, it is usually not clear what exactly the "true" eigenvalues should be. The above usage is only intended to illustrate the methodology.

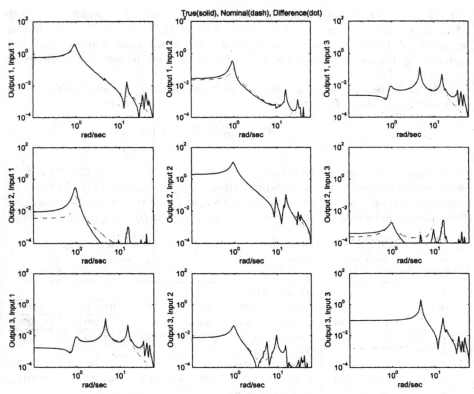

Figure 20. Frequency response magnitude comparison of flexible structure.

The perturbed model is analogous to a not totally accurate higher order model derived either from an analytical/mathematical modeling approach or empirically constructed from system identification algorithms (see for example [27], which discuss the identification problem for various types of flexible structures of interest to NASA).

Uncertainty model. Since a nominal model of the flexible structure is derived from a higher order perturbed model, its frequency and damping values differ from the true values even before an order reduction. This motivates choosing uncertainties about a subset of nominal eigenvalues to reflect inevitable errors in these frequencies and damping values. Of course this choice of uncertain parameters is almost natural for lightly damped flexible structures because they represent the dominant resonances in structural vibrations. Additionally, they are invariant to similarity transformations, which means that these parameters are independent of coordinate selection, clearly an attractive property in modeling physical systems. For these reasons, we select a subset of eigenvalues from the nominal model as candidates for parametric uncertainties and make use of additive uncertainties to cover the neglected 24 higher frequency modes, as illustrated in Figure 21. That this choice in the uncertainty structure is sufficient for a flexible structure is discussed in Section

4.2.2. An additional advantage of using additive uncertainty for the higher frequency modes is that it will also cover "missing" structural modes, regardless of its actual number, typical of FEM models when compared to empirically derived models at higher frequencies.

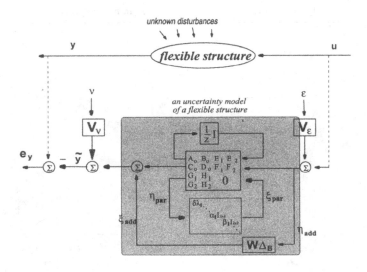

Figure 21. Mathematical model used in uncertainty bound determination.

A comparison of the magnitude gains of the true system versus the nominal model and the corresponding differences are shown in Figure 20. Although the magnitude gains of the nominal model response (dashed line) appears to closely approximate the true model response (solid line), a plot of their difference shows that large differences occur around structural resonance frequences due to phase errors. The figure also shows that the nominal model tries to capture four groups of low frequency structural resonances near .9, 4.6, 9, and 15 rad/sec while many other higher frequency secondary resonances, say beyond 20 rad/sec, are ignored. The two largest discrepencies between the true and nominal model frequency response occurs at resonances near .9 and 4.6 rad/sec.

Smallest model validating unmodeled dynamics. Table II shows the four cases simulated which are based on varying levels of eigenvalue uncertainty allowances, but with the same equivalent output noise allowances. Only Case 1, with no parametric uncertainty allowance, led to a convex optimization problem and the remaining cases were solved by using a Sequential Quadratic Programming algorithm found in Optimization Toolbox [29]. Figure 22 shows the computed smallest model validating additive uncertainties and the reference response from the truth (solid) and nominal (dash) models. These uncertainty bounds (circle-line) were curve fitted with low order stable rational transfer functions (dash-dot).

162

TABLE II. Cases simulated for smallest unmodeled dynamics.

Case	Unmodeled dynamics	Eigenvalue uncertainty allowance
1	Additive	none
2	Additive	.001 in first 3 modes
3	Additive	.002 in first 3 modes
4	Additive	.004 in mode 3 only

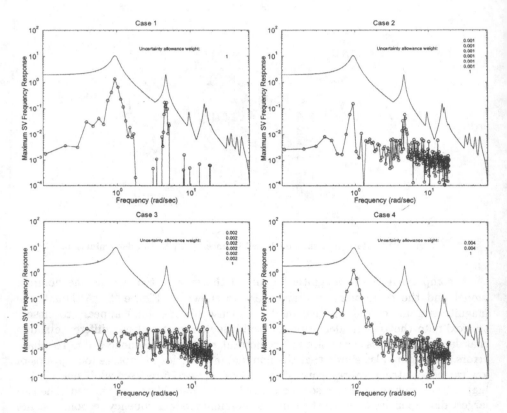

Figure 22. Smallest model validating additive uncertainties; Data (circle-line), Fitted uncertainty (dash-dot), True minus nominal (dot), True (solid), Nominal (dash).

In Case 1, the model validating additive uncertainty (circle-line) recovers the true output errors (dotted line) caused by a combination of eigenvalue errors in the nominal model (see Table I) and truncated higher frequency modes (see Figure 20). In Case 2, by allowing an eigenvalue allowance of .001 in the real and imaginary eigenvalue components of the first three modes (associated with resonant peaks at .9 and 4.6 rad/sec), the smallest model validating additive uncertainty reduces by an order of magnitude. Case 3 shows what happens to smallest additive uncertainty if eigenvalue uncertainty allowance is increased two fold to .002. Notice from Table I that an allowance of .001 used in Case 2 is just under the true parametric error while an allowance of .002 used in Case 3 is slightly larger than the true parametric error. Hence, the Case 3 additive uncertainty shown in Figure 22 is flat over frequency and does not show any sign of resonant dependent error, in contrast to Case 2 which shows a slight resonant dependent additive uncertainty (likely due to Mode 1 which is missing and Mode 4 which has a slighty larger error than .001, see Table I). In Case 4, only Mode 4 near 4.6 rad/sec is considered uncertain and as expected, the parametric uncertainty contribution to the additive uncertainty near this mode disappeared. Without any parametric uncertainty allowance for the modes near .9 rad/sec for Case 4, the additive uncertainty fully covered these parametric uncertainty errors for these modes. Doubling the parametric allowance to .004 did not noticably decrease the additive uncertainty. For brevity sake, the important effects of noise allowance in determining the smallest unmodeled dyanamics for model validation is not discussed and the interested reader is referred to [19].

4. Robust control of a large flexible structure

4.1. LARGE FLEXIBLE STRUCTURE TESTBED

Figure 23 shows a large flexible structure laboratory testbed at NASA to evaluate active control laws for vibration supression and fine pointing. The main truss is approximately 50 feet long with two vertical and two horizontal booms connected to the main truss. Cable attachements at the ends of the horizontal booms, which are connected to the ceiling by soft springs, suspend the entire structure to allow a general six degrees of freedom translation and rotation. A 16 foot diameter mirror, attached to a dish-like reflector located at one end of the main truss, is designed to reflect a laser beam source originating from the end of a vertical boom, to an optical detector array, located approximately 60 feet above the structure. More details of this Control-Structure Interaction Evolutionaty Model (CEM) testbed are available in [30].

Fine pointing for this CEM structure consists of regulating an impinging laser beam on a target plane as illustrated in Figure 24. A performance measure considered for the CEM system is the worst case error in the line-of-sight (LOS) pointing (for more details see [31]) in the presence of external disturbances and model uncertainties. For this task, six pairs of pneumatic thrusters are used for feedback control and two pairs of thrusters are used to simulate exogenous disturbances. Six accelerometers attached to the structure are used for sensing the vibration and pseudo rigid body (pendulum) motions and for feedback control. These thrusters and accelerometers are spatially distributed and configured to effectively control all six degrees of freedom and are approximately collated for translational degrees of freedom. These approximately collocated sensors and actuators, assuming perfect integration of accelerometer readings, limits the use of simple yet robust schemes such as Direct Velocity Feedback [32].

Dynamic characteristics. Figure 25 shows the acceleration responses in three axis corresponsing to sensors 5, 1, and 2, when actuated from thruster 2. Notice that the high modal density coupled with low damping makes the active control problem

164

Figure 23. Large flexible structure testbed at NASA.

very interesting since the controller bandwidth will likely roll-off through flexible
modes. This fact underscores a need for an accurate reckoning of the model, at least
in the neighborhood of loop gain crossover frequencies [33]. The analysis of stability,
performance, and its robustness, and subsequent compensator design may be a an
easy task for single-loop configuration, but not so for multi-loop high performance
system with questionable dynamical models. In short, although very detailed and
large mathematical models can be generated by current technology, yet it is difficult
to realize a highly reliable dynamical model of the CEM,[10] as expected in a typical
large flexible structure in space. This is mostly due to the complexity in the details
of the physical system, which limits the accuracy and reliability of the mathematical
representation. Efforts in model refinement using measurement data are also limited
due to its generally ad-hoc nature and limitations to the extent of testing possible.

In summary, due to the coupled multi-axis dynamics, desired level of disturbance
attenuation ruling out passive or structural redesign alone, and the uncertainties in
mathematical and empirical models and unknown disturbances, an application of
multivariable robust control is of significant interest.

---

[10] Careful experimental modal analysis, which can sometimes accurately estimate the resonant
frequencies and even damping ratios for a subset of significant modes, is usually not sufficient for
generating mathematical models that can accurately predict dynamic responses.

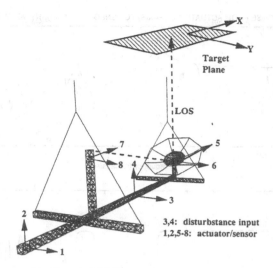

Figure 24. Line of sight control of a large flexible structure.

Design goal. The design goal is to maintain LOS pointing by rejecting disturbances over a control bandwidth of at least 2 Hz under the following uncertainties [34]: (i) frequency and damping, (ii) higher frequency truncated modes, (iii) structural mode shapes, and (iv) electro-pneumatic servovalve actuator dynamics. The disturbance rejection is quantified by specifying a factor of attenuation (say 10) for the dominant structural peaks which is equivalent to a corresponding increase in the open-loop damping in the modes contributing to LOS motion. Note also that the target bandwidth for the CEM structure is significantly faster than attitude controllers for rigid spacecraft having, say, a bandwidth of .1 Hz.

4.2. ROBUST CONTROL DESIGN MODEL

As note earlier, a robust control design model is intended to capture discrepencies in the mathematical model representation of physical systems, due to inherent uncertainties. Figure 26 shows an actual model mismatch in CEM between an estimated and predicted spectra for thruster 8 to accelerometer 8. At low frequencies, parametric errors in the resonant frequencies and damping factors are noted, while at higher frequencies, large errors which may be considered unmodeled dynamics are observed. It was found that system identification was helpful but was clearly inadequate in resolving model accuracy issues at higher frequencies.

4.2.1. *Nominal model*
Consider a NASTRAN finite element model with 638 finite element grid points. In physical coordinates, the linear motion is written as

$$M\ddot{\xi} + Q\dot{\xi} + K\xi = Eu; \quad y_{LOS} = F\xi; \quad y_a = H\ddot{\xi} \tag{61}$$

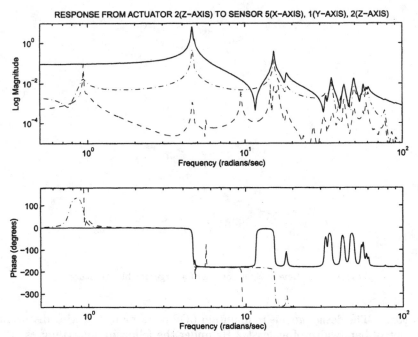

Figure 25. Predicted response from a 52 mode model of the CEM.

where y_{LOS} and y_a corresponds to the LOS displacements and acceleration outputs, respectively. Since ξ denotes the finite-element displacement vector, the nonzero entries in the corresponding columns and rows of E and H, denote the location and direction of the actuation, u, and sensor, y, while F is determined by the geometrical configuration and relative orientations of the LOS laser source, mirror and target (see [31] for more details). Using the matrix of structural mode shape vectors Ψ available from a finite-element model, we can rewrite the motion equation in structural modal coordinates

$$\ddot{\eta} + \text{diag}(2\zeta\omega)\dot{\eta} + diag(\omega^2)\eta = \Psi^T Eu \tag{62}$$

Typically, the designer includes in Ψ mode shapes corresponding to significant modes for a given dynamic environment. A conservative but simple approach is to include all modes that lie within the expected controller bandwidth. For example, for the CEM LOS control, the controller bandwidth is expected to be around 2 Hz, and we include all structural modes which falls below, say, 10 Hertz. This results in a truncated modal model consisting of 25 modes. Using a sampling rate of 125 Hz and assuming zero-order hold at all inputs, we can rewrite the motion equations in modal state space form

$$x_{k+1} = Ax_k + Bu_k, \quad y_k = Cx_k + Du_k \tag{63}$$

where $A = \text{diag}(\Lambda_1, \cdots, \Lambda_n)$, and $\Lambda_i := \begin{bmatrix} \alpha_i & \beta_i \\ -\beta_i & \alpha_i \end{bmatrix}$, for $i = 1, \ldots, n_{modes}$. Table III shows the frequencies and damping of a twelve mode nominal model. The first six

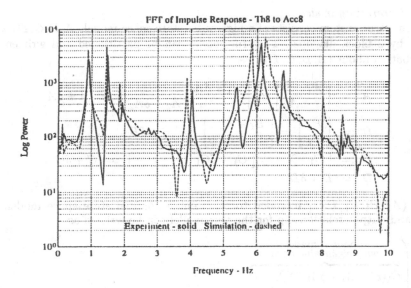

Figure 26. Typical mismatch between estimated spectra and predicted: thruster 8 to accelerometer 8.

corresponds to the suspension dependent pendulum motion which can be viewed as pseudo rigid-body modes.

TABLE III. Frequency and damping of 12 mode nominal model; boxed frequency and damping values indicate parametric uncertainty candidate.

Mode	Freq(r/s)	Damp	Description (based on NASTRAN mode shapes)
1	.924	.057	Translation in Y-axis
2	.936	.075	Translation in X-axis
3	.975	.075	Translation in Z-axis
4	4.59	.007	Rotation about Y-axis
5	4.70	.007	Rotation about Z-axis
6	5.49	.002	Rotation about X-axis
7	9.25	.002	First torsional about X-axis
8	10.9	.002	complex coupled
9	11.8	.002	complex coupled
10	14.4	.001	First antenna bending
11	15.9	.001	First bending about Y-axis
12	17.7	.001	First bending about Z-axis

4.2.2. *Uncertainty model*

Suppose a true system consists of a high order state space model (A, B, C, D) where A is 2 by 2 block diagonal. The true model can be partitioned and written as a perturbation about $(\bar{A}, \bar{B}, \bar{C}, \bar{D})$

$$A := \begin{bmatrix} A_1 & \\ & A_2 \end{bmatrix} := \begin{bmatrix} \bar{A}_1 + \delta A_1 & \\ & \bar{A}_2 + \delta A_2 \end{bmatrix} := \bar{A} + \delta A$$

$$B := \begin{bmatrix} B_1 \\ B_2 \end{bmatrix} := \begin{bmatrix} \bar{B}_1 + \delta B_1 \\ \bar{B}_2 + \delta B_2 \end{bmatrix} := \bar{B} + \delta B$$

$$C := \begin{bmatrix} C_1 & C_2 \end{bmatrix} := \begin{bmatrix} \bar{C}_1 + \delta C_1 & \bar{C}_2 + \delta C_2 \end{bmatrix} := \bar{C} + \delta C$$

$$D := \bar{D} + \delta D$$

where $(\bar{A}_1, \bar{B}_1, \bar{C}_1, \bar{D})$ is intended to represent a nominal state space model. The input and output relationship for the true model can be written as

$$y = \left(\overbrace{\bar{C}_1(zI - \bar{A}_1 - \delta A_1)^{-1}\bar{B}_1 + \bar{D}}^{\substack{\text{nominal model} \\ \text{with eigenvalue error}}} \right.$$

$$\left. + \underbrace{\delta C_1(zI - A_1)^{-1}B_1}_{\text{modal output error}} + \underbrace{\bar{C}_1(zI - A_1)^{-1}\delta B_1}_{\text{modal input error}} + \overbrace{C_2(zI - A_2)^{-1}B_2}^{\text{unmodeled dynamics}} + \underbrace{\delta D}_{\substack{\text{feedthrough} \\ \text{gain error}}} \right) u$$

The point here is that any unknown state-space system can be represented by an additive uncertainty structure about a lower order nominal model with eigenvalue uncertainties. The remaining issues include how large are the uncertainty balls about the nominal, and whether to treat modal output and input errors as parametric errors or combine them with unmodeled additive uncertainties. Notice that the latter issue involves the question of uniqueness in uncertainty structure and models, open issues in general. Nevertheless, the uncertainties in modal outputs and inputs appear easier to model as nonparametric uncertainties at the inputs and/or outputs. This is because δC_1 and δB_1 are coordinate dependent (unlike δA_1), and therefore difficult to quantify their physical significances based on their numerical values, and because they possibly avoids a large increase in the number of unknown parameters.

Damping and frequency uncertainties in state space. For a general linearly perturbed state space model

$$\left\{ \begin{array}{c} x_{k+1} \\ y_k \end{array} \right\} = \left(\begin{bmatrix} A_o & B_o \\ C_o & D_o \end{bmatrix} + \sum_{i=1}^{m} \delta_i \begin{bmatrix} A_i & B_i \\ C_i & D_i \end{bmatrix} \right) \left\{ \begin{array}{c} x_k \\ u_k \end{array} \right\} \tag{64}$$

where $\begin{bmatrix} A_i & B_i \\ C_i & D_i \end{bmatrix}$ denotes the i-th perturbation matrix associated with real scalar perturbation δ_i, Morton and McAfoos [35] demonstrated that this general linearly perturbed state space model can be represented as an LFT in

$$\left\{ \begin{array}{c} x_{k+1} \\ y_k \end{array} \right\} = F_l(M, \Delta) \left\{ \begin{array}{c} x_k \\ u_k \end{array} \right\} \tag{65}$$

as shown in Figure 27. The matrices E_i, F_i, G_i, and H_i can be computed from the

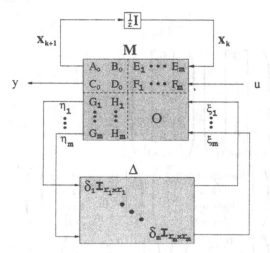

Figure 27. General linearly perturbed state space model in LFT form.

singular value decomposition

$$\left[\begin{array}{cc} A_i & B_i \\ C_i & D_i \end{array} \right] = \underbrace{U_{i_1} \Sigma_{i_1}^{\frac{1}{2}}}_{} \underbrace{\Sigma_{i_1}^{\frac{1}{2}} V_{i_1}^T}_{} := \left[\begin{array}{c} E_i \\ F_i \end{array} \right] \left[\begin{array}{cc} G_i & H_i \end{array} \right], \quad \Sigma_{i_1} > 0 \tag{66}$$

The order of the multiplicity, r_i, for the ith scalar parameter depends on the matrix rank of the corresponding perturbation matrix given by equation 66. Physically, this rank depends on the choice of the state variables and the parameterization of its perturbations in the state space matrices. In particular, for frequency and damping uncertainties in a modal state space model, one can represent them as linear perturbations in the A matrix,

$$\left\{ \begin{array}{c} x_{k+1} \\ y_k \end{array} \right\} = \left[\begin{array}{cc} A_o + \Delta A & B_o \\ C_o & D_o \end{array} \right] \left\{ \begin{array}{c} x_k \\ u_k \end{array} \right\} \tag{67}$$

where the nominal A_o matrix is given by

$$A_o := \left[\begin{array}{ccccccc} \lambda_1 & & & & & & \\ & \ddots & & & & & \\ & & \lambda_r & & & & \\ & & & \Lambda_1 & & & \\ & & & & \ddots & & \\ & & & & & \Lambda_n \end{array} \right], \quad \lambda_i \in \mathcal{R}, \quad \Lambda_i := \left[\begin{array}{cc} \alpha_i & \beta_i \\ -\beta_i & \alpha_i \end{array} \right] \in \mathcal{R}^{2 \times 2}$$

and the linear perturbations as

$$\Delta A := \begin{bmatrix} \delta\lambda_1 & & & & & \\ & \ddots & & & & \\ & & \delta\lambda_r & & & \\ & & & \delta\Lambda_1 & & \\ & & & & \ddots & \\ & & & & & \delta\Lambda_n \end{bmatrix}, \ \delta\lambda_i \in \mathcal{R}, \ \delta\Lambda_i := \begin{bmatrix} \delta\alpha_i & \delta\beta_i \\ -\delta\beta_i & \delta\alpha_i \end{bmatrix} \in \mathcal{R}^{2\times2}$$

$$= \sum_{i=1}^{r} \delta\lambda_i \left(\begin{bmatrix} 0 \\ \vdots \\ 1 \\ \vdots \\ 0 \\ [0_{n_y \times 1}] \end{bmatrix} \right) \left(\begin{bmatrix} 0 \\ \vdots \\ 1 \\ \vdots \\ 0 \\ [0_{n_u \times 1}] \end{bmatrix} \right)^{T}$$

$$+ \sum_{i=1}^{n} \delta\alpha_i \left(\begin{bmatrix} 0_{[r+2(i-1)]\times2} \\ I_{2\times2} \\ 0_{[n-2i]\times2} \\ [0_{n_y \times 2}] \end{bmatrix} \right) \left(\begin{bmatrix} 0_{[r+2(i-1)]\times2} \\ I_{2\times2} \\ 0_{[n-2i]\times2} \\ [0_{n_u \times 2}] \end{bmatrix} \right)^{T}$$

$$+ \sum_{i=1}^{n} \delta\beta_i \left(\begin{bmatrix} 0_{r\times2} \\ 0_{2(i-1)\times2} \\ \begin{bmatrix} 0 & 1 \\ -1 & 0 \end{bmatrix} \\ 0_{[n-2i]\times2} \\ [0_{n_y \times 2}] \end{bmatrix} \right) \left(\begin{bmatrix} 0_{r\times2} \\ 0_{2(i-1)\times2} \\ \begin{bmatrix} 0 & 1 \\ -1 & 0 \end{bmatrix} \\ 0_{[n-2i]\times2} \\ [0_{n_u \times 2}] \end{bmatrix} \right)^{T}$$

where $\delta\lambda_i$ denotes the purely real eigenvalue perturbation while $\delta\alpha_i$ and $\delta\beta_i$ denotes the real and imaginary components of the ith complex conjugate eigenvalue perturbation, respectively. Using the above expressions, singular value decompositions are not even necessary for their representation in LFT form.

In the robust control designs outlined in the next section, constant parametric uncertainties of 5 % were assumed for the first three modes and 1 % for the next three, i.e., $|\delta\alpha_i| \leq .05$, $|\delta\beta_i| \leq .05$, for $i = 1, 2, 3$, and $|\delta\alpha_i| \leq .01$, $|\delta\beta_i| \leq .01$, for $i = 4, 5, 6$.

Unmodeled high frequency modes. As noted earlier, it is generally difficult to accurately and reliably predict high frequency structural modes of large flexible structures either by black-box/experimental modeling or by finite element analysis. This limitation is mainly due to the simplifying assumptions which are often necessary in first principle modeling of the structure, on detail issues such as imprecise material properties, joint and boundary conditions, and imperfect alignment, length, and thickness of structural members, etc. A consequence is that mathematical models can often completely miss the presence (or absence) of a resonant peak, as seen for example in Figure 26.

For the CEM structure, the analytical model (with refinements based on experimental modal analysis) provided reasonably confident natural frequency values to within 2% and damping values to within 10%, for modes below 2 Hz. However, higher frequency modes, beginning with the 12th mode near 3 Hz, are not accurately known. With no reliable knowledge of the gain and phases of these higher frequency resonant modes, the standard approach in robust control is to gain stabilize them by rolling off the loop gain. This is efficiently handled by prescribing an uncertainty in a robust control design model, in contrast to an iterative loop-shaping procedure to trade-off with other performance goals. Hence, thirteen truncated higher frequency modes, ranging from about 3 Hz to 12 Hz, are accounted for by unstructured additive uncertainty. Figure 28 shows the maximum singular value frequency response for the LOS error from all thrusters to all accelerometers. The additive uncertainty block is represented by a fully populated complex 6×6 matrix. For flexible structures, the necessary assumption of stable perturbations always holds since all truncated higher frequency structural modes are dissipative in nature. A 4th-order stable rational

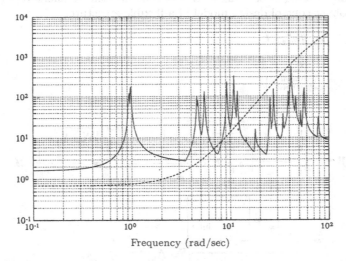

Frequency (rad/sec)

Figure 28. Maximum singular value frequency response from all thrusters to all accelerometers, additive uncertainty weight (dashed line).

filter is used as an upper bound fit of the additive uncertainty. Notice that all higher frequency modes, say, above 3 Hertz is covered by this additive uncertainty.

Other model errors. In the CEM, each actuator actually consists of a pair of pneumatic thrusters acting in opposite directions. As expected, the dynamics of each pneumatic thruster is not identical and resembles a first order filter with a sufficiently large break frequency of about 43 Hertz. Since this is about 20 times the control bandwidth, only its DC gain is factored into design and the uncertainties in the acutator dynamics are modeled as constant percentage uncertainty in each input channel, i.e., multiplicative uncertainty.

The inaccuracy in the FEM also causes mode shape uncertainties in the model. The mode shapes are by definition constants and hence it is appropriate to model

172

them as constant scaling errors at inputs or outputs. Hence, mode shape uncertainties and actuator uncertainties are combined into a sinlge multiplicative uncertainty at the plant input. A 5 % constant multiplicative input uncertainty is assumed for each input channel. Notice that higher levels with frequency weighting can be used instead if one seeks to curb control effort in the design.

Unknown exogenous disturbances. Exogenous disturbances and noises which are generally unknown are inevitable in practice and the question usually involves their influence path and size. As described earlier, in robust control all unknown exogenous disturbances are modeled as unknown but bounded. This is equivalent to a stochastic viewpoint of a bounded uniformly distributed noise. Since disturbances and noises can enter the control system in any number of ways, for convenience in later analysis and design, an equivalent noise at output is defined (for more details see [19]). Since flexible structures are intrinsically stable, their free responses can be recorded and the noise spectrum at the output estimated. This can then be normalized to unity 2-norm and any scaling or coloring filter absorbed into the augmented plant. Figure 29 shows a robust control design model of a flexible structure.

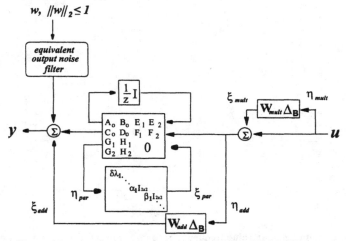

Figure 29. A robust control design model for CEM LOS control.

4.2.3. *Inherent modal sensitivity for lightly damped structure.*
Figure 30 illustrates the inherent sensitivity in a flexible structure mode when it is very lightly damped. For example, a 1 % error in frequency for a very lightly damped mode at $\zeta = .005$, will lead to huge errors in the gain and phase while a 5 % error will mean missing the entire contribution of the mode itself due to frequency shift and narrow resonance. From a robust control standpoint, this means that the degree of robustness attainable will be very sensitive to the size of uncertainty in the frequency for a mode which is lightly damped. This means that the resonant frequencies of very lightly damped modes should be accurately known because of its inherent model sensitivity and that modest levels of added passive damping should be effective in reducing this inherent sensitivity.

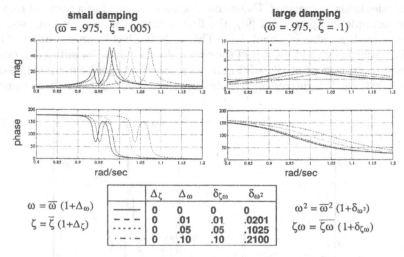

	$\Delta\zeta$	Δ_ω	$\delta_{\zeta\omega}$	δ_{ω^2}
——	0	0	0	0
– – –	0	.01	.01	.0201
······	0	.05	.05	.1025
· – · –	0	.10	.10	.2100

$$\omega = \overline{\omega}\,(1+\Delta_\omega)$$
$$\zeta = \overline{\zeta}\,(1+\Delta_\zeta)$$

$$\omega^2 = \overline{\omega}^2\,(1+\delta_{\omega^2})$$
$$\zeta\omega = \overline{\zeta\omega}\,(1+\delta_{\zeta\omega})$$

Figure 30. Sensitivity in mode 3 of CEM model due to frequency and damping errors.

4.3. DESIGN OF LOS CONTROLLER

The interconnection structure for the LOS control of CEM is shown in Figure 31. A constant diagonal LOS weight was used while an appropriately scaled first order

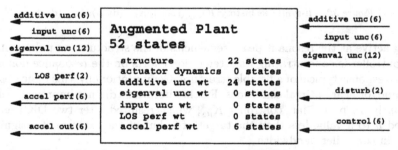

Figure 31. Interconnection structure for the LOS control of CEM.

low pass filter was used to specify the attenuation at the acceleration outputs. The controller was designed by μ synthesis via DK-iteration, i.e.,

$$\min_{K, D \in \mathcal{D}} \|DF_l(P,K)D^{-1}\|_\infty$$

Initially, by setting $D = I$, an H_∞ control design (see for example Chapter 14 of [9]) produced a controller, K_{H_∞}, which gave $\|F_l(P,K)\|_\infty = 22.5$. Although this H_∞ norm appears large, μ-analysis[11] indicated that $\mu = 1.90$ (see Figure 32) which means that roughly half the target performance can be guaranteed for half the size of the

---

[11] A mixed μ analysis [14] indicated that the complex scalar uncertainty assumption was not significantly conservative.

uncertainties initially specified. Using the D scaling obtained as a result of μ-analysis (see for example Chapter 10 of [9]), the first μ controller, K_{μ_1}, is obtained which achieves a peak μ value of about 1.15. This μ design has improved about 40 % in both the performance and the size of the uncertainty set over the initial H_∞ design, a significant improvement predicted. Figure 32 shows the upper bound μ values during the DK iterations. It is interesting to note from the μ plot that three peaks come close

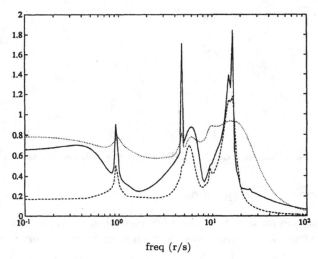

freq (r/s)

Figure 32. μ-synthesis history: K_{H_∞} (solid), K_{μ_1} (dash), K_{μ_2} (dot).

to being active at the resonant peak frequencies of the structure, an indication that the μ controller is concentrating its attenuation efforts on the resonance dominated dynamic response typical of a flexible structure. Computationally, this indicates good scaling and/or near optimal conditions. Further iterations did not improve the design significantly. In particular, controller K_{μ_2}, which resulted after two DK iterations improved μ to a value less than the target of 1, but at the expense of a significant increase in controller bandwidth.

4.4. SIMULATION AND EXPERIMENTAL RESULTS

The predicted open and closed loop LOS response to a unit band-limited white noise disturbance is given in Figure 33 for both X and Y axes. Notice that the modes beyond 3 Hz are essentially undamped, which is due to controller roll-off and the open loop value being below the target value. Table IV shows that the nominal closed-loop response satisfies the target response well. For example, in the X-axis LOS motion, the open-loop H_∞ norm is 3.13 as compared to 0.46 for the closed-loop response, which is less than the target norm of 0.625. As a result of optimizing the disturbance rejection from the disturbance inputs to the predicted LOS response, only modes that significantly participated in the LOS output were damped significantly, corresponding to the first six of the twelve modes present in the control design model. Although significant damping was added to these six modes,

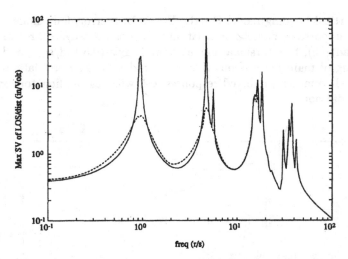

Figure 33. Predicted nominal LOS performance for K_{μ_1}; maximum singular value frequency response open loop(solid), closed loop(dash).

TABLE IV.   Comparison of LOS response for controller K_{μ_1}.

	LOS-X (H_∞ norm)	LOS-Y (H_∞ norm)
Open Loop	3.13	14.5
Target	0.625	2.0
Closed Loop	0.46	1.47

the root-mean-square of the LOS error was only reduced from an open-loop value of 1.14 inch to 0.75 inches. This is expected because the μ design is driven by the worst case response to guarantee robustness and does not seek optimality in an average sense, i.e., it is not an H_2-based design.

A parametric study involving various uncertainties modeled confirmed that robust performance was most sensitive to the size of the frequency uncertainty specified. For instance, the improved robust performance noted in the control design by reducing the actuator uncertainty to 2.5%, was insignificant when compared to the improvements due to halving frequency uncertainties. This observation is consistent with the earlier prediction that a small percentage error in the natural frequency for a very lightly damped mode can lead to significant errors in its frequency response.

Figure 34 shows a comparion between measured and predicted open loop responses from wide band disturbances at thrusters 3 and 4. For the X-axis LOS error, the dominant mode was the first vertical bending of the main body about Y-axis (16 rad/sec) which simulation model (dashed line) also predicts reasonably well (notice however that only gain comparisons are shown and discrepencies in the phases near the resonances may be significant for low damping). Other significant modes that contribute to LOS error includes translation in X-axis (.9 rad/sec) and rotation about Y-axis (4.5 rad/sec), which do not appear to match predicted responses as well as the first vertical bending mode. For the Y-axis LOS error, four modes appeared to be equally important, namely, translation in Y-axis (.9 rad/sec), rotation about X-axis (5.5 rad/sec), first torsional mode about X-axis (9.5 rad/sec), and the first lateral bending of main body about Z-axis (17 rad/sec). The first lateral bending mode appeared to match predicted responses well whereas the first torsional mode about X-axis did not.

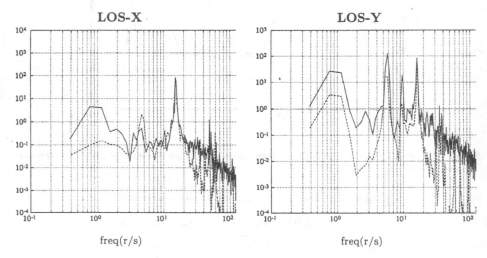

Figure 34. Measured (solid) and predicted (dash) open loop LOS responses due to wide band disturbances at thrusters 3 and 4.

Figure 35 shows a comparion between measured closed loop (solid lines) and open loop (dash lines) responses due to wide band disturbances at thrusters 3 and 4. For the X-axis LOS error, significant attenuation is observed in the dominant first vertical bending of the main body about Y-axis (16 rad/sec) and in the translational motion in X-axis (.9 rad/sec). A moderate level of attenuation in the rotation about Y-axis (4.5 rad/sec) is also noted. For the Y-axis LOS error, significant attenuation is observed in the rotational motion about X-axis (5.5 rad/sec) and in the first lateral bending of the main body about Z-axis (17 rad/sec). However, only a moderate level of attenuation is noted in the translation motion in Y-axis (.9 rad/sec) while the first torsional mode about X-axis (9.5 rad/sec) is practically undamped. This latter result is consistent with the significant mismatch noted earlier in Figure 34.

Figure 36 shows a comparison between measured and predicted closed loop responses from wide band disturbances at thrusters 3 and 4. The model appears to match the observed closed loop response reasonably well although significant areas for improvement exists, particularly in the Y-axis LOS response involving the translational motion in Y-axis (.9 rad/sec) and the first torsional mode about X-axis (9.5 rad/sec). As a historical note, the above discrepencies and limited predictability in the closed loop responses for the CEM testbed, despite the application of robust control laws and analysis (based on ad-hoc uncertainty bounds), partly motivated this author to develop a more methodical uncertainty bound determination procedure based on measurement data, as outlined in Section 3.

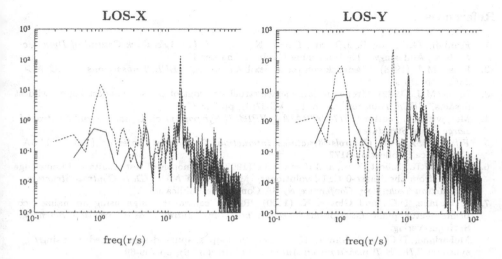

Figure 35. Measured closed loop (solid) and open loop (dash) LOS responses due to wide band disturbances at thrusters 3 and 4.

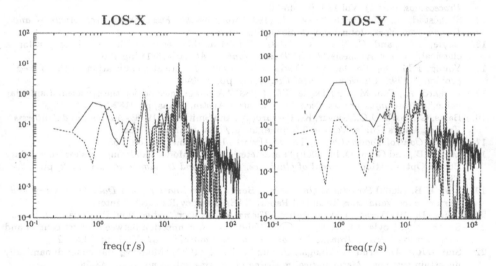

Figure 36. Measured (solid) and predicted (dash) closed loop LOS responses due to wide band disturbances at thrusters 3 and 4.

Acknowledgement

The author appreciates the useful comments from his colleague Dr. Daniel P. Giesy of NASA Langley Research Center, which helped improve this paper.

References

1. Franklin, G.F., Powell, J.D., and Emami-Naeini, A.E. (1986)*Feedback Control of Dynamic Systems,* Addison-Wesley Publishing Co, Inc., Chapter 1.
2. Balas, M.J. (1978) "Feedback control of flexible systems," *IEEE Transactions on AC,* AC-23(4), pp 673-9.
3. Balas, M.J. (1982) "Trends in large space structures control theory: Fondest hopes, wildest dreams," *IEEE Transactions on AC,* AC-27(3), pp 522-35.
4. Meirovitch, L., editor (1977-1993)*AIAA/VPI&SU Symposium on Dynamics and Control of Large Space Structures.*
5. *Fifth NASA/DoD Controls-Structures Interaction Technology Conference* (1992), NASA Conference Publication 3177.
6. Belvin,W.K., Elliott, K.B., and Horta, L.G. (1992) "A synopsis of test results and knowledge gained from the Phase-0 CSI Evolutionary Model,"*Fifth NASA/DoD Controls-Structures Interaction Technology Conference,* NASA Conference Publication 3177.
7. McFarlane, D.C., and Glover, K. (1990) "Robust controller design using normalized co-prime factor plant descriptions," *Lecture notes in Control and Information Sciences,* Springer-Verlag.
8. McFarlane, D.C., and Glover, K. (1992) "A loop shaping design procedure usingH_∞ synthesis," *IEEE Transactions on Automatic Control,* 37(6), pp 759-69.
9. Zhou, K. (1998) *Essentials of Robust Control,* New Jersey: Prentice Hall, Inc.
10. Doyle, J.C., Francis, B.A., and Tannenbaum, A.R. (1992)*Feedback Control Theory,* Macmillan Publishing Company, New York.
11. Doyle, J.C. (1982) "Analysis of feedback systems with structured uncertainties,"*IEE Proceedings,* Part D, Vol 133, pp. 45-56.
12. Skogestad, S., and Postlethwaite, I. (1997)*Multivariable Feedback Control: Analysis and Design,* New York: John Wiley & Sons, Chapter 8.
13. Doyle, J.C., and G. Stein, G. (1981) "Multivariable feedback design: Concepts for a classical/modern synthesis," *IEEE Transactions on AC,* AC-26(1), pp 4-16.
14. Young, P.M., Newlin, M., and Doyle, J.C., (1992) "Practical computation of the mixed problem," *1992 American Control Conference,* pp. 2190-4.
15. Packard, A., Fan, M., and Doyle, J.C. (1988) "A power method for the structured singular value," *27th Conference on Decision & Control,* Austin, Texas, pp.2132-37.
16. Balas, G., Doyle, J.C., Glover, K., Packard, A., and Smith, R. (1998)*μ-Analysis and Synthesis Toolbox,* User's Guide, Version 3 (online),http://www.mathworks.com
17. *Robust Control Toolbox* (2001), The Mathworks Inc, Online Version 2.08
18. Lim, K.B., and Giesy, D. P. (2000) "Parameterization of Model Validating Sets for Uncertainty Bound Optimization", *Journal of Guidance, Control, and Dynamics,* vol. 23, no. 2, pp. 222-230.
19. Lim, K.B. (2002) *Structured Uncertainty Bound Determination from Data for Control and Performance Validation,* Technical Paper, NASA Langley Research Center.
20. Ljung, L. (1999) *System Identification: Theory for the user,* 2nd edition, Prentice Hall, p. 32.
21. Smith, R.S., Doyle, J.C. (1992) "Model Validation: A connection between robust control and identification," *IEEE Transactions on Automatic Control,* vol 37, No 7, pp.942-952.
22. Smith, R.S., Dullerud, G., Rangan, S., and Poolla, K. (1997) "Model validation for dynamically uncertain systems," *Mathematical modeling of systems,* vol. 3, no. 1, pp. 43-58.
23. Poolla, K., Khargonekar, P.P., Tikku, A., Krause, J., and Nagpal, K. (1994) "A time-domain approach to model validation,"*IEEE Transactions on Automatic Control,* v.39, n.5, pp. 951-9.
24. Callafon, R.A. de, and Hof, Paul M.J. van den (2000) "Closed-loop model validation using coprime factor uncertainty models," SYSID 2000*Symposium on System Identification,* Santa Barbara, California.
25. Astrom, K.J., and Soderstrom, T. (1971) "System Identification: A survey," *Automatica,* v.7, pp.123-62.
26. Soderstrom, T., and Stoica, P. (1989) *System Identification,* Hempstead, Prentice Hall International.
27. Juang, J-N. (1994) *Applied System Identification,* Prentice Hall.
28. SYSID 2000: *Symposium on System Identification,* Santa Barbara, California, June 21-23.
29. T. Coleman, M. A. Branch, and A. Grace (1999)*Optimization Toolbox,* User's Guide, Version 2 (online). http://www.mathworks.com

30. Belvin, K.W., et.al (1991) "Langley's CSI evolutionary model: Phase 0,"*NASA TM 104165*.
31. Lim, K.B., and Horta, L.G. (1990) "A line-of-sight performance criterion for controller design for flexible structures," *1990 AIAA Dynamics Specialists Conference*, AIAA Paper No. 90-1226.
32. Balas, M.J. (1979) *Journal of Guidance and Control*, vol 2, no 3, pp 252-3.
33. Balas, G.J., and Doyle, J.C. (1994) "Control of lightly damped, flexible modes in the controller crossover region," *Journal of Guidance, Control, and Dynamics*, Vol. 17, No. 2, pp 370-7.
34. Lim, K.B., and Balas, G.J. (1992) "Line-of-sight control of the CSI evolutionary model μ-control," *1992 American Control Conference*.
35. Morton, B., and McAfoos, R. (1985) "A Mu-test for real parameter variations,"*1985 American Control Conference*, pp,135-8.

FINITE ELEMENT MODELS FOR PIEZOELECTRIC CONTINUA

PAOLO GAUDENZI
Università di Roma "La Sapienza",
Dipartimento di Ingegneria Aerospaziale e Astronautica
Via Eudossiana 16, 00184 Roma - Italy

1. Introduction

In recent years the development of new active materials and their use for structural applications have induced the need of new numerical modeling tools capable of simulating the complicated constitutive behaviour of these materials. A particularly prominent area of research has been the area of adaptive structure technology in which focus is given on the possibility of designing and building active structures with advanced distributed actuating and sensing capabilities [1]. To this end, several types of materials such as magnetostrictive materials, piezoelectric and electrostrictive materials, and shape memory alloys are being tested in order to identify their possible use in engineering practice as part of an "intelligent structure".

Piezoelectric materials have been particularly taken into consideration for this kind of applications because of their capability of producing in a relatively easy way both sensing and actuation.

Piezoelectric effects were discovered in 1880 by J. and P. Curie. When a mechanical force is applied to a piezoelectric material an electrical voltage is generated; this phenomenon is referred to as the direct piezoelectric effect. On the contrary, the converse piezoelectric effect is observed when the application of an electrical voltage produces strain in the material. In these cases an energy transfer is observed from mechanical to electrical energy or vice versa.

Piezoelectric materials have been widely used in transducers in several applications like strain gages, pressure transducers and accelerometers. The development of new products like lead zirconate titanate piezoelectric ceramics and low modulus piezoelectric films makes possible new kinds of structural engineering applications. In fact with these new piezoelectric materials it is possible to build composite structures in which the piezoelectric material is perfectly bonded (or even embedded in the case of fiber reinforced plastic composite laminates) to the passive (traditional) structure [2]. In such an assembly both the actuating function (by means of the converse effect) and the sensing function (by means of the direct effect) can be performed, provided that appropriate locations and geometries are chosen for the piezoelectric part.

A. Preumont (ed.), Responsive Systems for Active Vibration Control, 181–205.
© *2002 Kluwer Academic Publishers. Printed in the Netherlands.*

However, due to the rather complex nature of the physical behaviour, analytical and numerical methods for the analysis of piezoelectric materials are not yet fully developed.

One of the principal issues in the modeling is of course the coupling of the mechanical response with the electrical response; namely, the equations of mechanical equilibrium and continuity are coupled through the constitutive equations with the corresponding electrical equations.

Also the physical behaviour of a piezoelectric material is frequently governed by a nonlinear constitutive equation so that a nonlinear mathematical model has to be established [3].

For piezoelectric materials, a sound mathematical continuum model for the linear case has been established [4, 5, 6], and several analytical studies have used this model in mechanical vibrations of rods, plates and shells [7].

Elements of a nonlinear theory have also been developed [8]. On the computational side, some finite element developments have been presented for the linear case [9, 10] based on the approach proposed by Allik and Hughes [11]. In this model the displacements and electric potential are used as nodal unknowns. Recently, several numerical and analytical studies were performed on composite structures in which the piezoelectric part is bonded or embedded in a traditional structure or in a fiber reinforced plastic laminate [2, 12, 13, 14, 15].

In the following a finite element procedure that can be used to model the electro-mechanical coupled behavior of piezoelectric continua and that can also be used in a general nonlinear incremental finite element solution is fully described [16, 17]. The technique is based on establishing separately the finite element equations for the mechanical response and the electrical field. In this way the response for the converse and for the direct piezoelectric effects are solved. In order to fully account for the electro-mechanical coupling an iterative procedure is used. The procedure is then demonstrated by the solution of some 2D and 3D problems, including the case of a composite material with piezolectric phases.

2. The governing equations of piezoelectricity

In this section the equations of the linear theory of piezoelectricity for the steady state case are briefly summarized [4].

Let us consider a piezoelectric body in three-dimensional space. The body occupies a region V bounded by a surface S with an outward unit normal vector with components n_i. The following equations have to be satisfied in V:

Mechanical equilibrium equations

$$\tau_{ij,j} + f_i^B = 0 \tag{1}$$

Strain-displacement relations

$$\epsilon_{ij} = \frac{1}{2}(u_{i,j} + u_{j,i}) \tag{2}$$

Maxwell's equations for the quasi-static electric field

$$D_{i,i} = 0 \tag{3}$$

$$E_i = -\phi_{,i} \tag{4}$$

Constitutive equations

$$\tau_{ij} = c_{ijkl}\epsilon_{kl} - e_{kij}E_k \tag{5}$$

$$D_i = e_{ikl}\epsilon_{kl} + \varepsilon_{ij}E_j \tag{6}$$

where u_i is the displacement vector, ϵ_{ij} the strain tensor, τ_{ij} the stress tensor, f_i^B the body force vector, ϕ the electric scalar potential, D_i the electric displacement vector, E_i the electric field vector, c_{ijkl} the elastic constitutive tensor, e_{kij} the piezoelectric (strain) tensor, ε_{ij} the dielectric permittivity tensor. All the indices range over 1,2,3.

The boundary conditions are:
Natural mechanical boundary condition on S_f

$$\tau_{ij}n_j = f_i^S \tag{7}$$

Natural electrical boundary condition on S_σ

$$n_i D_i = \sigma^S \tag{8}$$

Essential mechanical boundary condition on S_u

$$u_i = u_i^* \tag{9}$$

Essential electrical boundary condition on S_ϕ

$$\phi = \phi^* \tag{10}$$

where f_i^S is the surface force and σ^S is the surface charge.

For the boundary surface S the following relations hold: $S_u \cup S_f = S$ and $S_u \cap S_f = 0$, $S_\phi \cup S_\sigma = S$ and $S_\phi \cap S_\sigma = 0$.

The constitutive equations are sometimes also written in the form

$$\epsilon_{ij} = F_{ijkl}\tau_{kl} + d_{kij}E_k \tag{11}$$

$$D_i = d_{ikl}\tau_{kl} + \varepsilon_{ij}E_j \tag{12}$$

where F_{ijkl} is the compliance tensor and d_{kij} is the piezoelectric tensor.

The following symmetries hold for the tensors that appear in the constitutive relations:

$$c_{ijkl} = c_{jikl} = c_{klij} \tag{13}$$

$$e_{kij} = e_{kji} \tag{14}$$

$$\varepsilon_{ij} = \varepsilon_{ji} \tag{15}$$

From equation (3) it is clear that no body density charge σ^B is assumed to be present in the piezoelectric material.

The complete set of governing differential equations consists of 22 equations (1)–(6) in 22 unknowns (u_i, ϵ_{ij}, τ_{ij}, ϕ, E_i, D_i), with the relevant boundary conditions (7)–(10).

A smaller set of four differential equations in four unknowns can be obtained from equations (1)–(6) in terms of u_i and ϕ:

$$c_{ijkl}u_{k,li} + f_j^B + e_{kij}\phi_{,ki} = 0 \qquad (16)$$

$$e_{kij}u_{i,jk} - \varepsilon_{ij}\phi_{,ij} = 0 \qquad (17)$$

Equation (16) is the three-dimensional equilibrium equation of the elastic body in terms of displacements with an additional term in which the piezoelectric tensor e_{kij} gives the electro-mechanical coupling. Equation (17) has the form of a divergence equation that describes field problems like heat flow or seepage in which, due to the coupling, a displacement dependent term is added.

If in equation (16) we assume that the electric potential is given throughout the body, the problem can be solved in analogy with the usual approach for linear thermoelasticity, by considering a new body force vector f_j^{B*} instead of f_j^B, with

$$f_j^{B*} = f_j^B + e_{kij}\phi_{,ki} \qquad (18)$$

A similar approach can be used for the electric field equation if u_i is assigned. The loading term in equation (17) is equivalent to a body charge density σ^{B*}, with

$$\sigma^{B*} = -e_{kij}u_{i,jk} \qquad (19)$$

3. The finite element equations for 3D solids

3.1. VARIATIONAL PRINCIPLES

From Equations (1) and (7) that describe the three-dimensional mechanical equilibrium of the body in the field V and on the boundary S for each time t, the *principle of virtual work* can be derived.

For every time t and for every possible choice of virtual displacements δu_i that are zero at and corresponding to the essential mechanical boundary conditions (9), the following relation holds [18]:

$$\int_{t_V} {}^t\tau_{ij}\,\delta_t\epsilon_{ij}d^tV = \int_{t_V} {}^tf_i^B\,\delta u_i d^tV + \int_{t_{S_f}} {}^tf_i^S\delta u_i^S\,d^tS_f \qquad (20)$$

where ${}^t\tau_{ij}$ is the stress at time t, $\delta_t\epsilon_{ij} = \frac{1}{2}(\frac{\partial \delta u_i}{\partial {}^tx_j} + \frac{\partial \delta u_j}{\partial {}^tx_i})$ is the virtual strain corresponding to δu_i, ${}^tf_i^B$ and ${}^tf_i^S$ are the body forces and the surface tractions at time t, ${}^tV$ and ${}^tS_f$ are the volume occupied at time t and the corresponding surface on which the tractions ${}^tf_i^S$ are prescribed and δu_i^S is the virtual displacement on ${}^tS_f$.

In analogy with the principle of virtual work a *principle of virtual electric potentials* can be stated as follows from equations (3) and (8). For every time t

and for every choice of virtual electric potential $\delta\phi$ that is zero at and corresponding to the essential electrical boundary condition (10), the following relation holds:

$$\int_{^tV} {}^tD_i \, \delta_t E_i d^t V = - \int_{^tS_\sigma} {}^t\sigma^S \delta\phi^S \, d^t S_\sigma \qquad (21)$$

where ${}^tD_i$ is the electric displacement vector at time t, $\delta_t E_i = -\frac{\partial\delta\phi}{\partial^t x_i}$ is the virtual electric field corresponding to $\delta\phi$, ${}^t\sigma^S$ is the surface charge density, and $\delta\phi^S$ is the virtual electric potential on the boundary surface ${}^tS_\sigma$.

In these variational principles no restrictions on constitutive relations are present [18]; consequently they can be applied also for the case of material nonlinearity. In equations (20) and (21) in which no dynamic effect is considered, time is used as a convenient variable which denotes different intensities of load applications and correspondingly different configurations.

3.2. THE CASE OF LINEAR CONSTITUTIVE RELATIONS

We now restrict the analysis to the case of linear constitutive relations. If we substitute equations (5) and (6) into equations (20) and (21) the following set of equations is obtained:

$$\int_V c_{ijkl}\epsilon_{kl} \, \delta\epsilon_{ij} dV - \int_V e_{kij} E_k \, \delta\epsilon_{ij} dV = \int_V f_i^B \delta u_i dV + \int_{S_f} f_i^S \delta u_i^S \, dS_f \qquad (22)$$

$$\int_V e_{ikl}\epsilon_{kl} \, \delta E_i \, dV + \int_V \varepsilon_{ij} E_j \, \delta E_i \, dV = - \int_{S_\sigma} \sigma^S \delta\phi^S \, dS_\sigma \qquad (23)$$

The same equations can be written also in matrix form:

$$\int_V \bar{\underline{\epsilon}}^T \, \underline{C} \, \underline{\epsilon} \, dV - \int_V \bar{\underline{\epsilon}}^T \, \underline{e} \, \underline{E} \, dV = \int_V \bar{\underline{u}}^T \, \underline{f}^B \, dV + \int_{S_f} \bar{\underline{u}}^{S\,T} \, \underline{f}^S \, dS_f \qquad (24)$$

$$\int_V \bar{\underline{E}}^T \, \underline{e} \, \underline{\epsilon} \, dV + \int_V \bar{\underline{E}}^T \, \underline{\varepsilon} \, \underline{E} \, dV = - \int_{S_\sigma} \sigma^S \bar\phi^S \, dS_\sigma \qquad (25)$$

3.3. FINITE ELEMENT DISCRETIZATION OF THE LINEAR PIEZOELECTRIC EQUATIONS

We now formulate the generic finite element equations for the variables $\underline{u}$ and ϕ. For every finite element m of the considered body we assume

$$\underline{u}^{(m)} = \underline{H}_u{}^{(m)}\hat{\underline{u}} \qquad \phi^{(m)} = \underline{H}_\phi{}^{(m)}\hat\phi \qquad (26)$$

where $\hat{\underline{u}}$ is the vector of nodal displacement and $\hat\phi$ is the vector of electric potential of the discretized body. By substituting equations (26) into equations (2) and (4) we obtain

$$\underline{\varepsilon}^{(m)} = \underline{B_u}^{(m)} \, \underline{\hat{u}} \quad \underline{E}^{(m)} = \underline{B_\phi}^{(m)} \hat{\phi} \tag{27}$$

We can then obtain, following the classical procedure of summing over all the elements:

$$\Sigma_m \int_{V^{(m)}} \underline{B_u}^{(m)T} \underline{C} \, \underline{B_u}^{(m)} \, dV^{(m)} \, \underline{\hat{u}} - \Sigma_m \int_{V^{(m)}} \underline{B_u}^{(m)T} \underline{e} \, \underline{B_\phi}^{(m)} \, dV^{(m)} \, \hat{\phi} =$$

$$\Sigma_m \int_{V^{(m)}} \underline{H_u}^{(m)T} \underline{f}^{B(m)} \, dV^{(m)} + \Sigma_m \int_{S_f^{(m)}} \underline{H_u}^{S(m)T} \underline{f}^{S(m)} \, dS_f^{(m)} \tag{28}$$

$$-\Sigma_m \int_{V^{(m)}} \underline{B_\phi}^{(m)T} \underline{e} \, \underline{B_u}^{(m)} \, dV \, \underline{\hat{u}} - \Sigma_m \int_{V^{(m)}} \underline{B_\phi}^{(m)T} \underline{\varepsilon} \, \underline{B_\phi}^{(m)} \, dV \, \hat{\phi} =$$

$$\Sigma_m \int_{S_\sigma^{(m)}} \underline{H_\phi}^{S\,(m)T} \sigma^{S(m)} \, dS_\sigma^{(m)} \tag{29}$$

These equation can be also written in a more compact form as follows:

$$\underline{k_{uu}} \, \underline{\hat{u}} + \underline{k_{u\phi}} \, \hat{\phi} = \underline{F_u}^B + \underline{F_u}^S \tag{30}$$

$$\underline{k_{\phi u}} \, \underline{\hat{u}} + \underline{k_{\phi\phi}} \, \hat{\phi} = \underline{F_\phi}^S \tag{31}$$

with:

$$\underline{k_{uu}} = \Sigma_m \int_{V^{(m)}} \underline{B_u}^{(m)T} \underline{C} \, \underline{B_u}^{(m)} \, dV^{(m)}$$

$$\underline{k_{u\phi}} = \underline{k_{\phi u}}^T = -\Sigma_m \int_{V^{(m)}} \underline{B_u}^{(m)T} \underline{e} \, \underline{B_\phi}^{(m)} \, dV^{(m)}$$

$$\underline{k_{\phi\phi}} = -\Sigma_m \int_{V^{(m)}} \underline{B_\phi}^{(m)T} \underline{\varepsilon} \, \underline{B_\phi}^{(m)} \, dV^{(m)}$$

$$\underline{F_u}^B = \Sigma_m \int_{V^{(m)}} \underline{H_u}^{(m)T} \underline{f}^{B(m)} \, dV^{(m)}$$

$$\underline{F_u}^S = \Sigma_m \int_{S_f^{(m)}} \underline{H_u}^{S(m)T} \underline{f}^{S(m)} \, dS_f^{(m)}$$

$$\underline{F_\phi}^S = \Sigma_m \int_{S_\sigma^{(m)}} \underline{H_\phi}^{S\,(m)T} \underline{\sigma}^{S(m)} \, dS_\sigma^{(m)}$$

where $\underline{k_{uu}}$ is the mechanical stiffness matrix, $\underline{k_{\phi\phi}}$ is the electrical permittivity matrix, $\underline{k_{u\phi}}$ is the piezoelectric matrix, $\underline{F_u^B}$ is the body force loading vector, $\underline{F_u^S}$ is the surface force loading vector, F_ϕ^S is the surface density charge vector.

3.4. AN ITERATIVE APPROACH FOR SOLUTION

Equations (30) and (31) are, respectively, the finite element expressions of the converse and the direct piezoelectric effects. If we assume that in equation (30) the potential is given throughout the body, we can move the expression $k_{u\phi}\hat{\phi}$ to the right hand side and consider it as an additional load vector. In this way the solution of equation (30) gives:

$$\underline{\hat{u}} = \underline{k_{uu}}^{-1}(\underline{F_u}^B + \underline{F_u}^S - \underline{k_{u\phi}}\hat{\phi})$$

Here the influence of the displacement field on the electric field has been neglected, but the problem has now a much simpler formulation.

The same process can be applied to equation (31) as far as the dispacement contribution is concerned. If the displacement field is known, an additional loading vector can be obtained moving $k_{\phi u}\hat{u}$ to the r.h.s. of the equation. The solution of the problem can be written as:

$$\hat{\phi} = \underline{k_{\phi\phi}}^{-1}(\underline{F_\phi}^S - \underline{k_{\phi u}}\hat{u})$$

Here the effect of the electric field on the displacement field has been neglected.

In some cases it is possible to deal with the direct and converse effects in a separate way but in general the coupling between equations (30) and (31) has to be fully taken into account. A possible way to obtain the fully coupled solution is to solve simultaneously equations (30) and (31) for $\underline{u}$ and ϕ. Alternatively, an iterative procedure can be adopted. Its steps can be briefly outlined as follows:

1. Solve equation (31) assuming that $\hat{\underline{u}} = 0$, thus obtain $\hat{\phi}$, $\underline{E}$, and $\underline{D}$;
2. Substitute the obtained value for $\hat{\phi}$ into equation (30) and solve for $\hat{\underline{u}}$, $\underline{\epsilon}$, $\underline{\tau}$;
3. Substitute the value obtained for $\hat{\underline{u}}$ into equation (31); again solve for $\hat{\phi}$, $\underline{E}$, and $\underline{D}$.
4. Compare the values of $\hat{\phi}$ obtained in step 3 with those obtained in step 1, by evaluating whether the following tolerance condition is satisfied:

$$(\|\underline{\hat{\phi}}^{(3)} - \underline{\hat{\phi}}^{(1)}\|)/\|\underline{\hat{\phi}}^{(1)}\| \le \beta_1$$

where β_1 is a tolerance parameter;
5. Substitute again the value obtained for $\hat{\phi}$, into equation (30) and solve for $\hat{\underline{u}}$, $\underline{\epsilon}$, $\underline{\tau}$;
6. Compare the values of $\underline{u}$ obtained in step 5 with those obtained in step 2, by evaluating whether the following tolerance condition is satisfied:

$$(\|\underline{\hat{u}}^{(5)} - \underline{\hat{u}}^{(2)}\|)/\|\underline{\hat{u}}^{(2)}\| \le \beta_2$$

where β_2 is a tolerance parameter. If the above two convergence conditions are not fulfilled continue with the analysis, repeating steps 3 to 6, until the two tolerance conditions are fulfilled.

A key advantage of this iterative approach is that already existing finite element programs that solve classical solid mechanics problems and field problems, like heat transfer, can directly be used in a reliable way providing that the constitutive law is modified.

This way both geometric and material nonlinearities can directly be included, and also a reduction in solution time may be accomplished. Of course the solution procedure assumes that convergence is reached in a reasonable number of iterations.

3.5. THE CASE OF NONLINEAR CONSTITUTIVE RELATIONS

The relations (20) and (21) express respectively the mechanical and electrical equilibrium at all times of interest. In order to establish a general solution scheme for nonlinear problems the development of incremental equilibrium equations is necessary.

We aim to establish a procedure that is both iterative, in the spirit of the discussion in Sec.3.4 , and also incremental, dealing at every iteration with the mechanical or electrical equilibrium.

Let us consider again equation (20). We assume that the conditions at time t have been calculated and that the displacements are to be determined for time $t + \Delta t$, where Δt is the time increment (note that in the steady–state case time is a dummy variable). As far as electrical variables are concerned, we use the last calculated values, that is to say, the mechanical displacements are the only primary unknown variables. Mechanical equilibrium is considered at time $t + \Delta t$ in order to solve for the displacements at time $t + \Delta t$:

$$\int_{t_V} c_{ijkl}\epsilon_{kl}\, \delta\epsilon_{ij}dV = {}^{t+\Delta t}R - \int_{t_V} {}^t\tau_{ij}\, \delta\epsilon_{ij}dV \tag{32}$$

where

$$^{t+\Delta t}R = \int_{t_V} {}^{t+\Delta t}f_i^B \delta u_i dV + \int_{t_{S_f}} {}^{t+\Delta t}f_i^S \delta u_i^S\, dS_f \tag{33}$$

In equation (32) ${}^t\tau_{ij}$ includes the last calculated value for the electric field, see eq. (5). Eq.(32) represents a linearization of the mechanical response and the first step of a Newton–Raphson iteration [18].

Whereas the proposed expression for the mechanical equilibrium equation allows to calculate the displacements for every time step, assuming that the value of the electric field is known, a similar expression can be written for the electrical incremental equilibrium, assuming that the stresses are known.

To that purpose let us consider now equation (21). We assume that the conditions at time t have been calculated and that the electrical displacements are to be determined for time $t + \Delta t$, where Δt is the time increment. As far as the mechanical variables are concerned, we use the last calculated values, so that the electric potential is the only unknown variable. Electrical equilibrium is considered at time $t + \Delta t$ in order to solve for the electric potential at time $t + \Delta t$:

$$\int_{t_V} \varepsilon_{ij} E_j \, \delta E_i \, dV = - \int_{t_{S_\sigma}} {}^{t+\Delta t}\sigma^S \delta\phi^S \, dS_\sigma - \int_{t_V} {}^t D_i \, \delta E_i \, dV \qquad (34)$$

In eq.(34) ${}^t D_i$ includes the last calculated mechanical effect, see eq.(6).

In order to obtain a fully coupled solution of eq.(32) and eq.(34) we proceed as follows: we first perform Newton-Raphson iterations on eq.(32) using the last calculated value for the electric field in the calculation of the stresses ${}^{t+\Delta t}\tau_{ij}{}^{(k)}$ until at the k–th iteration

$$ {}^{t+\Delta t}R - \int_{t_V} {}^{t+\Delta t}\tau_{ij}{}^{(k)} \, \delta\epsilon_{ij} dV = 0 \qquad (35)$$

We then move to eq.(34) and we use the last calculated value the mechanical effect in the calculation of the electrical displacements and we perform Newton–Raphson iterations on eq.(34). At this stage an updated value for ${}^{t+\Delta t}E_k$ is available for solving again eq.(32). This procedure is continued until convergence is reached for both the mechanical equilibrium and the electrical equilibrium at time ${}^{t+\Delta t}$.

4. The governing equations of a 3D dielectric body with piezoelectric inclusions

Let us consider a dielectric body that occupies a region V bounded by a surface S with an outward unit normal vector with components n_i of a three-dimensional space. Let us assume that the body consists of a dielectric matrix of volume $V^{(m)}$ bounded by a surface $S^{(m)}$ and several inclusions each occupying the volume $V^{(i)}$ bounded by a surface $S^{(i)}$.

The three-dimensional equations that govern the problem of the linear electrical and mechanical response of the continuum body with N inclusions can be summarized as follows: In the volume occupied by the dielectric matrix $V^{(m)}$:

Mechanical equilibrium equations

$$\tau_{ij,j} + f_i^B = 0 \qquad (36)$$

Strain-displacement relations

$$\epsilon_{ij} = \frac{1}{2}(u_{i,j} + u_{j,i}) \qquad (37)$$

Maxwell's equations for the quasi-static electric field

$$D_{i,i} = 0 \qquad (38)$$

$$E_i = -\phi_{,i} \qquad (39)$$

Mechanical constitutive equations

$$\tau_{ij} = c_{ijkl}\epsilon_{kl} \qquad (40)$$

Electrical constitutive equations

$$D_i = \varepsilon_{ij} E_j \tag{41}$$

In the volume $V^{(i)}$ occupied by the i–th inclusion:

Mechanical equilibrium equations

$$\tau_{ij,j}{}^{(i)} + f_i^B = 0 \tag{42}$$

Strain-displacement relations

$$\epsilon_{ij}{}^{(i)} = \frac{1}{2}(u_{i,j}{}^{(i)} + u_{j,i}{}^{(i)}) \tag{43}$$

Maxwell's equations for the quasi-static electric field

$$D_{i,i}{}^{(i)} = 0 \tag{44}$$

$$E_i{}^{(i)} = -\phi_{,i}{}^{(i)} \tag{45}$$

Constitutive equations

$$\tau_{ij}{}^{(i)} = c_{ijkl}{}^{(i)} \epsilon_{kl}{}^{(i)} - e_{kij}{}^{(i)} E_k{}^{(i)} \tag{46}$$

$$D_i{}^{(i)} = e_{ikl}{}^{(i)} \epsilon_{kl}{}^{(i)} + \varepsilon_{ij}{}^{(i)} E_j{}^{(i)} \tag{47}$$

where u_i is the displacement vector, ϵ_{ij} the strain tensor, τ_{ij} the stress tensor, f_i^B the body force vector, ϕ the electric scalar potential, D_i the electric displacement vector, E_i the electric field vector, c_{ijkl} the elastic constitutive tensor, e_{kij} the piezoelectric (strain) tensor, ε_{ij} the dielectric permittivity tensor. All the indices range over 1,2,3. The superscript $^{(i)}$ denotes the variables relevant to the i–th inclusion.

The boundary conditions are:

Natural mechanical boundary condition on the boundary of the composite continuum where tractions are prescribed S_f

$$\tau_{ij} n_j = f_i^S \tag{48}$$

Natural electrical boundary condition on the boundary of the composite continuum where surface charges are prescribed S_σ

$$n_i D_i = \sigma^S \tag{49}$$

Essential mechanical boundary condition on the boundary of the composite continuum where displacements are prescribed S_u

$$u_i = u_i^* \tag{50}$$

Essential electrical boundary condition on the boundary of the composite continuum where electric potentials are prescribed S_ϕ

$$\phi = \phi^* \tag{51}$$

where f_i^S is the surface force and σ^S is the surface charge.

For the boundary surface S the following relations hold: $S_u \cup S_f = S$ and $S_u \cap S_f = 0$, $S_\phi \cup S_\sigma = S$ and $S_\phi \cap S_\sigma = 0$.

Mechanical boundary condition on the boundaries between the matrix and the i-th piezoelectric inclusion

$$\tau_{ij} n_j = \tau_{ij}^{(i)} n_j^{(i)} \tag{52}$$

$$u_i = u_i^{(i)} \tag{53}$$

Electrical boundary condition on the boundaries between the matrix and the i-th piezoelectric inclusion

$$n_i D_i = n_i^{(i)} D_i^{(i)} \tag{54}$$

$$\phi = \phi^{(i)} \tag{55}$$

The following symmetries hold for the tensors that appear in the constitutive relations (see [6]):

$$c_{ijkl} = c_{jikl} = c_{klij} \tag{56}$$

$$e_{kij} = e_{kji} \tag{57}$$

$$\varepsilon_{ij} = \varepsilon_{ji} \tag{58}$$

From equation (38) it is clear that no body density charge σ^B is assumed to be present in the piezoelectric material.

In presence of N inclusions the complete set of governing equations consists of 22x(N+1) equations (36)–(47) in 22x(N+1) unknowns (u_i, ϵ_{ij}, τ_{ij}, ϕ, E_i, D_i for the dielectric matrix and the corresponding ones for each of the inclusions), with the relevant boundary conditions (48)–(55).

From Equations (36) and (48) that describe the three-dimensional mechanical equilibrium of the dielectric matrix in the field $V^{(m)}$ and on the boundary $S^{(m)}$ for each time t, and from equations (42) and (52) that describe the three-dimensional mechanical equilibrium of the i-th inclusion in the field $V^{(i)}$ and on the boundary $S^{(i)}$, the *principle of virtual work* can be derived.

For every time t and for every possible choice of virtual displacements δu_i and $\delta u_i^{(i)}$, relevant respectively to the dielectric matrix and to all the piezoelectric inclusions $i = 1, 2, \ldots, N$, that satisfy the essential mechanical boundary conditions (50) and (53), the following relation holds:

$$\int_{tV^{(m)}} {}^t\tau_{ij}\, \delta_t\epsilon_{ij} d^tV + \sum_{i=1}^{N} \int_{tV^{(i)}} {}^t\tau_{ij}{}^{(i)}\, \delta_t\epsilon_{ij}{}^{(i)} d^tV =$$

$$\int_{tV^{(m)}} {}^tf_i^B\, \delta u_i d^tV + \sum_{i=1}^{N} \int_{tV^{(i)}} {}^tf_i^B\, \delta u_i{}^{(i)} d^tV + \int_{tS_f} {}^tf_i^S \delta u_i^S\, d^tS_f \qquad (59)$$

where ${}^t\tau_{ij}$ is the stress at time t, $\delta_t\epsilon_{ij} = \frac{1}{2}(\frac{\partial \delta u_i}{\partial^t x_j} + \frac{\partial \delta u_j}{\partial^t x_i})$ is the virtual strain corresponding to δu_i, ${}^tf_i^B$ and ${}^tf_i^S$ are the body forces and the surface tractions at time t, ${}^tV$ and ${}^tS_f$ are the volume occupied at time t and the corresponding surface on which the tractions ${}^tf_i^S$ are prescribed and δu_i^S is the virtual displacement on ${}^tS_f$. The notation used allows to distinguish between the variables relevant to the dielectric matrix and to the i-th piezoelectric inclusion.

In analogy with the principle of virtual work a *principle of virtual electric potentials* can be stated as follows from equations (38), (49), (44) and (52). For every time t and for every choice of virtual electric potentials $\delta\phi$ and $\delta\phi^{(i)}$ that verify the essential electrical boundary condition (51) and (55), the following relation holds:

$$\int_{tV^{(m)}} {}^tD_i\, \delta_t E_i d^tV + \sum_{i=1}^{N} \int_{tV^{(i)}} {}^tD_i{}^{(i)}\, \delta_t E_i{}^{(i)} d^tV =$$

$$-\int_{tS_\sigma} {}^t\sigma^S \delta\phi^S\, d^tS_\sigma \qquad (60)$$

where ${}^tD_i$ is the electric displacement vector at time t, $\delta_t E_i = -\frac{\partial \delta\phi}{\partial^t x_i}$ is the virtual electric field corresponding to $\delta\phi$, ${}^t\sigma^S$ is the surface charge density, and $\delta\phi^S$ is the virtual electric potential on the boundary surface ${}^tS_\sigma$.

The governing finite element equations can easily be obtained in this case in an analogous way to the one described in Sec.3.

5. Numerical results

In this section the results of two simple electroelastic analyses for the linear case and for the case of nonlinear constitutive equations are given in order to demonstrate the capability of performing a coupled electro-mechanical analysis of 2D continua using the procedure described in Sec. 3.

5.1. ANALYSIS OF A TWO-DIMENSIONAL BEAM UNDER ELECTRICAL AND MECHANICAL LOADING

Consider a rectangular strip of piezoelectric material occupying the region $\mid x \mid \le l$, $\mid z \mid \le h$ of a two-dimensional space, as shown in figure 1.

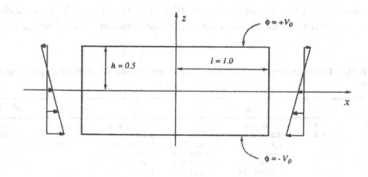

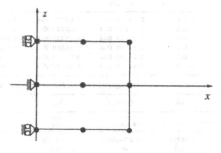

Figure 1.

TABLE I.   Data used in Solution of Problem in Figure 1

s_{11} mm^2/N	s_{13} mm^2/N	σ_0 N/mm^2	σ_1 N/mm^3	V_0 V	h mm	d_{31} mm/V	d_{33} mm/V
0.1944444 E-4	-0.083333 E-4	-5.0	2.0 E+1	1.0 E+3	5.0 E-1	-1.8 E-7	3.6 E-7

The material has been polarized along the thickness, that is along the z direction, and is assumed to be transversely isotropic.

The governing equations of the two–dimensional plane–stress problem are given in Appendix A.

The following boundary conditions apply for the problem:

At $z = \pm h$

$$\phi = \pm V_0 \quad \sigma_z = 0 \quad \tau_{xz} = 0$$

At $x = \pm l$

$$D_x = 0 \quad \sigma_x = \sigma_0 + \sigma_1 z \quad \tau_{xz} = 0$$

The analytical solution of the problem is given in Appendix A.

Only one half of the structure is considered in the finite element modeling due to symmetry with respect to the z axis and the data in Table I are used.

TABLE II. Convergence of displacements, electric potentials, electric displacements and stresses.

x	z	w_{exact}	$w^{(2)}$	$w^{(5)}$	$w^{(2)}/w_{ex}$	$w^{(5)}/w_{ex}$
1.0	0.0	-1.72916 E-4	-1.94444 E-4	-1.72916 E-4	1.1244	1.0000
1.0	0.5	-5.22152 E-4	-5.54444 E-4	-5.22152 E-4	1.0618	1.0000
1.0	-0.5	1.56181 E-4	1.23889 E-4	1.56181 E-4	0.7932	1.0000
0.5	0.0	-0.43229 E-4	-0.48611 E-4	-0.43229 E-4	1.1244	1.0000
0.5	0.5	-3.92464 E-4	-4.08611 E-4	-3.92465 E-4	1.0411	1.0000
0.5	-0.5	2.85868 E-4	2.69722 E-4	2.85868 E-4	0.9435	1.0000
		u_{exact}	$u^{(2)}$	$u^{(5)}$	$u^{(2)}/u_{ex}$	$u^{(5)}/u_{ex}$
1.0	0.0	2.62778 E-4	2.62778 E-4	2.62778 E-4	1.0000	1.0000
1.0	0.5	4.35694 E-4	4.57222 E-4	4.35694 E-4	1.0494	1.0000
1.0	-0.5	0.89861 E-4	0.68333 E-4	0.89861 E-4	0.7604	1.0000
0.5	0.0	1.31389 E-4	1.31389 E-4	1.31389 E-4	1.0000	1.0000
0.5	0.5	2.17847 E-4	2.28611 E-4	2.17847 E-4	1.0494	1.0000
0.5	-0.5	0.44930 E-4	0.34166 E-4	0.44930 E-4	0.7604	1.0000
		ϕ_{exact}	$\phi^{(1)}$	$\phi^{(3)}$	$\phi^{(1)}/\phi_{ex}$	$\phi^{(3)}/\phi_{ex}$
1.0	0.0	29.9003	0.0	29.9004 E-4	0.0	1.0000
x	z	$\sigma_{xx\ exact}$	$\sigma_{xx}^{(2)}$	$\sigma_{xx}^{(5)}$	$\sigma_{xx}^{(2)}/\sigma_{xx\ ex}$	$\sigma_{xx}^{(5)}/\sigma_{xx\ ex}$
0.887298	-0.387298	-12.745966	-12.7460	-12.7460	1.0000	1.0000
0.887298	0.0	-5.0	-5.0	-5.0	1.0000	1.0000
0.887298	+0.387298	2.745966	2.74597	2.74597	1.0000	1.0000
		$D_{z\ exact}$	$D_z^{(1)}$	$D_z^{(3)}$	$D_z^{(1)}/D_{z\ ex}$	$D_z^{(3)}/D_{z\ ex}$
0.887298	-0.387298	-0.292 E-4	-0.301 E-4	-0.292 E-4	1.0308	1.0000
0.887298	0.0	-0.292 E-4	-0.301 E-4	-0.292 E-4	1.0308	1.0000
0.887298	+0.387298	-0.292 E-4	-0.301 E-4	-0.292 E-4	1.0308	1.0000

Both electrical and mechanical simulations use a single 9 node element. The iterative procedure converges to the exact solution in five steps, since the values computed for stresses at step 5 are equal to the values calculated at step 2. Some calculated values for displacements, stresses, electrical potential, and electrical displacements at the different steps are reported in Table II. In the same table the exact solution is also given.

5.2. ANALYSIS OF THE PIEZOELECTRIC ACTUATION OF AN ALUMINUM STRUCTURE

We now consider the experiments performed on an aluminum cantilever by Anderson and Crawley [19]. Two 2024 aluminum beam specimen are examined. They have thicknesses of 3.21 mm and 1.59 mm, each has length of 356 mm and width of 51 mm. Two G–1195 piezoceramic actuators with dimensions $63.5 \times 25.4 \times 0.254$ mm were bonded to each of the upper and lower surfaces of the beam at a distance

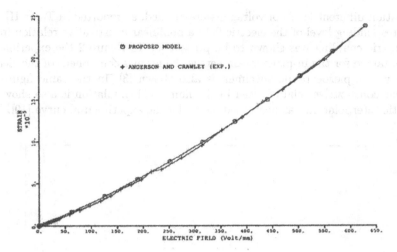

Figure 2.

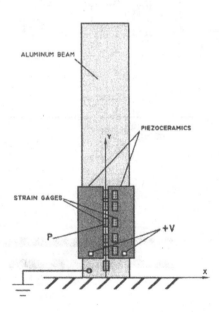

Figure 3.

of 25.4 mm from the clamped end of the cantiliver. A small strip was left free in order to enable strain measurements.

The beam specimen were statically deformed by applying an electric field in a direction normal to the middle plane of the actuators. In order to produce bending the actuators on the top of the beam were subjected to a field equal in value but opposite in sign with respect to the actuators on the bottom. During the static deformation different levels of voltages were applied, as reported in Table III. For the corresponding level of the electric field a nonlinear constitutive relation in the piezoelectric coupling was shown to be present [3]. In Figure 2 the experimental curve obtained for the in-plane strain versus the normal component of the electric field of a free piezoceramic specimen is also shown [3]. In the same figure the nonlinear constitutive relation used in the numerical simulation is also shown. A quadratic interpolation has been used to model the experimental curve [19].

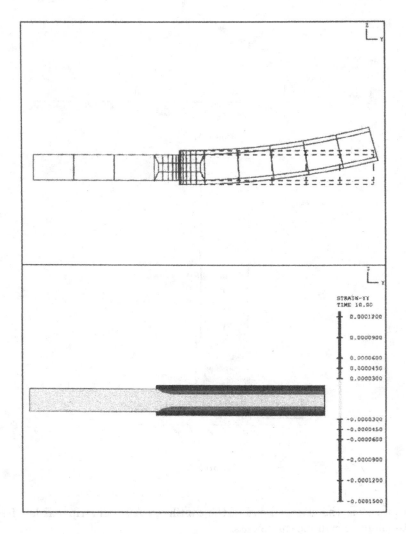

Figure 4.

A two-dimensional finite element analysis was performed in the bending plane using 2D elements both to represent the aluminum structure and the actuators.

A portion of the mesh close to the end of the actuators is shown in Figure 4 a. Perfect bond was assumed between the actuators and the structure.

Table III shows a comparison between the numerical prediction obtained with 10 time steps and the experimental results for the bending strain measured at point P (see Figure 3 on the top of the structure). Different loading cases are reported for the two considered thicknesses. In Figure 4 b the distribution of in-plane strain along the thickness is reported for the case of beam thickness 1.59 mm and applied potential $\phi = 100$ V.

TABLE III.   Comparison between experimental and numerical results (ϵ_{11} is evaluated at P; see Figure 3)

thickness mm	ϕ V	E_3 V/mm	microstrain experimental	microstrain numerical
3.21	50	197.9	14.7	15.54
	100	393.7	32.6	32.48
	150	590.6	52.4	50.90
1.59	50	197.9	26.8	24.26
	100	393.7	60.1	52.03

5.3. ACTIVE COMPOSITE WITH A PIEZOELECTRIC INCLUSION

In this section the simple case of one inclusion only has been selected for performing several loading cases of mechanical and electrical nature in presence of different boundary conditions.

The geometry of the inclusion, illustrated in figure 5 is representative of a 3D volume of a composite medium infinitely long in one direction with a cylindrical inclusion, also infinitely long in the same direction. Such geometry can be viewed as the elemental representative volume of a piezoelectric fiber composite material body.

The numerical study was performed for a quarter only of the volume by assuming the relevant symmetry conditions. The sides A,B,C,D,E of the volume were considered to remain planar in all the studied cases. The side F was considered in both cases of planar constrain and in the free condition.

The first group of cases deals with electrical loading: the electrical potential is set equal to zero for $z = 0$ and to $\phi = 100$ V for $z = t_z$. In figure 6 the response in terms of electric potential is shown for four different cases. The ratio between the permittivity of the fiber and the permittivity of the matrix $P = \epsilon_f/\epsilon_m$ is set equal to $60, 30, 6, 0.1$ for the different cases. As it is clear from the picture, the bigger drop of the potential occurs in the part of the composite with the lower permittivity. Since this constant is in general higher in the fiber than in the matrix,

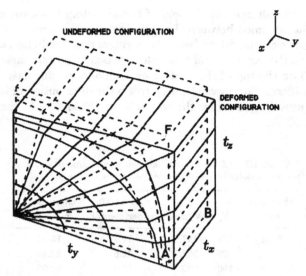

Figure 5. Mesh of the representative volume of the composite in the undeformed and deformed configuration.

as a result the electric field will be in general higher in the matrix than in the fiber. This could be a problem for the activation of the composite since in order to get high piezoelectrically induced strains the electric field in the piezoelectric inclusion should be as high as possible.

Figures 7, 8 and 9 all refer to the same electrical loading (applied potentials on the faces at $z = 0$ and $z = t_z$. Figure 7 is relevant to the case with all the planes assumed to stay plane and shows the response in terms of electric potentials and in terms of electric displacements and the deformed configuration (solid line). In figure 8 the normal stresses in x and in z directions and the displacements in y and z directions are presented.

In figure 9 three cases of mechanical loadings are illustrated. In figure 9 a a uniform displacement in the x direction is assumed at the edge at $x = t_x$. The normal stress component in the same direction is aslo shown.

Figures 9 b and 9 c show the cases with uniform displacements in y and z direction at the sides, respectively, at $y = t_y$ and $z = t_z$. The distribution of the corrisponding normal stress components is shown as in figure 9 a.

As it is clear from the band-plots, the values of the various variables are not subjected to any smoothing at the boundaries of the elements before being plotted.

The results of these analyses could be used for the characterization of some of the engineering constant of the composite materials, considered in this case at a macroscopic case as an homogeneous medium with appropriate elastic constants.

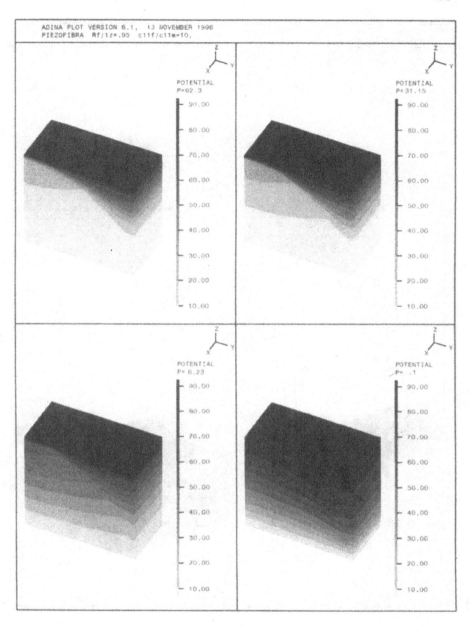

Figure 6.

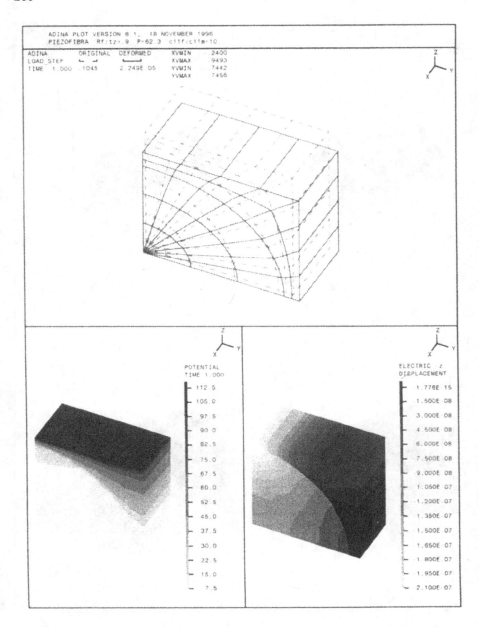

Figure 7.

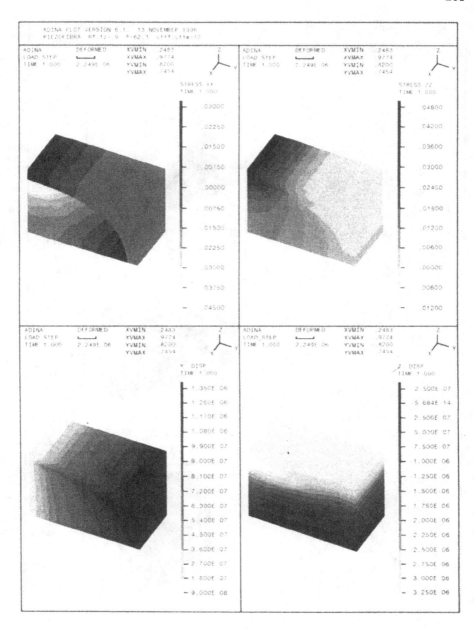

Figure 8.

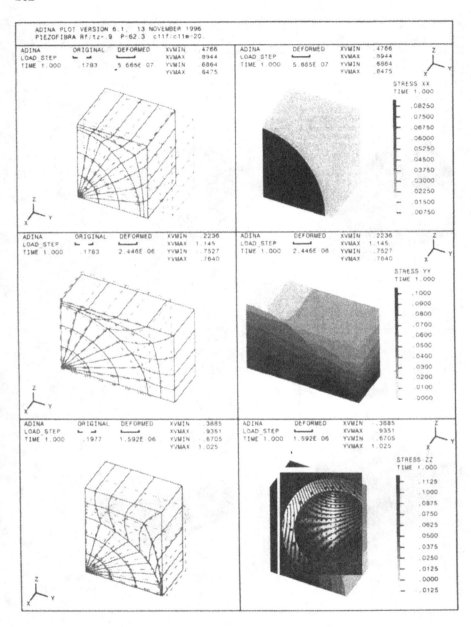

Figure 9.

Acknowledgments

Fig.2 and fig.4 and parts of sec. 2 and 3 reprinted by permission of Sage Publication Ltd from P. Gaudenzi, K.J. Bathe, An iterative finite element procedure for the analysis of piezoelectric continua, *Journal of Intelligent Material Systems and Structures*, **6**, pp. 266–273, ©Sage Publication Ltd 1995.
Fig.5,6,7,8,9 and parts of sec. 4 and 5, reprinted from *Computers & Structures*, **65(2)**, P. Gaudenzi, On The Electromechanical Response Of Active Composite Materials With Piezoelectric Inclusions, pp. 157–168, copyright 1997, with permission from Elsevier Science.

References

1. Wada B.K., Fanson J.L., Crawley E.F., (1989) Adaptive Structures, in *Adaptive structures*, editor B.K.Wada, pp. 1–8, ASME, New York.
2. Crawley E.F., de Luis J., (1987) Use of Piezoelectric Actuators as Elements of Intelligent Structures, *AIAA Journal*, **25(10)**, pp. 1373–1385.
3. Crawley E.F., Lazarus K.B., (1989) Induced Strain Actuation of Isotropic and Anisotropic Plates, *AIAA Journal*, **29(6)**, pp. 944–951.
4. Tiersten H.F., (1969) *Linear Piezoelectric Plate Vibration*, Plenum Press.
5. Mindlin R.D., (1972) Elasticity, piezoelectricity and cristal lattice dynamics, *Journal of Elasticity*, **2**, pp. 217–282.
6. Institute of Electrical and Electronics Engineers Inc. (The), (1987) *IEEE Standard on Piezoelectricity*, New York, 1987.
7. Mindlin R.D., (1972) High frequency vibrations of piezoelectric crystal plates, *International Journal of Solids and Structures*, **8**, pp. 895–906.
8. Maugin G.A., (1988) *Continuum mechanics of electromagnetic solids*, North-Holland.
9. Tzou H.S., Tseng C.I., (1991) Distributed vibration control and identification of coupled elastic/piezoelectric systems: finite element formulation and applications, *Mechanical Systems and Signal Processing*, **5(3)**, pp. 215–231.
10. Lerch R., (1990) Simulation of Piezoelectric Devices by Two– and Three–Dimensional Finite Elements, *IEEE Transaction on Ultrasonics, Ferroelectrics and Frequency Control*, **37(2)**, pp. 233–247.
11. Allik H., Hughes T.J.R., (1970) Finite element method for piezoelectric vibration, *International Journal for Numerical Methods in Engineering*, **2**, pp. 151–157
12. Im S., Atluri S.N., (1989) Effects of a Piezo-Actuator on a Finitely Deformed Beam Subjected to General Loading, *AIAA Journal*, **27(12)**, pp. 1801–1807.
13. Wang B.T., Rogers C.A., (1991) Modeling of Finite-Length Spatially-Distributed Induced Strain Actuators for laminate Beams and Plates, *Journal of Intelligent Material Systems and Structures*, **2**, pp. 38–58.
14. Hagood N.W., von Flotov A., (1991) Damping of Structural Vibrations with Piezoelectric Materials and Passive Electrical Network, *Journal of Sound and Vibration*, **146(2)**, pp.243–268.
15. Ha S.K., Keilers C., Chang F.K., (1992) Finite Element Analysis of Composite Structures Containing Distributed Piezoceramics Sensors and Actuators, *AIAA Journal*, **30(3)**, pp. 772–780.
16. Gaudenzi P., Bathe K.J., (1995) An iterative Finite Element Procedure for the Analysis of Piezoelectric Continua, *Journal of Intelligent Material Systems and Structures*, **6**, pp. 266–273.
17. Gaudenzi P., (1997) On The Electromechanical Response Of Active Composite Materials With Piezoelectric Inclusions, *Computers & Structures*, **65(2)**, pp. 157–168.
18. Bathe K.J., (1996) Finite Element Procedures, Englewood Cliffs, Prentice-Hall
19. Anderson E.H., Crawley E.F., (1989) Piezoceramic actuation of one– and two–dimensional structures, Massachusetts Institute of Technology, Space Systems Laboratory, SSL 5-89, Cambridge, MA 02139.

20. Parton V.Z., Kudryavtsev B.A., Senik N.A., (1989) Electroelasticity, *Applied Mechanics: Soviet reviews. Volume 2: Electromagnetoelasticity*, editors G.K. Mikhailov and V.Z. Parton, pp. 1–58, Hemisphere Publishing Corporation.

A. A 2D closed form solution

Consider a rectangular strip of piezoceramics occupying the region $\mid x \mid \leq l, \mid z \mid \leq h$ of a two-dimensional space, as shown in fig.1. The piezoelectric material has been polarized along the thickness, that is along z direction.

The governing equations can be written as follows ([20]):

$$D_{x,x} + D_{z,z} = 0 \tag{61}$$

$$\sigma_{x,x} + \tau_{xz,z} = 0 \quad \tau_{xz,x} + \sigma_{z,z} = 0 \tag{62}$$

$$E_x = -\frac{\partial \phi}{\partial x} \quad E_z = -\frac{\partial \phi}{\partial z} \tag{63}$$

$$\epsilon_x = \frac{\partial u}{\partial x} \quad \epsilon_z = \frac{\partial w}{\partial z} \quad \gamma_{xz} = \frac{\partial u}{\partial z} + \frac{\partial w}{\partial x} \tag{64}$$

$$D_x = \varepsilon_{11} E_x + d_{15} \tau_{xz} \quad D_z = \varepsilon_{33} E_z + d_{31} \sigma_x + d_{33} \sigma_z \tag{65}$$

$$\epsilon_x = s_{11}\sigma_x + s_{13}\sigma_z + d_{31}E_z \quad \epsilon_z = s_{13}\sigma_x + s_{33}\sigma_z + d_{33}E_z \quad \gamma_{xz} = s_{55}\tau_{xz} + d_{15}E_x \tag{66}$$

Here equations (65) and (66) are the constitutive relations derived for the plane stress case in the form of equations (11) and (12) of Sec.2.

Consider now the following boundary conditions:

At $z = \pm h$

$$\phi = \pm V_0 \quad \sigma_z = 0 \quad \tau_{xz} = 0 \tag{67}$$

At $x = \pm l$

$$D_x = 0 \quad \sigma_x = \sigma_0 + \sigma_1 z \quad \tau_{xz} = 0 \tag{68}$$

For this problem the exact solution was given by Boriseiko et al. [20]:

$$u = s_{11}(\sigma_0 - \frac{d_{31}V_0}{s_{11}h})x + s_{11}(1 - k_{31}^2)\sigma_1 xz \tag{69}$$

$$w = s_{13}(\sigma_0 - \frac{d_{33}V_0}{s_{13}h})z + s_{13}(1 - k_s^2)\sigma_1 \frac{z^2}{2} - s_{11}(1 - k_{31}^2)\sigma_1 \frac{x^2}{2} \tag{70}$$

$$\phi = V_0 \frac{z}{h} - \frac{d_{31}\sigma_1}{2\varepsilon_{33}}(h^2 - z^2) \quad E_z = -\frac{V_0}{h} - \frac{d_{31}\sigma_1}{\varepsilon_{33}}z \quad D_z = -\varepsilon_{33}\frac{V_0}{h} + d_{31}\sigma_0 \quad (71)$$

$$\sigma_x = \tau_{xz} = E_x = D_x = 0 \tag{72}$$

with

$$k_{31}^2 = \frac{d_{31}^2}{s_{11}\varepsilon_{33}} \quad k_s^2 = \frac{d_{33}d_{31}}{s_{13}\varepsilon_{33}} \tag{73}$$

LATEST TRENDS IN THE DEVELOPMENT OF PIEZOELECTRIC MULTI-DEGREE-OF-FREEDOM ACTUATORS/SENSORS

R. BANSEVIČIUS
Kaunas University of Technology
Donelaicio 73, LT-3006 Kaunas, Lithuania

Abstract

The concept of *Piezomechanics* (Piezoelectricity + Mechanics + Control) describing a complex interaction of dynamic effects and precision-engineered devices is presented. *Piezomechanics* is considered as part of the more broad philosophy of *Mechatronic* devices. New structure units of precise mechanisms - *active kinematic pairs* are described; in them forces or torque are generated in the contact zone between components. A lot of schematics of piezoelectric multi-degree-of-freedom actuators/sensors are presented; they include the actuators with separated power and control systems and actuators, in which transformation of high frequency resonant oscillations into continuous motion is taking place. It is shown that functional ability of the mechanisms can be increased by using hybrid composite systems, made from two or more alternating layers of piezoceramic and magnetostrictive plates or films.

1. Introduction

Recent advances in the development, theory and applications of new smart materials, structures and devices, including materials with extremely high piezoelectric or magnetostrictive properties *(e.g.,* Terfenol-*D* or *PMN* - lead magnesium niobate) have extended the area of *Mechatronics,* providing new systems with very high levels of integration and multifunctionality. The classical approach to engineering (i.e., interactions of mass and energy) is performed on several levels and, as in *Mechatronics,* the factor of information is present at all levels. The route to a better understanding of these processes indicates the need to introduce the concept of *Piezomechanics* (Piezoelectricity + Mechanics + Control Systems), describing the holistic complex interaction of dynamic effects, energy transformations and devices based on them, while the physics involved in piezoelectric theory may be regarded as the coupling between Maxwell's equations of electromagnetism and the elastic stress equations of motion [19,31].

A. Preumont (ed.), Responsive Systems for Active Vibration Control, 207–238.

Piezomechanics is considered as a subsystem of Mechatronics, representing a new facet of engineering and design, yielding some devices that are revolutionary in nature. Some concepts, e.g. actuators with an infinite number of degrees of freedom, transmission of energy to actuators through some distance, active bearings, etc., could be described as solutions looking for a problem. In some cases the introduction of piezomechanical systems creates a new synergistic effect and, as in all integrated systems, the problem of rnaximum interaction between subsystems is the key to optimum design. Although the introduction of new piezoactive materials, which have been finding application in the areas of actuators, transducer technology, energy transformers, control devices, etc., has been very intensive in recent years [1,18,25,29], the main concepts, ideas and effects are relatively unknown to design engineers.

2. The Object and Structure of Piezomechanics

In most integrated piezoelectric devices a very important feature is the availability of various feedback systems: adaptive, bilateral, cross, position, rate, torque/force, sensory, even visual. Almost every one of the known piezoelectric effects (piezoelectricity, piezomagnetism - stress dependence of magnetic properties, piezo-optical effect - the change produced in the index of refraction of a light-transmitting material by externally applied stress, piezoresistive effect, etc.) could be used in the design of precision systems and devices. As a rule both direct and reverse piezoelectric effects of piezoactive materials are used, very often simultaneously, in the same transducer. It enables the possibility of adaptive feedback systems (derivative, force, multiple sensory or position) to control stresses, acceleration, velocity and the position of moving links.

The main nonlinear effects, which are of great importance to Piezomechanics, are:

1) control of friction forces down to *zero* value;
2) transformation of high frequency mechanical oscillations into continuous or stepped motion;
3) structure control of kinematic pairs and mechanisms:
4) generation of static or quasistatic displacements or stresses in elastic bodies;
5) touch, slip and inner force sense for sensory-controlled devices;
6) nonlinear and solitary waves in piezoelectric media;
7) hysteresis damping in piezoactive materials:
8) periodic shocks (impacts).

An important quality of piezomechanical systems is the potential to control the structure of the system, including the number of degrees of freedom of the mechanical components (e.g. kinematic pairs); this makes it into very powerful design tool for Precision Engineering. Another important factor is that structure of a system could be changed in very short time, very often depending only on the natural frequencies of the dynamic system. The possibility of controlling the structure is related to the following two properties, which are found only in piezomechanical systems:

*(i) The variable topology of the driving forces - the po*sition of excitation forces or torque *[F$_i$(X$_i$)]*, *i* = 1, 2, ..., *m* in piezomechanical systems could be controlled simply by connecting *m* electrodes in the corresponding areas of transducer to signal generator. This also permits change in the types and forms of oscillations, to generate static or dynamic deformations of deformable bodies (e.g., mirrors - in adaptive optics), linear and nonlinear waves, etc.

(ii) Variable (in *space*) *vector of polarization* gives the opportunity to generate multi-component motions, 3-*D* trajectories, non-sinusoidal oscillations, etc.

Fig. 1 presents examples of devices with structure control and variable vector of polarization. In the mechanism, shown in Fig. 1*a*, the following modes of operation are possible:

• The application of constant potential to some of the electrodes causes elliptical deformations of a cylindrical transducer (Fig. 1*c*); thus the joint becomes a single degree of freedom system with the possibility of rotation around the axis *A'-A'* (lying in the *x"y"* plane, where *x"* and *y"* axes are parallel to the corresponding *x* and *y* axes). The position of the *A'-A'* axis in the plane *x"y"* is controlled by applying a constant potential to certain groups of electrodes.

• Translational movement along z and rotation around z and any axes lying in the *x"y"* plane (4 degrees of freedom) can be performed by generating resonant oscillations of the transducer, the type, form, and phase of which determines the direction and type of movement. The oscillations generate oblique impacts in the contact zone between transducer and sphere, which are transformed into continuous motion of the sphere. According to the hypothesis of *viscous friction* [1], the tangential component of the impact pulse is independent of the magnitude of its normal component and is defined by the coefficient of instantaneous friction at impact. The latter depends on the properties and condition of the surfaces in the contact zone.

According to the hypothesis *of dry friction,* the tangential component of dry friction is proportional to its normal component and the coefficient of proportionality is equal to the coefficient of dry friction. Rotation around the *z*-axis is performed when radial oscillations of the travelling wave type are being generated. The transformation of *travelling* wave oscillations *into* continuous motion is effected via the frictional interaction of the travelling wave motion of the actuator and the driven component.

The application of a constant potential to all electrodes fixes the sphere in the transducer and makes the number of degrees of freedom equal to 0. But it should be stressed here that piezoelectric actuators could be considered as having two levels of degree of mobility: the first as described above and dealing with relatively large displacements (from ∞ to ~ 0.1 μm); the second level deals with small deformations of

210

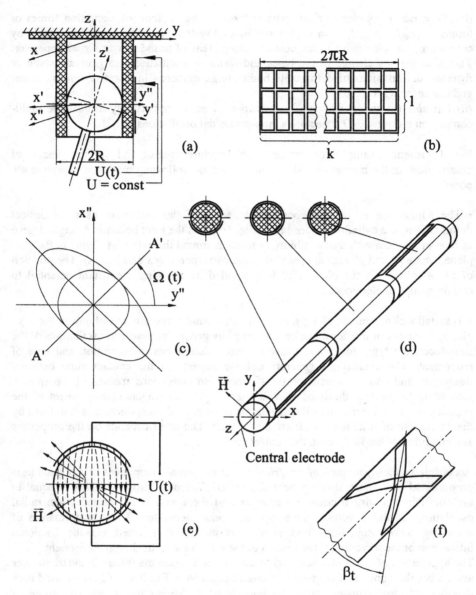

Figure 1. Structure control in *Piezomechanics*: (*a*) schematic of a piezoelectric actuator with 4 degrees of freedom; (*b*) the development of the outer electrode of the transducer; (*c*) formation of axis A'-A' with controllable position Ω (t) in $x''y''$ plane; (*d*) transducer with variable vector of polarisation H and the form of electrodes; (*e*) direction of potential in a cross-section of the transducer; (*f*) the form of electrode in the case of torsional oscillations
($\beta_t = \pi/4$).

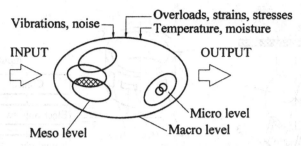

Figure 2. Structure of a piezomechanical system.

the piezoelectric transducer in the range of 0 to ~10 µm, including subnanometer behaviour [25], as in a piezoelectric micromanipulator, used for example, in a tunnel microscope. In this concept, even in the clamped position the actuator has 6 degrees of mobility for small displacements, which could be generated by applying potential to corresponding electrodes. This mode of operation could be called *two channel control* and can be performed simultaneously or in sequence, the first movement being transformation of high-frequency mechanical oscillations to continuous motion; the second - generation of small deformations in the transducer, leading to corresponding displacements or rotation of the gripper in all 3 directions.

One of the applications of *variable vector of polarization* is presented in Fig. 1*d*. A cylindrical rod made from *PZT* is initially polarized radially using a central electrode (in the form of a wire) and the electrode on the outer cylindrical surface of rod. After the poling is complete, the outer electrode is divided into three sections. The application of an electric potential to the right and left groups of electrodes results in a field that is only fully aligned with the direction of polarization near the electroded surface. It gives the possibility to generate four types of motion in the rod:

• longitudinal displacements or oscillations along the z axis;
• flexural displacements or oscillations in the xz plane;
• flexural displacements or oscillations in the yz plane;
• torsional displacements or oscillations (changing central electrode in a way shown in Fig. 1*f*);
• any combination of all of them.

Piezomechanics could be described as a complex system consisting of subsystems of different levels (Fig. 2). At the *macro* level the complete system is interacting with the environment (outer excitation), changing its structure or parameters when some external conditions (or parameters) are being altered. At the *meso* level it is the interaction between mechanical, electrical, optical, control, etc., systems; partial decomposition of the whole system into individual functions and interactions at that level could be realized. At the *micro* level linear or nonlinear processes of energy, forces, or motion transformation are taking place.

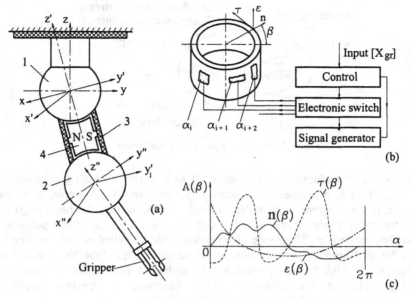

Figure 3. Piezoelectric robot.

The energy and information transformation at different levels could be illustrated by a piezoelectric robot (Fig. 3a), consisting of two spheres *1* and *2*, made from passive material (e.g. steel), with a piezoelectric transducer *3* between them. The contact forces are ensured by the help of a permanent magnet *4*. Consisting of two spherical joints (two kinematic pairs), the robot has 6 degrees of freedom. The motions (rotations) in all directions are achieved by generating high frequency resonant mechanical oscillations (of ultrasonic frequency), which at the contact zone are transformed into continuous motion [1,10,18].

At the *macro level* the input signal (the required trajectory of the gripper), the mass of the object and the position of the gripper define the vector of the moments, acting

$$[M_1(x),\ M_2(y'),\ M_3(z'),\ M_4(z'),\ M_5(x''),\ M_6(y_1')]^{-1}$$

along the axes $x,\ x'',\ y',\ z',\ y_1$ that must be generated in the contact zones between the links *1, 2* and *2, 3*. At the *meso level* the dynamic structure of the transducer is defined, determining the zones of local excitation, corresponding to the electrodes, α_i, α_{i+1}, ... on the outer surface of the transducer (Fig. 3b). At this level oscillations at the contact zones are transformed into continuous rotations about the axis, defined by the character of the distribution of three mutually orthogonal component oscillations (Fig. 3c), which are determined at the lowest - micro level - transformation of electrical energy from the signal generator to mechanical oscillations of the transducer.

The piezoelectric robot shown in Fig. 3 is just one of many applications of *Piezomechanics*, but it clearly indicates that the holistic integration of various systems, with the aim of optimum product design, is a complicated problem due to complexity of energy conversion processes in piezomechanical systems.

3. Energy and Signal Transformation

Energy conversion in piezomechanical devices plays a vital role. In new piezoactive materials, e.g., Terfenol-*D*, conversion efficiency from electrical to mechanical energy is very high, with magneto-mechanical coupling factors being greater than 0.7. That, together with rapid response time, flexible control possibilities and the facility with which they can be fabricated into complex forms, structures and arrays, make the piezomechanical devices ideal for Precision Mechanics, where incompatible demands for accuracy, high speed, dexterity, efficiency and reliability pose the greatest problems for designers.

The emergence of new "smart" materials (e.g., with a photostrictive effect - a phenomenon by which, in electrooptic ceramics, a constant electromotive force is generated with exposure to light) made it possible to widen the classical triangle of energy conversion adding such energy sources as thermal, radiant, chemical, etc.

It should be noted that piezomechanical devices usually use more than one energy conversion, thus piezoelectric actuators (ultrasonic motors, vibromotors) use 2 energy conversions:

Electrical energy $\Rightarrow$ Mechanical oscillations $\Rightarrow$ Continuous motion.

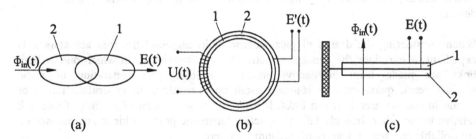

Figure 4. Transducers - sensors, using both piezoelectric (*1*) and magnetostrictive (*2*) materials: (*a*) energy conversion; (*b*) combined piezoelectric transducer with low input and high output impedance; (*c*) magnetic field sensor.

An optically controlled piezoelectric robot uses 3 conversions, including the coupling of photo voltaic and inverse piezoelectric effects. In [3] a new approach to sensor, piezoelectric, and actuator design is given - systems, consisting of coupled piezoelectric and magnetostrictive converters; in this case the generation of oscillations or static displacements is being carried out by both magnetic $\Phi_m(t)$ and electric $U_E(t)$ fields (Fig. 4). It enables the transfer of energy to the device from some distance, e.g., the control of movement of micro manipulators inside the human body with the power source (signal generator) outside it.

By using this approach it is possible to combine in one piezoelectric system a low impedance input with high impedance output and vice versa. As a non-linear system, transducers could transform the frequency, amplitude, and phase of the input signal, which could be an electrical signal or mechanical parameters, e.g., stresses, displacements, acceleration, temperature, etc. In the simplest case, conversion of magnetic to electrical energy, a double energy conversion is performed.

It should be noted here that some universities are including the elements of Piezomechanics into more general Mechatronics teaching module. In Kaunas University of Technology this subject is being presented at Master's level as a separate module from 1995. The content of the module, which is unique in its kind, is given in *Supplement 1*.

4. Active Kinematic Pairs

Recently new publications devoted to the research and development of piezoelectric actuators with one and several degrees of freedom (DOF) have been published [2,8,9], introducing new class of adaptive (or "intelligent") mechanisms, capable to control their parameters or kinematic structure in response to the changes in environment or the variation of dynamic parameters. As a rule they use both direct and inverse piezoelectric effects and are characterized by multi-functionality, combining several features in one device.

When considering the design of piezoelectric multi-degree-of-freedom actuators it is expedient to introduce the concept of active kinematic pair [8], in which one or both links are made from piezoactive material, enabling the generation of static displacements, quasi-static or resonant oscillations, resulting in generating forces or torque in contact area between links. It leads to the generation of motion of one link relative to the other. In such definition, active kinematic pair is a kinematic pair with a controllable number of degrees of freedom W, where

$$W = 6 - s(\zeta) \tag{1}$$

Here s is the number of constraint conditions, restricting the relative motion of each kinematic pair ($1 \leq s \leq 6$); ζ is a set of the control parameters, changing the constraint conditions. In particular cases, it may be that $s = s(t)$ when the structure of the mechanism is programmable; $s = s(dx/dt)$, $s = s(d^2x/dt^2)$ when the structure of the mechanism in controlled, depending on speed and the acceleration of generalized coordinates x_i; and, in a general case, also depending on constrain reaction magnitudes F_i:

$$s = s \ (dx_i/dt, d^2x_i/dt^2, x_i, F_i, ..., t) \qquad (2)$$

A number of constrain conditions s can be varied in different ways. The simplest is control of the friction, acting in the pair, usually when the elements of the pair are closed by force. Here either the friction coefficient can be varied, or the magnitude of the force executing the closure. This in achieved by the excitation of high frequency tangential or normal vibrations in the contact zone of the pair. The electrorheological and magnetorheological liquids, in which viscosity can be varied in a wide range, can also be used successfully. At $s = 6$, the number of degrees of freedom is equal to zero and the pair becomes stationary.

Active kinematic pair in characterized by:

(i) Control of number of degrees of freedom W;
(ii) Generation of forces or torque in the contact area between links.
(iii) Additional features: self-diagnostic, multi-functionality, self-repair, self-adaptation.
(iv) Two levels of degrees of freedom, where the first level is related to big displacements (transformation of resonant frequency oscillations into continuous motion), the second - small displacements (static and quasistatic deformation of active link, generated by using inverse piezoeffect and sectioned electrodes of transducer).

The most important feature of these is the possibility to control the number of degrees of freedom. Kinematic structure control must satisfy certain relationships, which may be developed using Ozol's [23] and modified Grubler's expressions:

$$W = f - 6k + \sigma; \ W = 3n - 2p_1 - p_2 \qquad (3)$$

Where f - total number of DOF of kinematic pairs, σ - number of repeated constrains; n - number of mobile links; p_1 - number of kinematic pairs with one DOF; p_2 - number of kinematic pairs with two DOF; k - number of closed loops. In case of planar mechanisms with no repeated constrains the following expressions for their synthesis can be derived from (3):

$$\Delta W = \Delta f - 3\Delta k; \ \Delta W = 3\Delta n - 2\Delta p_1 - \Delta p_2 \qquad (4)$$

The development of active materials made it possible to control and modify most of the mechanical parameters, characterizing the performance of the mechanisms. Usually three types of control mechanism are involved:

(i) The use of electrorheological and magnetorheological fluids (ERF, MRF), possessing the unique ability to change from liquid to solid in a fraction of a second

when subjected to an electric or magnetic field. The degree of solidification can be controlled by varying the strength of the applied field.

(ii) The generation of multi-component high frequency oscillations in the contact area between the links reduces the friction up to zero value (in case of squeeze film); the same control mechanism can be used to control the effective mass or moment of inertia, coefficient of restitution in systems with impacts, etc.

(iii) The control of static and quasi-static stresses in piezoactive materials. This control mechanism enables control of the natural frequencies, damping and stiffness of typical links.

Actuation, sensing and signal processing in intelligent mechanisms are the main functions, embedded in which are energy conversion and information transfer systems. Such mechanisms can modify their behavior in response to change in the dynamics of the process and the disturbances.

An important feature of mechanisms containing links made from active materials (e.g. piezoceramics) is that very often various functions (those of actuators, sensors, oscillators, generators of static and quasi-static displacements, etc.) can be performed by the same transducer, enabling the development of methods for designing adjustable or adaptive mechanisms, so that multiple tasks can be performed by the same mechanism, including the ability to accommodate manufacturing and part-positioning errors in order to widen applicability.

Fig.5 shows three levels of adaptivity of *active kinematic pair* with W degrees of freedom.

The lowest - *first level* - is characterized by the ability of the mechanism to change its parameters in real time to enable the end-effector of a given mechanism configuration to accomplish different tasks, depending on the value of the adjustable parameter. It is understood that some prior information regarding the mechanism transfer function is available. The *second level* takes into account full or partial information on external ant state-dependent disturbances ant the correction of the program is taking place. On the *third level* such functions as self-diagnosis and self-repair are introduced and multifunctionality of mechanisms with active links are being exploited.

The schematic of active kinematic pair - piezoelectric actuator with three possible levels of adaptivity is shown in Fig.6. The generation of resonant traveling wave oscillations in spherical piezoelectric transducer *2* leads to the rotation of link *1* around the corresponding axes x', y' or z'. The position of the axis of rotation is defined by the topology of electrodes, connected to n-phase ($n \geq 3$) electrical signal generator.

Here, on the first level of adaptivity, the control system defining the actuating torque to produce the desired motion of link *1* has a feedback providing it with information about its parameters. Compensation of disturbances $\xi(t)$ - second level - will include change of state of the contact between links *1* and *2* by changing the diameter of outer sphere

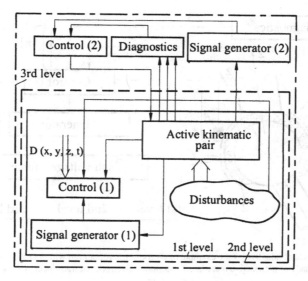

Figure 5. Three levels of adaptivity in active kinematic pairs

(applying high static voltage to certain electrodes) and corresponding change of the control algorithm.

On the third level system diagnostics is included; here an adaptation process, based on full information of mechanism state (including wear, aging of piezoactive material, etc.) and errors readjust the parameters in the nonlinear model until the position and velocity errors of link *l* along the nominal trajectory are minimized.

There is a strong correlation between *intelligent* and *adaptive* mechanisms, which can be divided into three groups:

(i) Mechanisms, in which outer excitation or control results in redistribution of strains, stresses, forces, reactions, etc. (as in self-aligning mechanisms).
(ii) Mechanisms with controllable parameters, such as mass, moment of inertia, damping, stiffness, friction, etc. Typical examples are mechanisms with redundant degrees of freedom (very often the efficiency of the system is related to the degree of redundancy), self-optimizing mechanisms and automatic balancing devices.
(iii) Mechanisms with controllable kinematic structure. This is the most advanced class of intelligent mechanisms, in which the kinematic structure of the system can be changed in a very short time, depending only on the natural frequencies of dynamic systems.

More advanced cases can also involve the change of algorithm, e.g. when changing the type of mechanism.

218

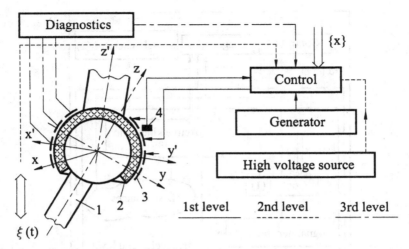

Figure 6. Three levels of adaptivity - the example of piezoelectric robot joint with 3 DOF: *1* - passive link; *2* - link- transducer, made of piezoactive material; *3* - sectioned electrodes of transducer; *4* - 3-*D* relative motion sensor.

5. Piezoelectric Multi-Degree-of-Freedom Actuators/Sensors

This class of actuators is based on *Active kinematic pairs*, in which oscillations in the contact zone of the links result in generation of the torque or forces, acting in contact zone of the links. The direction of these forces or torque can be controlled by introducing spacial phase shift of oscillations, e.g. by activating specific sectioned electrodes of the transducer. The most recent example is Robot's eye - one of the prototypes of EC Copernicus project CP941109 SPA "Smart piezoelectric multi-degree-of-freedom actuators/sensors" (Fig. 7), in which miniature CCD camera *1* is fixed in the passive sphere *2*, contacting with the piezoelectric ring *3*. There is a constant pressure in the contact zone, realized with the help of permanent magnet *4*. All system represents kinematic pair of the 3rd class having 3 DOF. The electrodes *5* of the piezoelectric ring are sectioned (in this case into 3 symmetrical parts); activating any of them with AC of resonant frequency results in the rotation of the sphere around axis, position of which can be controlled by changing the activated electrode. Generation of travelling wave oscillations of the ring (by applying 3 phase AC to all three electrodes) results in the rotation of the sphere around the axis of the ring. The resolution of such actuator in every direction is approx. 2 angular seconds, which exceeds the requirements for Robot vision systems.

The existing methods to control types and forms of resonant oscillations make it possible to design mechanisms, in which the same active link is used in two kinematic pairs - to increase redundancy of the system (Fig.8). This system was already mentioned in Fig.3; here the form of sectioned electrodes and the distribution of the three component oscillations in the contact zone are given. The first kinematic pair comprises frame *1* and piezoelectric transducer *2*, the second - transducer *2* and link *3*. Spring *4* ensures constant between all links; both contacting surfaces are spherical [7].

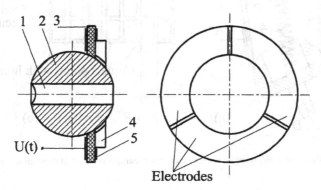

Figure 7. The schematic of piezoelectric Robot's eye.

Generation of high frequency multi-component oscillations of the contact points by activating certain electrodes ("*A*") of link *2* leads to the rotation of link *2* in relation to link *1*. Changing the position of oscillation pattern by $\pi/3$ (it leads to the change of position of vibration nodes in contact points) results in the rotation of link *3* in relation to link *2*. Total number of DOF of mechanism is $W = 3 + 3 = 6$.

Using direct piezoeffect it is possible to extract additional information (with the help of electrodes "*B*") on forces and torque, acting on link *2* and the state of contacting surfaces. This information can be used to reduce the positioning errors and correct the trajectory of motion.

Typical example of this class of new devices in given in Fig.9 as a laser scanning device, where *1* is the mirror, *2* - piezoceramic disc with sectioned electrodes, *3* - the spring. The mirror has three DOF; the direction and velocity of rotation of mirror depends on the distribution in the contact area of the three vector components of high-frequency oscillations, which are controlled by activating certain zones of the transducer [1,7].

Similar system can be effected for the precise positioning of the objects on the plane. Generation of three component oscillations in the active link of the kinematic pair, shown in Fig.10 leads to the displacement of transducer *2* in the direction, defined by the type and amplitude of oscillations. By using the phase shift of spacial oscillations distribution in the contact zone between piezoelectric ring and passive plane (3rd class kinematic pair), positioning system on the plane can be realized. In this case the electrodes of the ring are sectioned into several parts; activation of the specific electrodes leads to the translation of the ring in direction, related to the position of the

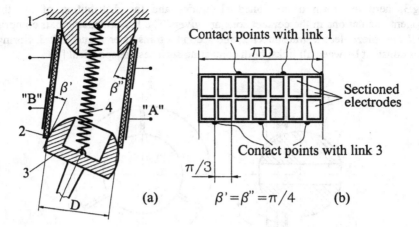

Figure 8. Manipulator with piezoelectric actuator/sensor, being a member of two active kinematic pairs.

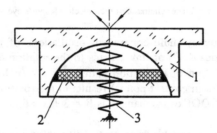

Figure 9. Piezoelectring laser scanning device

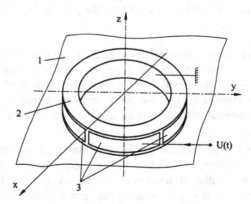

Figure 10. Positioning device: 1 - plane, 2 - piezoelectric transducer with axial poling, 3 - sectioned electrodes.

activated electrode. Activation of all electrodes (minimum number of electrodes is 3) with a phase shift results in the rotation of the ring around its axis. So the number of DOF in this case is 3.

Combination of active kinematic pairs with several degrees of freedom allows the design of microrobots and micromanipulators, characterized by very high resolution.

It is worth to note that it is possible to realize non-traditional systems with $W = 6$, in which piezoelectric transducer is contacting with liquid. A hollow piezoelectric sphere (Fig.11a) is made with two groups of hexagonal electrodes, inner and outer [3]. The outer electrode is segment into a matrix of symmetrical hexagonal electrodes, covering the entire surface. Rotations around axes x, y, z (Θ_x, Θ_y, Θ_z) are controlled by transforming the sphere to an ellipsoid (Fig.11b) and generating the travelling-wave-type oscillations. These would be enabled by connecting the corresponding electrodes to a multiphase signal generator:

$$R = R_0 \ [1+\xi\sin (n_1\varphi-\omega t)]$$

Where ξ is the relative oscillation amplitude, n_1 is the wave number and φ is the angular coordinate of the considered cross-section. The tangential ε_t and the radial ε_r displacements of points on the neutral surface of the sphere are then related by

$$\frac{d\varepsilon_r}{dt} = -\varepsilon_r \frac{d\varphi}{dt}$$

(5)

Here $d\varphi/dt$ is the angular velocity of the wave which equals (at $n_1=2$) the angular frequency of the generator. The interaction of this traveling-wave-type deformation of the sphere with the liquid, which is a mechanism similar to peristaltic motion, results in the rotation of the sphere around the corresponding axes.

Translational motions of the sphere in the direction of the axes x, y, z (X, Y, Z) are controlled by generating non-harmonic (asymmetric) oscillations $\Psi_x(t)$, $\Psi_y(t)$ or $\Psi_z(t)$ of the front zone in the sphere and higher mode harmonic oscillations of the rear part of the sphere. In this case the operation relies on the nonlinear character of damping in the liquid, which is independent of the absolute velocity of surface oscillation.

Research and development of piezoelectric motors (or vibromotors) in Vibration Centre of Kaunas University of Technology has been initiated in 1969. In *Supplement 2* the main milestones of this extremely interesting R&D area are presented. Based on the results of R&D, a new course for post-graduate students "Piezoelectric (Ultrasonic) Actuators" has been developed (*Supplement 3*); from 1996 this course is being read at Kaunas University of Technology. In 1997 this course has been presented for the post-graduate students of Aalborg University, Denmark.

222

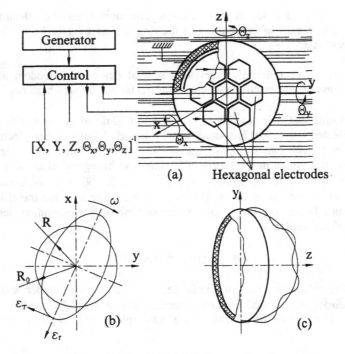

(a)   Hexagonal electrodes

(b)   (c)

Figure 11. "Piezofish" with six degrees of freedom

6. Active Bearings, Supports and Slides

A classical approach to reduce or eliminate static and dynamic errors of bearings, supports and guides is to increase the accuracy and stiffness of system elements. This sharply increases the costs of the devices. Further gains in accuracy of high precision elements is gradually reaching its economically acceptable limits. The mechatronic approach to this problem is to integrate mechanical system with electronics and control. Further to reducing the cost and increasing the final accuracy, this alternative introduces new properties in the existing systems.

The term *Active bearing* here is related to a kinematic pair with W degrees of freedom ($1 \leq W \leq 5$), in which torque or forces are generated in the contact zone between elements, leading to the relative motion. As in existing devices one element is fixed, allowing the transmission of forces to the other element.

The development of active bearings was made possible through the latest advances in creating new piezoactive materials [31] and the development of various methods to transform high frequency multi-component mechanical oscillations into continuous or step motion [1,25]. The integration of unique properties of piezoactive transducers and actuators (high resolution, low time constant, easy control of forms, types and parameters of oscillations, possibility to generate multi-component static, quasi-static and resonant displacements) with control system made it possible to sharply reduce or even fully eliminate most errors of bearings, supports and guides, used in high precision measuring devices.

At a fundamental level the structures resemble slide bearings with the main difference being that one or both elements are made from piezoactive material or constitute part of a piezoelectric transducer. To achieve rotation or translation motion, high frequency oscillations are generated in the contact zone between two elements.

There are three basic methods to transform high frequency resonant oscillations into continuous motion, which greatly influence the construction and design of active bearings:

(*i*) The transformation of *high frequency oblique impacts*. The method is based on the superposition of tangential components of the impact pulse and involves the use of two-component oscillations, generated by the special configuration transducer. The example of active bearing of this type is given in Fig.12a, where at every contact zone two component (longitudinal and flexural) resonant oscillations of the transducer excite periodic oblique impacts. If the fundamental mode of longitudinal oscillations and the second form of flexural oscillations are used, the direction of tangential components of oblique impact pulses is the same for both contact zones. This is achieved by using sectioned electrodes of special configuration, being connected to the multi-phase harmonic signal generator. For practical design purposes it is important that there is one common node for both types of oscillations, which can be used to fix transducer to the housing.

Oblique impacts can also be excited by using two contacting transducers, perpendicular to one another, in which one-component resonant oscillations are generated (Fig.12b). In this schematic any lower form of longitudinal oscillations can be used; the use of second form of longitudinal resonant oscillations with two nodes simply solves the problem of fixing the transducer to the housing.

(*ii*) The transformation of *travelling wave oscillations* into continuous motion through frictional interaction of the travelling wave motion of the transducer and driven component. If one active component of the bearing is made in the form of a hollow cylinder, for travelling wave we have:

$$R = R_0 [1 + \zeta \sin(n_1 \varphi - \lambda t)] \qquad (6)$$

Where ζ is the relative oscillation amplitude; n_1 is the wave number and φ is angular coordinate of the considered cross section. Then the tangential ε_t and the radial e_r displacement of points on the neutral surface of the active element are described by:

$$d\varepsilon_t/dt = -\varepsilon_r \, d\phi/dt \qquad (7)$$

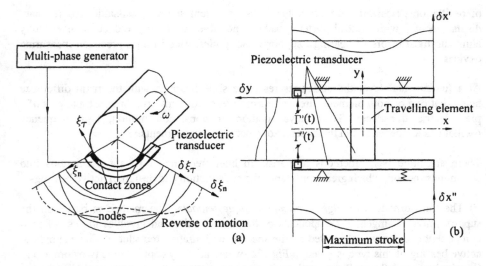

Figure 12. The schematics of active bearings and supports with oblique impacts generated by: (*a*) - curvilinear transducer ($\delta\xi_\tau$ and $\delta\xi_n$ - the distribution of the amplitudes of tangential and normal components of oscillation); (*b*) - the superposition of two one-component oscillations in transducers, perpendicular to one another

Here $d\varphi/dt$ is the angular velocity of the wave, equal (at $n_1 = 2$) to the angular frequency of the harmonic signal. Angular speed of the rotor ω depends on a value of the interference fit between components and applied external moment, but in any case it is less than $1/R \times d\varepsilon_n/dt$ due to frictional interaction between both components.

Though this type of actuator is well studied [7,26,29], still exist a problem of the absence of nodes in ring type transducer, in which travelling wave radial oscillations are generated. One possible solution of this problem is to use large length to diameter ratio of the active element, with an amplitude phase shift of π/n_1 of oscillatory motions in the planes forming the ends of the active element (Fig.13). If cross section of the transducer $Q_1(x) = $ constant, the radius of the neutral surface R can be expressed as:

$$R(x, \varphi) = R_0 + \varepsilon_n/h \times x \cos(n_1\varphi - \lambda t) \tag{8}$$

When $x = -h$, $R(\varphi; x = -h) = R_0\varepsilon_n \cos(n_1\varphi - \lambda t)$; where ε is the amplitude of normal displacement in cross section $x = -h$. When $x = 0$, $R(x,\varphi) = R_0$ and in this case the modal plane is situated in the middle of the transducer and can be used for fixing it to the housing (Fig.13).

(*iii*) The transformation of non-harmonic oscillations into continuous motion using non-linear dependence of friction force upon velocity in contact zone. Saw-tooth oscillations are the most desirable since they produce the highest velocity in relation to the contact zone. The disadvantage is the absence of common nodes for fundamental and first modes of oscillation. This causes some difficulty when fixing the transducer to the housing. In this case the use of travelling transducer is a possible solution.

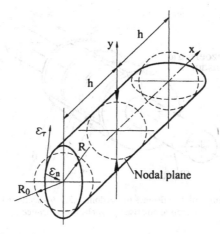

Figure 13. Nodal plane in cylindrical transducer and corresponding form of oscillations

Multi-degree-of-freedom active bearings and supports consist of kinematic pairs with several degrees of freedom W ($1 \leq W \leq 5$), in which one or both elements are made from piezoactive material with predetermined excitation zones.

Below are two examples of active supports for precision engineering devices, based on the use of piezoceramic transducers and frictional interaction of the elements of a kinematic pair. It is easy to notice, that in all schematics such traditional errors as radial or axial play, backlash and dead zones are minimised. As in active bearings with $W = 1$, the resolution of all movements is very high (it lies in the range of 0.1 to $0.01 \mu m$) and time constant is low (in the order of 0.5 to 5 ms), depending on the mass and moment of inertia of element. In the schematic, shown in Fig.14, the number of axial n and radial m electrode sectors are: $n = m = 3$.

Active supports with $W = 3$ (three rotations around axes x,y,z) can be performed on a basis of a kinematic pair with 3 degrees of freedom. The example of such a device is given in Fig.15.

Here spherical transducer *1* with piezoelectric element *2* is in contact with the component *3* with exponential internal surface, activated by piezoelectric element *4*.
The use of non-linear effects, kinematic structure control and both direct and inverse piezoeffects allows creating devices, having new qualities as compared with traditional bearings, supports and guides. The devices become multi-functional by taking advantage of the following features:

(*i*) They act as a support and high stiffness physical restraint for a component under investigation.

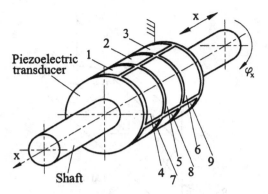

Figure 14. Active slide bearing with two degrees of freedom: 1,2,...,9 - sectioned electrodes (excitation zones) of a radial polarised piezoelectric transducer.

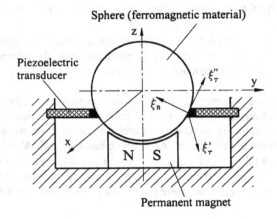

Figure 15. Active support with W = 3.

(*ii*) The controlled forces or torque are generated in the contact zone resulting in the relative movement of travelling element.

(*iii*) Piezoelectric actuators used in active bearings are characterised by thermo-, mass-, tenso- and gyro-sensitivity, which can be effectively applied to measure these parameters by measurement of the electrical parameters (voltage, phase or frequency), sourced from the specially sectioned electrodes.

(*iv*) It is possible to generate two types of motion simultaneously: transformation of high frequency resonant oscillations into continuous or step motion (resolution lies in the range of 0.1 to 0.01 μm) and static or quasi-static displacements of piezoelectric transducers under the applied electric field (resolution lies in the nanometre range). This feature enables the effect of two channel motion control, e.g. in *adaptive active bearing*, in which the position of the rotor in space can be controlled by applying high voltage to the specially selected electrodes of piezoelectric transducer.

(*v*) Forces or torque generated in the contact zone of active bearing are directly related to the friction forces. It is well established (e.g. in [1,13,17]) that the value of friction force can be controlled by mechanical oscillations. This phenomenon allows the control of forces or torque, acting on travelling component of active bearing. The frequency of oscillations controlling the friction must be 3 to 10 times greater than the frequency of oscillations being transformed into continuous motion.

(*vi*) The dependence of forces or torque in the contact zone on friction force can be used to design *centring devices*, in which mass m (Fig.16) will be automatically centred about specified z-axis.
The oblique impacts in the contact areas are generated; centering is completed when $N1 = N_2$. The device can be transformed into active bearing, when the phase of oscillations of one of the vertical transducers is shifted by π.

(*vii*) *Multi-functionality* of active bearings enables a simplification of the design of traditional precision devices, in which actuators and sensors are separated. Using direct and inverse piezoeffect it is possible to integrate both function into one transducer, thus achieving higher resolution and reducing the time constant of the whole device.

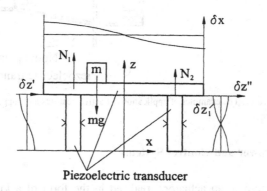

Piezoelectric transducer

Figure 16. Active support working as centering devices

The main area of application of active bearings is precision component surface and profile measuring systems, in which scanning of the surface is being performed. One example of such problem is the error evaluation in high precision ball bearings. Here simultaneous measurements of profile and surface is achieved by the rotation of the component. This is made possible by the use of piezoelectric transducers (Fig.17*a*), contacting with the component in two zones, having the same pattern of oscillation distribution and phase shift between normal and tangential components of oscillations. As external forces are absent, it is evident that errors caused by the torque generated in the contact zone are negligible.

228

In schematic shown in Fig.17*b*, in a waveguide the travelling waves are generated; nodal plane here coincides with the diaphragm, separating vacuum and atmospheric pressure.

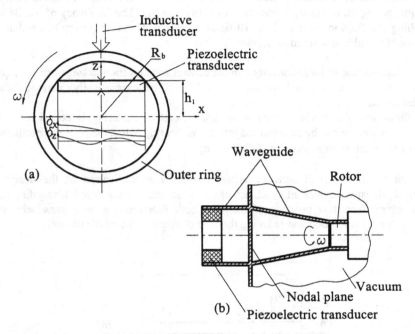

Figure 17. The examples of applications of active bearings and supports.

7. Separation of Power and Control Systems

The simplest structure of an actuator, realised in the form of a kinematic pair with separated power and control systems, is shown in Fig. 18. Here $\{P_j\}$ - vector of outer excitation (forces or torque, applied to the travelling link; in most cases $j=1$); $\{Q_i\}$ - vector of control signals; $\{X_k\}$ - vector of displacements, in which l usually corresponds to the number of degrees of freedom of the kinematic pair.

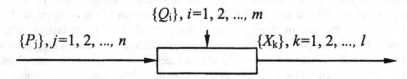

Figure 18. The structure of actuators with separated power and control systems

The concept of power and control systems separation allows to realise actuators with two or more DOF. Two examples of such systems are presented in Fig. 19 and 20. In both cases the power system consists of dynamically (Fig. 19) or statically (Fig. 20) unbalanced rotor, mounted in a travelling link. Resonance oscillations in piezoelectric transducers (Fig. 19), which control the friction between sphere and link, are generated by the signal generator. The function of the control system is to generate oscillations at the right moments and for a certain duration in every transducer, depending on the desirable trajectory of the link 2.

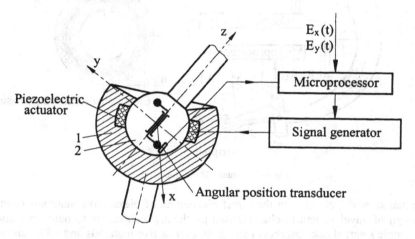

Figure 19. Spherical joint with two degrees of freedom

In Spherical joint (Fig. 19) while in the fixed position the friction force moment between links *1* and *2* exceeds the moment, generated by centrifugal forces. Generating oscillations in the contact zones between links gives the opposite effect.

In Trunk-like manipulator (Fig. 20) the change of the structure is being effected by the control of viscosity of electro-rheological fluid, situated between two flexible hoses. The trajectory of the grip in this case is defined by phase and duration of periodic high voltage electric pulses, applied to the electrodes on the hoses.

8. The Application of Piezoactive Composite

Electro- or magneto-rheological fluids (ERF, MRF) change their viscosity properties when high voltage or magnetic fields are applied to them. This phenomenon allows control of the stiffness and damping of the structure, which can be changed dynamically by applying high voltage. Several methods of vibration and structure dynamic control have been developed [1,3,25]; in most of cases the peaks of resonance are moved to

230

higher frequencies due to the increase of the total stiffness and peak amplitudes are decreased due to the increase of the total damping of the structure.

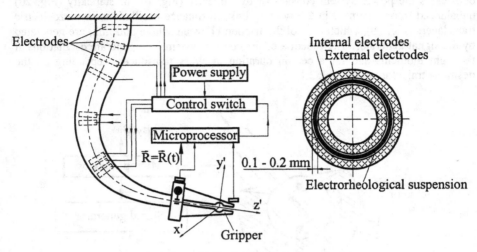

Figure 20. Trunk - like manipulator

Combination of ER effect with the direct piezoeffect of piezoactive materials enables the design of novel structures characterised by the *self-adaptation* to outer excitation. For example smart shock absorbers using both piezoactive materials and ERF can react to loads and re-adjust itself to optimise deceleration and damping. It is proposed that new sensors, consisting of coupled piezoelectric and magnetostrictive converters can be used to advantage. Such sensors enable the generation of oscillations or static displacement both by magnetic $\Phi_m(t)$ and electric $U_E(t)$ fields. Using both energy conversion methods it is possible to transfer energy to the device remotely, e.g. controlling the movement of a micro manipulator within the human body.

8.1. PIEZOELECTRIC/MAGNETOSTRICTIVE SENSORS/ACTUATORS

As it is shown in Fig. 4*b*, the typical energy transformation in this type of sensor/actuator is as follows: magnetic field $\Rightarrow$ mechanical stresses $\Rightarrow$ electrical charge. Fig.21 shows an example of this type of sensor/actuator; a piezoceramic ring with axial poling is mounted in a larger ring, made of a magnetostrictive material (e.g. nickel, Terfenol-*D*, etc.). In this case conversion between magnetic and electric energy is performed. Depending on the input and output signals, the device can be used as:

(*i*) *a low frequency (non-resonant) piezo-electric voltage transformer.* In this case

$$U_1(t) = U_0 + U_A \cos\omega t;$$
$$U_2(t) = U_B \cos(\omega t + \lambda); \qquad U_0 \geq U_A,$$

where $U_B/U_A \gg 1$ and $\lambda \approx 0$. Such transformers combine low impedance input with high impedance output and vice-versa. The frequency range in a non-resonant case lies between (0.1-0.5) Hz and $0.3\omega_0$, (zone "A", Fig.21b), where ω_0 is the lowest resonant frequency of radial oscillations of the transducer;

(*ii*) *a high voltage (resonant) transformer.* In this case $\omega = \omega_0$ and ratio U_B / U_A depends on the internal damping of the transducer;

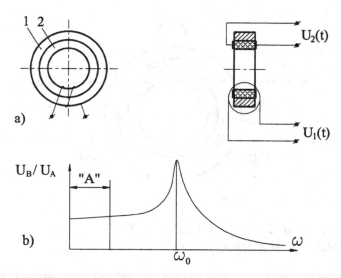

Figure 21. Schematics of composite piezoelectric/magnetostrictive actuator (*a*) and its frequency characteristic (*b*).

Both in resonant and non-resonant cases high impedance input and low impedance output can be realised using the inverse effect of change in magnetisation, when the dimensions of a transducer are altered by the application of U_2 (*t*) to the electrodes of the piezoceramic ring. In this case $U_B \gg U_A$; evident applications of such systems include transformers for very small power supply units.

(*iii*) *frequency multiplier by 2.* This feature is realised when $U_0 = 0$:

$U_1 (t) = U_A \cos \omega t$;
$U_2 (t) = U_B \cos(2\omega t + \lambda_1)$; $\lambda_1 = \lambda_1 (\omega)$.

(*iv*) *an ultrasonic transducer/actuator with two channel control.* In this case it can perform the role of an active link, in which the combination of two types of oscillations are realised, including the generation of non-harmonic oscillations. The actuator shown in Fig.21 can be used for the generation of resonant radial oscillations; but more complicated cases involving the sectioning of electrodes and separating of excitation zones - as in the actuator/sensor, shown in Fig.22, in which in addition to radial oscillations, two types of flexural radial travelling waves can be generated simultaneously.

$U_1(t) = U_A \cos \omega t$;
$U_2(t) = U_B \cos(2\omega t + \lambda_1)$; $\lambda_1 = \lambda_1(\omega)$.

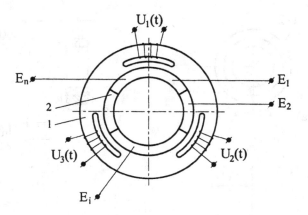

$U_j(t) = U_0 + U_a \cos [\omega t + 2(j-1)\pi/p]$; $j=1,2,...,p$; $p \geq 3$
$E_i(t) = U_T \cos(\chi t + 2\pi/ik)$; n= 3k; i = 1,...n

Figure 22. An actuator generating two radial travelling waves with multiple frequencies (*k* - coefficient, defining the mode of radial resonant oscillations).

Several energy conversions are also used in multi-layer piezoelectric/magnetostrictive (P/M) sensors (Fig.23). With the development of new technology for depositing thin Terfenol films, it is possible to develop P/M films, using as a basis flexible piezoactive materials (Polyvinylidene fluoride polymer PVDF, Piezorubber PZR, etc.). In Fig.24 a comparison is shown between the Hall effect transducer as a magnetic field sensor, and a composite P/M magnetic field sensor. In the Hall effect sensor, when a magnetic field is applied across a current-carrying conductor, an electric voltage will be developed orthogonally to both the current and the magnetic field, as shown schematically in Fig.24*a*. In the composite P/M sensor dimensional changes in the magnetostrictive film are induced by a magnetic field, leading to the generation of electrical charges on the electrodes of the piezoactive film.

If compared with the Hall effect sensor, composite P/M sensor has the following advantages:

(*i*) no voltage or power supply is required;

(*ii*) minimum thickness of a typical Hall effect sensor is approx. 1mm, while for a composite P/M sensor it can be from 0.1 to 0.01mm;

(*iii*) one of the electrodes of the PVDF film can be sectioned into a matrix of electrodes, thus allowing for the measurement of the distribution of alternating magnetic field in some specific area (e.g. in a gap between stator and rotor of an electric drive).;

(*iv*) the temperature range of composite P/M sensors can be extended up to Curie temperature, at which the material changes crystalline phase or symmetry. For typical PZT compositions it lies between 250 and 450 ^{0}C.

It is evident that the Hall effect is independent of frequency (down to DC), whilst the lowest frequency of a composite P/M sensors is related to the input impedance of a charge amplifier.

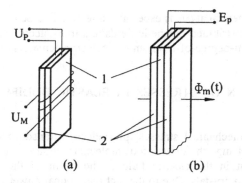

(a)   (b)

Figure 23. Bimorph (*a*) and symmetrical (*b*) composite P/M sensors/actuators: *1* - layer made from piezoceramic material; *2* - layer from magnetostrictive material.

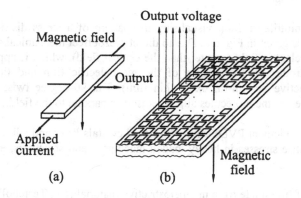

Figure 24. The schematics of a magnetic field sensors based on the Hall effect (*a*) and P/M composites (*b*).

For the theoretical analysis of combined piezoelectric/magnetostrictive transducers existing methods of finite element analysis can be applied, taking into account the equations both for magnetostriction and piezoelectric transformations:

$$S_1 = c_{11}{}^H \varepsilon_1 - e_{m11} H_1;$$
$$B_1 = e_{m11} \varepsilon_1 + \mu_0 \mu_{m11}{}^\varepsilon H_1;$$
$$S_1 = c_{11}{}^E \varepsilon_1 - e_{p31} E_3;$$
$$D_3 = e_{p31} \varepsilon_1 + z_{p33}{}^\varepsilon E_3,$$

Where S - mechanical strain; H, B - magnetic field strength and flux density; ε - deformation; c - tensor of stiffness constants when magnetic and electric fields are of constant values; e - tensor of dynamic coefficients of both transducers; μ - magnetic permeability; E, D - electric field strength and density; z - dielectric permeability.

The use of finite element analysis is especially effective in case of sectioned electrodes, with non-uniformly distributed magnetic fields and transducers with variable vector of poling as used in multi-degree-of-freedom piezoelectric actuators and intelligent mechanisms.

8.2. VISUALISATION OF STRESSES IN ELASTIC BODIES AND MAGNETIC FIELDS

When subjected to mechanical stresses, piezoelectric transducers generate sufficient electrical charge to change the direction of molecular alignment in liquid crystals. The process of alignment in the electric field is the result of the anisotropic constant characteristic of liquid crystals. Due to the fact that optical power is supplied externally (by sunlight or room lighting), liquid crystals need supply only the relatively minute amount of power (microwatts per square centimetre) required to change a device's reflective optical properties.

A schematic of multilayer film, visualising the pattern of a stress distribution in the elastic structure, is given in Fig.25a. Due to direct piezoeffect mechanical stresses in the structure induce electric field (greater than the critical field), which is applied between matrix-like electrodes on the surface of the piezoelectric film and the transparent electrode (conductive indium-tin oxide thin film). The 90-degree twist in the liquid crystal is destroyed as the molecules align themselves parallel to the field.

Using as an active element PVDF film with liquid crystals layer and adhesive base, it is possible to visualise stresses in heavy loaded machinery and structures, e.g. bridges or cranes.

Adding a layer of film made from magnetostrictive material (e.g. Terfenol) it is possible to visualise the distribution of alternating magnetic field strength in a specific area or location (Fig.25b).

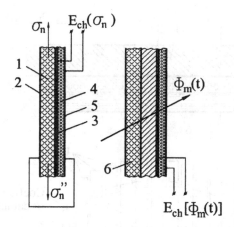

Figure 25. Schematics of a multi-layer film visualising stresses (*a*) and magnetic fields (*b*): *1* - PVDF film; *2* - continuous electrode; *3* - matrix-like electrode; *4* - twisted nematic cell; *5* - transparent electrode, connected to the continuous electrode; *6* - magnetostrictive material.

8.3. MULTI-LAYER STRUCTURES AND FILMS FOR ADAPTIVE DAMPING

Such systems are capable of introducing damping into vibrating systems in such a way that damping will be directly related to the maximum value of the stresses in the structure without applying an external electric field. The stiffness and damping of the structure can be changed dynamically by applying to the ERF layer high voltage, generated due to direct piezoeffect by the piezoceramic transducer, and thus effect active control of its dynamic behaviour.

ERF undergo significant reversible changes when subjected to a high voltage electric field - mass redistribution, stiffness and dissipative characteristics. These features are implemented in a multi-layer structure, consisting of piezoactive layer with sectioned electrodes and the layer of ERF (Fig.26); here high voltage on the electrodes of piezoactive layer influences the flow resistance of ERF. The frequency range of damping depends on the applied ERF.

The system presented in Fig.26 possesses the features both of passive and active damping systems. Passive systems have simple mechanical structures and no external power supply is needed. In classical systems passive dampers can only be tuned best for a relatively small frequency range, whereas an adaptive damper shown in Fig. 26 is characterised by a wide frequency range. Active dampers are much more elaborate since they form a closed loop; in a piezoelectric/ERF damper viscosity is directly related to the stresses induced in piezoactive materials by oscillations.

236

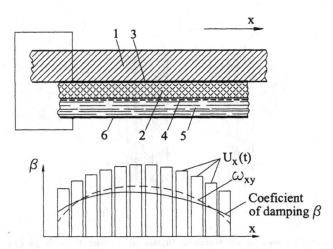

Figure 26. Damping the vibrations of structures by combined piezoelectric/ERF dampers: 1 - vibrating structure; 2 - piezoactive layer; 3 - continuous electrode; 4 - sectioned electrode; 5 - layer of ERF; 6 - electrode, in contact with the ERF layer; U_x (t) - the distribution of electric charge on the sectioned electrodes; σ_{xy} - stresses in the active layer.

Despite short impulse time, piezoelectric/ERF dampers can be effectively used for shock damping, especially in the case of periodic excitation. As the shock excitation in piezoelectric transducers generates high voltage, the fluid viscosity changes can be of high order.

References

1. Bansevicius, R. and Ragulskis, K. (1981) *Vibromotors*, Mokslas, Vilnius (in Russian).
2. Bansevicius, R., Dalay, H., and Knight, J. (1996) Piezomechanics as a Component of Mechatronics, *The Intern. Journal of Mechanical Engineering Education* 24, 73-87.
3. Bansevicius, R., Parkin, R., Jebb, A., and Knight, J. (1996) Piezomechanics as a Sub-system of Mechatronics: Present State of the Art, Problems, Future Developments, *IEEE Transactions on Industrial Electronics* 43, 23-30.
4. Bansevicius, R., and Ragulskis, K. (1971) Vibrational drive mechanisms for flexible tapes, *Proc. 3rd World Congr. Theory Mach.& Mechanisms*, Kupari, Yugoslavia, vol. A, 218-225.
5. Bansevicius, R. and Milner, J.H. (1994) Transducers, piezoelectric, *AIAA J.*, 364-367.
6. Bansevicius, R., Baskutis, S., and Sakalauskas, S. (1989) Sensorisierte schwingungskopfe fur koordinaten-messmaschinen zur messung von geometrischer und mechanischer charakteristik der teile, *Int. Wiss. Koll. TH Ilmenau*, vol. 34.
7. Bansevicius, R. and Ragulskis, K. (1979) Piezoelectric drives with several degrees of freedom for micromanipulators, *Proc. 5th World Congr. Theory Mach, & Mechanisms*, Montreal, Canada, 827-830.
8. Bansevicius, R., Tolocka, T., and Knight, J. (1995) Adaptive mechanisms: Trends and Some Applications, *Mechanika* No.1, Technologija, Kaunas, 47-52.
9. Bansevicius, R., Tolocka, T., and Knight, J. (1995) Intelligent Mechanisms in Mechatronics: Parameters, Design, Application, *Proceed. of the 9th World Congress on the Theory of Machines and Mechanisms*, Milano, Italy, vol.4, 3053-3057.
10. Bansevicius, R., Jebb, A., Knight, J., and Parkin, R. (1994) Piezomechanics: a New Sub-system in Mechatronics and its Applications in Precision Mechanics, *Proceedings of Joint Hungarian-British International Mechatronics Conference*, Budapest, 21-23, September, 361-366.
11. Chang, S.H. and Wang, H.C. (1990) A high speed impact actuator using multilayer piezoelectric ceramics, *Sensors and Actuators (A-Physical)* 24, 239-244.
12. Crawley, E. F. and De Luis, J. (1987) Use of piezoelectric actuators as elements of intelligent structures, *AIAA J.* 25, 1373-85.
13. Flynn, A.M., Tavrow, L.S., Bart, S.F., Brooks, R.A., Ehrlich, D.J., Udayakumar, K.R., and Cross, L.E. (1992) Piezoelectric micromotors for microrobots, *J. Micromech. Syst.* 1, 44-51.
14. Gotz, B., Martin, T., and Wieduwilt, T. (1994) Optimization of piezoelectric actuators, *Actuator '94 (Proc. 4th Int. Conf. New Actuators)*, Axon Technologie Consult GmbH, Bremen, Germany, June 15-17, 123-128.
15. Harb, S., Smith, S.T., and Chetwynd, D.G. (1992) Subnanometer behavior of a capacitive feedback, piezoelectric displacement actuator, *Rev. Sci. Instrum.* 63, 1680-1689.
16. Hellebrand, H., Cramer, D., Probst, L., Wolff, A., and Lubitz, K. (1994) Large piezoelectric monolithic multilayer actuators, *Actuator '94 (Proc. 4th Int. Conf. New Actuators)*, Axon Technologie Consult GmbH, Bremen, Germany, June 15-17, 119-122.
17. Howald, L., Rudin, H., and Guntherodt, H.-J. (1992) Piezoelectric inertial stepping motor with spherical rotor, *Rev. Sci. Instrum.* 63, 3909-3912.
18. Jebb, A. and Bansevicius, R. (1991) Actuator/sensor mechanisms for microrobots, *Proc. 8th World Congr. Theory Mach., Mechanisms*, Prague, Czechoslovakia, 875- 878.
19. Joshi, S.P. (1991) Non-linear constitutive relations for piezoceramic materials, *Proc. ADPA/AIAA/ASME/SPIE Conf. Active Mater. Adaptive Structures*, G. J. Knowles, Ed., Alexandria, VA, 217-225.
20. Knight, J., Bansevicius, R., and Gaivenis, A. (1995) Multi-degree-of-freedom Actuators with High Resolution for Mechatronics Applications, *International Journal of Intelligent Mechatronics: Design and Production* 1, 61-68.
21. Koyama,T. and Asada, K. (1991) Design of an arm with double actuators for high speed and high accuracy manipulation, *Proc. Amer. Contr. Conf.*, vol. 2, 1435-1437.
22. Kurosawa, M. and Ueha, S. (1989) High speed ultrasonic linear motor with high transmission efficiency, *Ultrasonics* 27, 39-44.
23. Ozol, O. (1979) *Design of Mechanisms*, Zvaigzne, Riga (in Russian).
24. Preumont, A., Dufour, J.-P., and Malkian, C. (1992) Active damping by a local force feedback with piezoelectric actuators, *J. Guidance, Contr, Dynamics* 15, 390-395.
25. Ragulskis, K., Bansevicius, R., Barauskas, R., and Kulvietis, G. (1988) *Vibromotors for Precision Microrobots, Hemisphere Publishing*, ISBN 0-9116054905, New York.
26. Sashide, T. and Kenjo, T. (1993) *An Introduction to Ultrasonic Motors*, Oxford Science Publ.,

Clarendon Press.
27. Smits, J. G. (1991) The effectiveness of a piezoelectric bimorph to perform mechanical work against various spring-type loads, *Ferroelec.* **120**, 241-252.
28. Toyama, S., Sugitani, S., Zhang Guogiang, Miyatani, Y., and Nakamura, K. (1995) Multi-degree-of-freedom Spherical Ultrasonic Motor, *Proceed. of the IEEE Intern. Conf. on Robotics and Automation*, vol.3, 2935-2940.
29. Ueha, S. and Tomikawa, Y. (1993) *Ultrasonic Motors. Theory and Applications*, Oxford Science Publications, Clarendon Press, Oxford.
30. Umeda, M., Ohnishi, K., Nakamura, K., Kurosawa, M., and Ueha, S. (1990) Dumbbell-shaped small size hybrid ultrasonic motor, *Jpn. J. Appl. Physics* **29**, pt. 1, S29-1, 191-193.
31. Ushino, K. (1992) Applied aspects of piezoelectricity, *Key Eng Mater.* **66-67**, 331-37.
32. Wallaschek, J. (1994) Piezoelectric Ultrasonic Motors, *Proceed. of the Second Intern. Conf. on Intelligent Materials*, ICIM'94, 1279-1290.

DAMPING CONTROL IN SYSTEMS ASSEMBLED BY SEMI-ACTIVE JOINTS

RAINER NITSCHE AND LOTHAR GAUL
Institute A of Mechanics, University of Stuttgart,
Pfaffenwaldring 9, 70550 Stuttgart, Germany

Abstract. Vibration properties of most assembled mechanical systems depend on frictional damping in joints. The nonlinear transfer behavior of the frictional interfaces often provides the dominant damping mechanism in a built-up structure and plays an important role in the vibratory response of the structure. For improving the performance of systems, many studies have been carried out to predict, measure, and/or enhance the energy dissipation of friction. To enhance the friction damping in joint connections a semi-active joint is investigated. A rotational joint connection is designed and manufactured such that the normal force in the friction interface can be influenced with a piezoelectric stack disc. With the piezoelectric device the normal force and thus the friction damping in the joint connection can be controlled. A control design method, namely semi-active control, is investigated. The recently developed LuGre friction model is used to describe the nonlinear transfer behavior of joints. This model is based on a bristle model and turns out to be highly suitable for systems assembled by joints. The semi-active method is well suited for large space structures since the friction damping in joints turned out to be a major source of damping. To show the applicability of the proposed concept to large space structures a two-beam system representing a part of a large space structure is considered. Two flexible beams are connected with a semi-active joint connection. It can be shown that the damping of the system can be improved significantly by controlling the normal force in the semi-active joint connection. Experimental results validates the damping improvement due to the semi-active friction damping.

Key words: Nonlinear systems, semi-active damping, friction modelling, nonlinear control and observer design

1. Introduction

Regardless of the degree of model uncertainty, semi-active elements have, like passive elements, the ability to dissipate system energy. Through implementation of an appropriate adaptive control law, semi-active elements are able to adapt to different vibration environments and/or system configurations. Another advantage over passive damping elements is their ability to utilize sensor information from other parts of the structure, a so-called non-collocated sensor/actuator architecture [8]. Typically, semi-active elements have low power requirements and are less massive than their active counterparts. Certainly, somewhat lower performance is obtained and complete controllability is lost. The system is stabilizable to the origin but is not controllable to other points in the state space.

A nonlinear, local control law is derived to maximize the energy dissipated by the

A. Preumont (ed.), Responsive Systems for Active Vibration Control, 239–251.

240

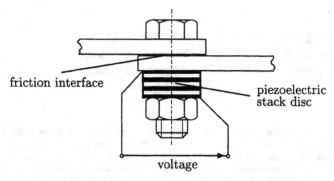

Figure 1. Semi-active joint.

frictional joint using the normal force as control input. This concept of varying the normal force in a frictional joint to enhance the dissipation of energy from a vibrating structure is investigated in the present paper.

2. Semi-active Joint Connection

In the following, a friction model with velocity dynamics is introduced. The model is used to describe the behavior of a semi-active revolute joint shown in Fig. 1 and Fig. 5. The friction torque couples two flexible Euler-Bernoulli beams as shown in Fig. 3.

In Fig. 1 a semi-active joint, patented by the first author under *DE 197 02 518 A1*, is realized with a piezoelectric disc plate. Applied voltage at the stack disc tries to thicken the piezoelectric material which, due to the constraint, results in increasing the normal force. In this paper the dynamic behavior of the piezoelectric material will be neglected, so the normal force is proportional to the input voltage. A modification of the dynamic friction model proposed by [5] is used to describe the friction characteristics. The starting point of the model derivation is the force caused by solid-to-solid contact, as shown in Fig. 2. At the microscopic level the surfaces are very irregular and make contact at a number of asperities which can be thought of as a contact between bristles. The bristles deflect like springs and give raise to the friction force, when a tangential force is applied. If the contact is sufficiently large some of the bristles deflect such that they slip off each other. Due to the irregular shape of the surfaces, contact phenomena are highly random [6]. Therefore the model is based on the average behavior of the bristles that make up the contact. This model was designed to reproduce all observed friction phenomena over a wide range of operating conditions. It is given by

$$M_{\mathrm{f}} = r\, F_{\mathrm{N}}\mu(\varphi,\dot\varphi,\dot\theta), \quad \mu = \sigma_0\varphi + \sigma_1\dot\varphi + \sigma_2\dot\theta \tag{1a}$$

$$\dot\varphi = \dot\theta - \sigma_0\,\frac{|\dot\theta|}{g(\dot\theta)}\,\varphi\,, \qquad \varphi(0) = \varphi_0 \tag{1b}$$

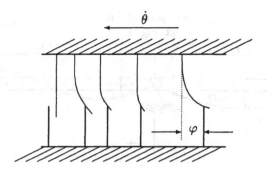

Figure 2. Model of the friction interface.

$$g(\dot{\theta}) = F_c + F_\Delta \exp\left(-\left(\dot{\theta}/\dot{\theta}_s\right)^2\right) \tag{1c}$$

where the friction moment is M_f, relative sliding velocity at the friction interface is $\dot{\theta}$, and the normal force F_N is the control input. The effective radius at which the friction force acts is denoted as r. The internal friction state φ represents the average deflection of the bristles as can be seen in Fig. 2. The parameter F_c is the Coulomb friction level while the sum $F_c + F_\Delta$ corresponds to the stiction force. The so-called Stribeck velocity $\dot{\theta}_s$ [2] determines how $g(\dot{\theta})$ varies within its bounds $F_c < g(\dot{\theta}) \le F_c + F_\Delta$. The stiffness of the bristles is described by σ_0, while the two parameters $\sigma_{1,2}$ control the dynamic dependence of friction on velocity. The variable μ, which is defined in (1a), can be interpreted as a state dependent friction coefficient.

With this model it is possible to describe presliding displacement[1], stick-slip motion including the Stribeck effect, and other rate dependent friction phenomena. Moreover the numerical efficiency and accuracy is very high as one do not have to detect zero velocity differing from static friction models. Extensive analysis of the model and its application can be found in [11].

The elastic bristles store energy while the system is sliding. Some of this energy is returned to the system when the sliding direction is reversed. This takes place just prior the direction reversal and is given by the value of φ.

3. Two-Beam Model

The system selected for study consists of two linear elastic beams connected by a semi-active rotational joint as shown in Fig. 3 which is similar to the one treated in [7, 9]. Each beam is simply supported, but resistive moments are transmitted from one beam to the other through the intermediate friction joint. In particular, the system consists of two beams and three pin joints, where the center joint is semi-active (cf. Fig. 1). Before any control system analysis and design can be performed, a mathematical model of the system is required. Since only

---

[1] Sometimes also called microslip.

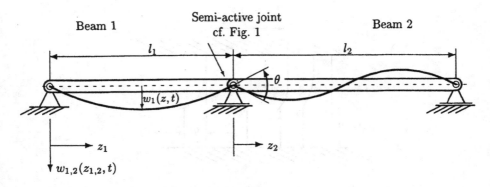

Figure 3. Two flexible beams connected by a semi-active joint. For illustration the first mode of the left and the second mode of the right beam is shown.

small flexural displacements of the beams are considered, the beams are modeled as Euler-Bernoulli beams. The equations of motion for the pinned-pinned beam system are derived using the assumed modes method. The flexural displacements of the two beams denoted by w_1 and w_2, are given by space time decomposition

$$w_i(z_i, t) = \sum_{k=1}^{N_i} \Phi_{ik}(z_i)\, w_{ik}^*(t)\,, \quad i = 1, 2 \tag{2}$$

where $\Phi_{ik}(z_i)$ is the k-th mode shape function, $w_{ik}^*(t)$ is the k-th modal coordinate, and N_i is the number of modes used to model the i-th beam deflection. The modes Φ_{ik} are chosen to be the eigenfunctions for pinned-pinned beams:

$$\Phi_{ik}(z_i) = A_{ik}\, \sin\left(\frac{k\pi z_i}{l_i}\right) = \sqrt{\frac{2}{l_i}}\, \sin\left(\frac{k\pi z_i}{l_i}\right) \tag{3}$$

with normalized A_{ik}, such that $\int_0^{l_i} \Phi_{ik}^2(z_i)dz_i \overset{!}{=} 1$. Note that the k-th natural frequency is given by

$$\omega_{ik} = \left(\frac{k\pi}{l_i}\right)^2 \sqrt{\frac{EI_i}{\mu_i}}\,, \quad i = 1, 2, \quad k = 1, \ldots, N_i \tag{4}$$

where EI_i is the flexural rigidity and μ_i is the mass per unit length of beam i. Using (3) and (4) yields the following equations of motion for the combined system in Fig. 3 while taking the modal damping ratios ζ_{ik} into account:

$$\mu_1 \ddot{w}_{1k}^* + 2\mu_1 \zeta_{1k}\omega_{1k}\dot{w}_{1k}^* + \mu_1 \omega_{1k}^2\, w_{1k}^* = \left.\frac{d\Phi_{1k}(z_1)}{dz_1}\right|_{l_1} M_{\mathrm{f}} \tag{5a}$$

$$k = 1, \ldots, N_1$$

$$\mu_2 \ddot{w}_{2k}^* + 2\mu_2 \zeta_{2k}\omega_{2k}\dot{w}_{2k}^* + \mu_2 \omega_{2k}^2 w_{2k}^* = -\left.\frac{d\Phi_{2k}(z_2)}{dz_2}\right|_{0} M_{\mathrm{f}}\,, \tag{5b}$$

$$k = 1, \ldots, N_2$$

where M_f is the moment (1) transmitted between the two beams. The relative angular displacement and angular velocity of the two beams at the active joint can be calculated from:

$$\theta = \sum_{k=1}^{N_i} w_{2k}^* \left. \frac{d\Phi_{2k}(z_2)}{dz_2} \right|_0 - \sum_{k=1}^{N_i} w_{1k}^* \left. \frac{d\Phi_{1k}(z_1)}{dz_1} \right|_{l_1}, \tag{6a}$$

$$\dot{\theta} = \sum_{k=1}^{N_i} \dot{w}_{2k}^* \left. \frac{d\Phi_{2k}(z_2)}{dz_2} \right|_0 - \sum_{k=1}^{N_i} \dot{w}_{1k}^* \left. \frac{d\Phi_{1k}(z_1)}{dz_1} \right|_{l_1}. \tag{6b}$$

With the state vector

$$\boldsymbol{\eta} = \left[w_{11}^*, \ldots, w_{1N_1}^*, w_{21}^*, \ldots, w_{2N_2}^* \right]^{\mathsf{T}}, \tag{7}$$

(5) can be written with (1) in matrix notation as

$$M\ddot{\boldsymbol{\eta}} + D\dot{\boldsymbol{\eta}} + K\boldsymbol{\eta} + \boldsymbol{b}M_f = 0, \quad \boldsymbol{\eta}(0) = \boldsymbol{\eta}_0 \tag{8}$$

where M, K, and $\boldsymbol{b}$ are easily identified. As system outputs the flexural displacements $y_1 = w_1^o$ and $y_2 = w_2^o$ at z_1^o and z_2^o are chosen for beam one and two, respectively (cf. Fig. 4):

$$y_1 = \sum_{k=1}^{N_1} \Phi_{1k}(z_1^o)\, w_{1k}^*, \quad y_2 = \sum_{k=1}^{N_2} \Phi_{2k}(z_2^o)\, w_{2k}^*. \tag{9}$$

These quantities can be easily measured using strain gages.

The two-beam system may be viewed as a general case of linear, distributed elastic systems coupled with a semi-active joint. The pinned-pinned configuration is chosen solely for the resulting form of the modal expansions. The modal formulation is applicable to arbitrary coupled elastic systems by simply replacing the modal eigenfunctions and natural frequencies with those of the actual system.

4. Controller Design

The intention is to design a 'damping controller' which additionally dissipates system energy due to friction in the semi-active joint. For this the two-beam system (8), (1) can be written as a control-affine system

$$\dot{\boldsymbol{x}} = [\dot{\boldsymbol{\eta}}^{\mathsf{T}}, \ddot{\boldsymbol{\eta}}^{\mathsf{T}}, \dot{\varphi}]^{\mathsf{T}} = \boldsymbol{f}(\boldsymbol{x}) + \boldsymbol{g}(\boldsymbol{x})u, \quad \boldsymbol{x} \in \mathbb{R}^n \tag{10}$$

while u is the normal force F_N in the friction interface of the semi-active joint and is the control input. The equilibrium $\boldsymbol{x} = 0$ of the uncontrolled part

$$\dot{\boldsymbol{x}} = \boldsymbol{f}(\boldsymbol{x}), \quad \boldsymbol{f}(0) = 0 \tag{11}$$

of the system (10) is stable and the task of the control is to provide additional damping which will render $\boldsymbol{x} = 0$ asymptotically stable. If a radially unbounded Lyapunov function $V(\boldsymbol{x})$ is known such that

$$L_f V(\boldsymbol{x}) \leq 0 \quad \forall \boldsymbol{x} \in \mathbb{R}^n \tag{12}$$

then it is tempting to employ $V(x)$ as a Lyapunov function for the whole system (10).

Remark 1.1 (Lie Derivative) *This paper does not require the formalism of differential geometry and employs the Lie derivative only for notational convenience: If $f : \mathbb{R}^n \to \mathbb{R}^n$ is a vector field and $V : \mathbb{R}^n \to \mathbb{R}$ is a scalar function, the notation $L_f V$ is used for $\frac{\partial V}{\partial x} \cdot f(x)$. It is recursively extended to*

$$L_f^k V(x) = L_f(L_f^{k-1} V(x)), \quad k > 1, \quad L_f^0 V(x) = V(x). \tag{13}$$

In view of $L_f V(x) \le 0$, the time-derivative of $V(x)$ for (10) satisfies

$$\dot{V}(x) \le L_g V(x) u. \tag{14}$$

This shows that $\dot{V}(x)$ can be made more negative with the control law

$$u(x) = -\kappa L_g V(x) \quad \forall \kappa > 0. \tag{15}$$

A suitable Lyapunov function for the two-beam system is a function representing the kinetic and potential energy of (8), that is:

$$V = \frac{1}{2} \left(\boldsymbol{\eta}^\top K \boldsymbol{\eta} + \dot{\boldsymbol{\eta}}^\top M \dot{\boldsymbol{\eta}} \right) . \tag{16}$$

The control law (15) using (16) yields

$$u = -\kappa L_g V(x) = \kappa \, \mu \, \dot{\theta} \tag{17}$$

where $\kappa > 0$ is the control parameter. This damping control law minimizes the return of the stored frictional energy to the mechanical system while it maximizes $M_f = u \, r \, \mu$ when it opposes the direction of relative sliding in the friction interface.

For the realization of the damping controller (17) the relative velocity $\dot{\theta}$ and the state dependent friction coefficient μ are required. Since only the flexural displacements y_1 and y_2 according to (9) are measured, an observer is needed to estimate the required quantities to implement the proposed feedback (17). For this an operating point observer will be designed next.

5. Friction Observer

For an implementation of (17) a state estimator is required as the friction state μ is not measurable. Simulation and experimental studies showed that the system (8) is sufficiently accurate described by *only the first mode* of each beam. With the state vector

$$z = [w_{11}^*, w_{21}^*, \dot{w}_{11}^*, \dot{w}_{21}^*, \varphi]^\top \tag{18}$$

and the already used replacement of F_N by u, the nonlinear system (8), (1) can be rewritten in a more general notation:

$$\dot{z} = \boldsymbol{\psi}(z, u), \quad z(0) = z_0, \quad z \in \mathbb{R}^5, \quad u \in \mathbb{R}^1 \tag{19a}$$

$$\boldsymbol{\zeta} = [w_{11}^*, w_{21}^*]^\top = h(z) \tag{19b}$$

while the modal coordinates represented in ζ are easily measurable using strain gauges and taking (9) into account. According to [4], the nonlinear observer

$$\dot{\hat{z}} = \underbrace{\psi(\hat{z}, u)}_{\text{simulation}} + \underbrace{G\left[\zeta - \hat{\zeta}\right]}_{\text{correction}}, \quad \hat{z}(0) = \hat{z}_0, \quad \hat{\zeta} = h(\hat{z}) \tag{20}$$

belongs to (19), with a nonlinear simulation and a correction-term which is a typical structure for a Luenberger observer.

Since only small displacements are regarded, the system is always situated in the neighborhood of the stationary solution

$$\psi(z_s, u_s) = 0, \quad \text{with} \quad z_s = 0, \ u_s = u_{\min}. \tag{21}$$

The gain matrix G in (20) is determined by regarding the linearized system around the operating point (21). Thus the differential equation of the observer error $\tilde{z} = \hat{z} - z$ is linear:

$$\dot{\tilde{z}} \quad = (A - GC)\,\tilde{z}, \quad \tilde{z}(0) = \hat{z}_0 - z_0,$$

$$A = \left.\frac{\partial \psi}{\partial z}\right|_{z_s, u_s}, \quad C = \left.\frac{\partial h}{\partial z}\right|_{z_s, u_s} = \begin{pmatrix} c_1^{\mathsf{T}} \\ c_2^{\mathsf{T}} \end{pmatrix}. \tag{22}$$

To verify operating point observability, one usually proves wether the observability matrix of the linearized n-th order system has full rank, i.e

$$\text{rank} \begin{pmatrix} C \\ CA \\ \vdots \\ CA^{n-1} \end{pmatrix} = n. \tag{23}$$

This is exactly the Kalman rank condition for observability [10]. Since the computer algebra system MATHEMATICA [12] is used for observer design, it is more advantageous to prove whether the *selection matrix*

$$Q_s = \begin{pmatrix} c_1^{\mathsf{T}} \\ c_1^{\mathsf{T}} A \\ \vdots \\ c_1^{\mathsf{T}} A^{\gamma_1 - 1} \\ c_2^{\mathsf{T}} \\ \vdots \\ c_2^{\mathsf{T}} A^{\gamma_2 - 1} \end{pmatrix} \quad \text{with } \gamma_1 = 3, \ \gamma_2 = 2 \tag{24}$$

is regular [3]:

$$\det Q_s = -\left(u_{\min} \, \sigma_0 \, \left. \frac{\mathrm{d}\Phi_{11}(z_1)}{\mathrm{d}z_1}\right|_{l_1} \right) \neq 0. \tag{25}$$

Note that the γ_i are the so-called *observability indices*, determining the dimension of the respective subsystem [10] and the sum of order numbers is given by

$$m_i = \sum_{j=1}^{i} \gamma_j, \quad i = 1, 2. \tag{26}$$

The parameters σ_0, $\left.\frac{d\Phi_{11}(z_1)}{dz_1}\right|_{l_1}$ in (25) are guaranteed to be always nonzero as well as the minimal normal force $u_{\min}$ for physical reasons. Hence the system is operating point observable.

By using a computer algebra system for system analysis, better predictions about the properties of the system, i. e observability or controllability have become possible. This is not the case by a pure numerical analysis.

The gain matrix G in (20) can be calculated recursively by

$$G = \quad \left\{ \left[p_{10}I + p_{11}A + \ldots + p_{1\gamma_1-1}A^{\gamma_1-1} + A^{\gamma_1} \right] s_1 , \right.$$
$$\left. \left[p_{20}I + p_{21}A + \ldots + p_{2\gamma_2-1}A^{\gamma_2-1} + A^{\gamma_2} \right] s_2 \right\} L^{-1}, \tag{27}$$

with the already defined observability indices $\gamma_1 = 3$ and $\gamma_2 = 2$. With the sums of order numbers (26) the gain matrix (27) can be evaluated with the vectors s_i representing the m_i-th column of the *inverse* selection matrix (24):

$$s_i = Q_s^{-1}[0 \ldots 0 \underset{\underset{m_i}{\uparrow}}{1} 0 \ldots 0]^{\mathsf{T}}, \qquad i = 1, 2. \tag{28}$$

The matrix L in (27) consists of the columns which do not vanish in the following matrix:

$$C \left[s_1, As_1, \ldots, A^{\gamma_1-1}s_1, \mid s_2, As_2, \ldots, A^{\gamma_2-1}s_2 \right] . \tag{29}$$

As the characteristic equation of the linearized observer error (20) is composed of both linearized observer subsystems, i. e

$$\prod_{j=1}^{2} \left(\lambda^{\gamma_j} + p_{j\gamma_j-1}\lambda^{\gamma_j-1} + \cdots + p_{j1}\lambda + p_{j0} \right) \overset{!}{=} \prod_{j=1}^{2} \prod_{i=1}^{\gamma_j} (\lambda - \lambda_{ji}) . \tag{30}$$

The design parameters p_{ji} in (27) can be determined by an eigenvalue assignment using (30). They are chosen such that the observer error dynamics is faster than the closed loop dynamics. The dimensioning rule (27) is consistent with the rule of Ackermann [1] for linear time-invariable observers.

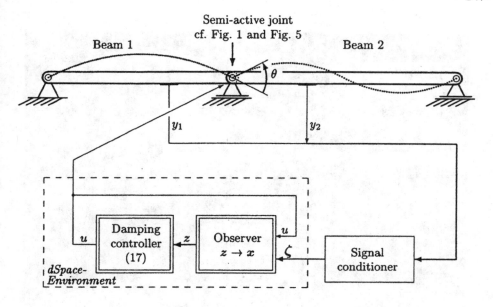

Figure 4. Schematic diagram of the setup of the two-beam experiment.

6. Closed Loop System

The closed loop system with observer can be seen in Fig. 4. The flexural displacements y_1 and y_2 at z_1^o and z_2^o can be measured for beam one and two, respectively. Using (9) and (19b) the vector of measurements ζ in (20) is easily identified. Of course the controller (17) uses the states calculated by the observer (20) i. e $\hat{\dot{\theta}}$ and $\hat{\mu}$ instead of $\dot{\theta}$ and μ, respectively.

7. Experimental Results

The experimental setup of the two-beam experiment is shown in Fig. 6. The hardware realization of the semi-active joint connecting the two beams is shown in Fig. 5. The paramters of the friction model (1) are determined by means of a nonlinear system identification method and, furthermore, the modal parameters of the two-beam system are identified by means of an experimental modal analysis. A discussion of these two topics would be prolix and thus cannot be done here.

To apply an initial deflections to the two-beam system, a 5 kg mass is attached to the left beam via a thread and a mechanism of idler pulleys. This is illustrated clearly in Fig. 6. A sudden release of the load is obtained by burning the thread which leads to a free vibration of the two-beam system.

The proposed feedback (17) and observer (20) are implemented on a dSpace real-time environment, as indicated in Fig. 4 and Fig. 6.

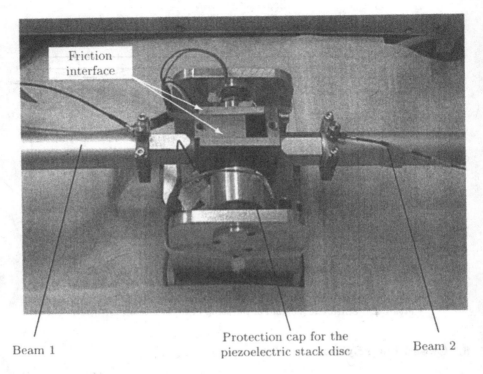

Figure 5. Semi-active rotational joint built into the two-beam experiment, cf. Fig. 6.

The temporal evolutions of the flexural displacements y_1 at $\frac{7}{10}l_1$ are shown in Fig. 7. In Fig. 7a the free vibration of the *passive* system with a constant normal force in the friction interface is shown. Sticking or microslip occurs in the joint. Hence the linear material damping is primarily responsible for the energy dissipation which is fairly small.

The semi-active system is realized by applying the proposed feedback (17) and observer (20) to the experimental setup of the two-beam system. The improvement in vibration damping by controlling the normal force in the friction interface of a semi-active joint is considerable as can be seen clearly in Fig. 7b.

8. Conclusion

The behavior of a system with dry friction damping and state-dependent normal contact force has been discussed. The presented results show that joints with a variable, controlled normal force can substantially improve the performance compared to joints with constant normal force.

A friction model with velocity dynamics has been used to describe the friction phenomena. A local control law based on an instantaneous maximization of energy

Left joint   Beam 1   Semi-active joint cf. Figure 5   Beam 2   Right joint

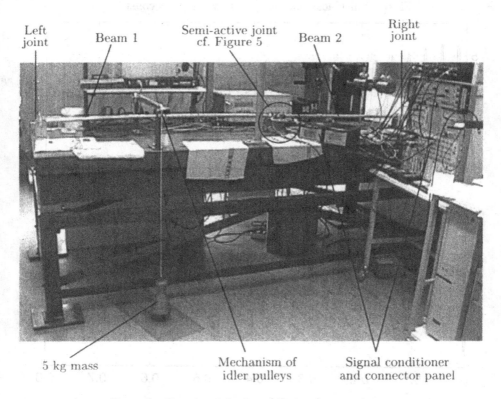

5 kg mass   Mechanism of idler pulleys   Signal conditioner and connector panel

Figure 6.   Experimental setup of the two-beam system.

dissipation rate has been designed. Moreover, the controller minimizes the return of the stored frictional energy to the mechanical system.

For the implementation of the controller, a state of the used friction model is required, which cannot be measured directly. Hence, a nonlinear operating point observer has been designed, allowing the realization of the introduced feedback.

Experimental results demonstrate the efficiency of such semi-active dry friction dampers in flexible structures.

As energy is always dissipated, the semi-active concept is more stable than a fully active one and is insensitive with respect to the spillover problem. It is also sometimes more effective than a purely passive approach.

The method presented can be applied to appropriate truss structure junctures at strategic locations e.g. throughout a large generic space structure. Therefore it can easily be extended to a whole class of large flexible structures to control vibration damping.

250

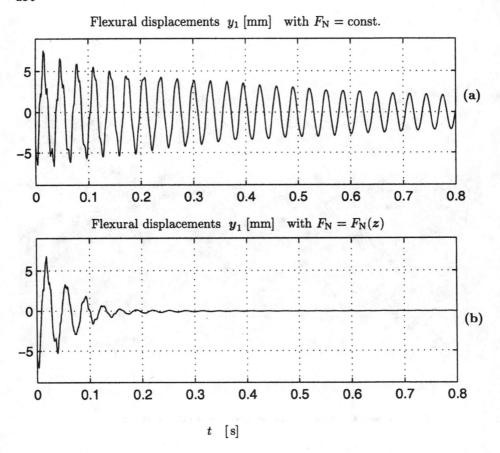

Figure 7. Experimental results of the two-beam experiment. Passive system with a *constant* normal force (a), semi-active system (b).

References

1. Ackermann, J.: 1983, *Abtastregelung*. Springer-Verlag.
2. Armstrong-Hélouvry, B., P. Dupont, and C. Canudas de Wit: 1994, 'A Survey of Models, Analysis Tools and Compensation Methods for the Control of Machines with Friction'. *Automatica* **30**(7), 1083–1138.
3. Birk, J.: 1992, 'Rechnergestützte Analyse und Lösung nichtlinearer Beobachtungsaufgaben'. Ph.D. thesis, VDI-Fortschritt-Berichte Nr.8/294, VDI-Verlag.
4. Birk, J. and M. Zeitz: 1988, 'Extended Luenberger observer for nonlinear multivariable systems'. *Int. J. Control* **47**, 1823–1836.
5. Canudas de Wit, C., H. Olsson, K. Åström, and P. Lischinsky: 1995, 'A New Model for Control of Systems with Friction'. *IEEE Trans. Automat. Control* **40**(3), 419–425.
6. Gaul, L. and R. Nitsche: 2001, 'The Role of Friction in Mechanical Joints'. *Applied Mechanics Reviews* **54**(2), 93–105.

7. Gaul, L., R. Nitsche, and D. Sachau: 1998, 'Semi-Active Vibration Control of Flexible Structures'. In: U. Gabbert (ed.): *Modelling and Control of Adaptive Mechanical Structures*. Düsseldorf, pp. 277–286.

8. Hyland, D., J. Junkins, and R. Longman: 1993, 'Active Control Technology for Large Space Structures'. *AIAA J. of Guidance, Control, and Dynamics* **16**(5), 801–821.

9. Lane, J., A. Ferri, and B. Heck: 1992, 'Vibration Control Using Semi-Active Friction Damping'. In: R. Ibrahim and A. Soom (eds.): *ASME-Publication DE-Vol 49*. pp. 165–171.

10. Nijmeijer, H. and A. van der Schaft: 1990, *Nonlinear Dynamical Control Systems*. John Wiley & Sons, Inc.

11. Nitsche, R.: 2001, 'Semi-Active Control of Friction Damped Systems'. Ph.D. thesis, VDI-Fortschritt-Berichte Nr.8/907, VDI-Verlag.

12. Wolfram, S.: 1991, *Mathematica – A System for Doing Mathematics by Computer*. Addison Wesley.

PIEZOELECTRIC STACK ACTUATOR:

FE-MODELING AND

APPLICATION FOR VIBRATION ISOLATION

UWE STÖBENER* AND LOTHAR GAUL**

*Halcyonics GmbH
Tuchmacherweg 12, 37079 Göttingen, Germany
**Institute A of Mechanics, University of Stuttgart
Pfaffenwaldring 9, 70550 Stuttgart, Germany

Abstract. The piezoelectric stack actuator is one of the most established actuator device for active systems. In this paper a finite element (FE) formulation for a piezoelectric rod is derived. The FE formulation provides a model of stack actuators by which the dynamic behaviour is calculated. To validate the FE formulation and to evaluate the performance of a vibration isolation system with stack actuator an experimental setup is designed and tested. In a first set of experiments a feedforward control concept is used and the results are compared with the FE simulation. Feedforward control requires a reference signal which is not always available in real applications. Therefore a second set of experiments is performed with a proportional feedback controller.

Key words: piezoelectric actuator, FEM, vibration isolation, active mount

1. Introduction

Piezoelectric stack actuators are manufactured by several companies and used in a variety of technical applications, such as for motion control and nano positioning. They are available in different sizes, for different voltages (HV=High Voltage 0-1000V, LV=Low Voltage 0-100V) and for different load levels. Nevertheless the design principle, which is shown in Fig. 1, is identical for all stack actuators. The stack consists of PZT layers and electrodes. The displacement of the complete stack is given by the sum of the displacements of all layers. The typical layer thickness is $0.5mm$ up to $1.0mm$ and the typical actuator displacement is $10\mu m$ up to $150\mu m$. The electrodes are alternating connected to the voltage supply and to the ground. Due to the polarization of the PZT material the piezoelectric effect is sensitive to the orientation of the electric field. To achieve a maximum actuator displacement all PZT layers are build into the stack with an alternating polarization direction. Therefore all layers change their lenght in the same direction.

By neglecting nonlinear material and hysteresis effects the stack can be replaced by a single PZT rod, whose lenght is identical to the thickness of the sum of all

A. Preumont (ed.), Responsive Systems for Active Vibration Control, 253–265.

layers. The driving voltage U_{act} has to be replaced by a 'virtual' voltage U_{act}^*.

$$U_{act}^* = n_{PZT} U_{act} \tag{1}$$

The voltage U_{act}^* is calculated by the product of the number of PZT layers n_{PZT} and the driving voltage U_{act}.

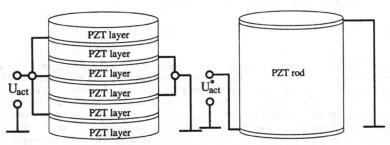

Figure 1. Piezoelectric stack and equivalent rod

2. FE Formulation for a Piezoelectric Rod

Earlier work of FE formulations for piezoelectric material was published by [1, 5]. In the nineties a lot of publications have been written about FE models of piezoelectric structures, e.g. [3, 4, 6] In the next chapters a FE formulation for a piezoelectric actuator in stacked design is derived. As mentioned above the stack is modeled by a rod.

2.1. FINITE ROD ELEMENT

Fig. 2 illustrates a 3-node finite element for a piezoelectric rod. The rod length is L and its cross section is A. The nodal forces are $\mathbf{N}^i$, $\mathbf{N}^j$, $\mathbf{N}^k$, the nodal displacements are $\mathbf{u}^i$, $\mathbf{u}^j$, $\mathbf{u}^k$, the nodal electric charges are q^i, q^j, q^k and the nodal electric potentials are φ^i, φ^j, φ^j. The superscripts i,j,k refer to the node numbers. The forces and displacements are vector quantities with components in cartesian coordinates (e.g. $\mathbf{u}^i = \mathbf{u}_x^i + \mathbf{u}_y^i + \mathbf{u}_z^i$) whereas the charges and potentials are scalar quantities.

2.2. VIRTUAL WORK

The equation of motion is derived from D'Alembert's generalization of principle of virtual displacements

$$\int\limits_V (\rho \delta u_i \ddot{u}_i + \delta \varepsilon_{ij} \sigma_{ij}) \, dV = \int\limits_V \delta u_i b_i dV + \int\limits_A \delta u_i t_i dA, \tag{2}$$

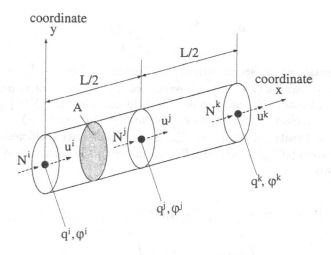

Figure 2. Three node finite element of a piezoelectric rod

with the volume V and the cross section A of the rod. The mass density is expressed by ρ, δu_i is the virtual displacement, $\ddot{u}_i$ is the acceleration, $\delta\varepsilon_{ij}$ is the virtual strain, σ_{ij} is the stress, b_i is the volume force and t_i is the traction. The electric flux conservation is described by

$$\int_V \delta E_i D_i\, dV + \int_V \delta\varphi q_V\, dV + \int_A \delta\varphi q_A\, dA = 0, \tag{3}$$

with D_i as the electric flux or electric displacement. δE_i is the virtual electric field, q_V is the charge per volume, q_A is the charge per area and $\delta\varphi$ is the virtual electric potential. Superposition of these equations leads to

$$\int_V (\rho\delta u_i \ddot{u}_i + \delta\varepsilon_{ij}\sigma_{ij})\, dV - \int_V \delta E_i D_i\, dV =$$

$$\int_V \delta u_i b_i\, dV + \int_A \delta u_i t_i\, dA + \int_V \delta\varphi q_V\, dV + \int_A \delta\varphi q_A\, dA. \tag{4}$$

2.3. CONSTITUTIVE EQUATIONS

In the following step the constitutive equations for the piezoelectric material are introduced. The nonlinear behaviour and hysteresis effects of piezoelectric materials are neglected in this paper. The assumption of linear material behaviour is valid for small electric field strenght and small mechanical stresses since significant hysteresis effects are related to high voltages and large mechanical loads. Under these conditions the following linear set of equations is used

$$\sigma_{ij} = C^E_{ijkl}\varepsilon_{kl} - e_{ijm}E_m, \tag{5}$$

$$D_m = e_{mij}\varepsilon ij + \xi^\sigma_{mn}E_n. \tag{6}$$

In these equations σ_{ij} are the stress coordinates ($[N/m^2]$), C^E_{ijkl} are the elasticity stiffness constants ($[N/m^2]$) measured for a constant electric field, ε_{kl} are the strain coordinates, E_m are the electric field coordinates ($[N/C]$ or $[V/m]$), e_{ijm} are the piezoelectric field stress coefficients ($[C/m^2]$ or $[N/(V m)]$), D_m is the electric flux density ($[C/m^2]$) and ξ^σ_{mn} is the permittivity coefficient measured for constant stress ($[C^2/(N m^2)]$ or $[F/m]$). Inserting the constitutive Eqs. (5,6) to Eq. (4) gives

$$\int_V \rho\delta u_i \ddot{u}_i dV + \int_V \delta\varepsilon_{ij} C^E_{ijkl}\varepsilon_{kl} dV - \int_V \delta\varepsilon_{ij} e_{ijm} E_m dV$$

$$- \int_V \delta E_m e_{mij}\varepsilon_{ij} dV - \int_V \delta E_m \xi^\sigma_{mn} E_n dV =$$

$$\int_V \delta u_i b_i dV + \int_A \delta u_i t_i dA + \int_V \delta\varphi q_V dV + \int_A \delta\varphi q_A dA. \tag{7}$$

Eq. 7 is the coupled electromechanical virtual work principle.

2.4. FINITE ELEMENT APPROXIMATION

The finite element method approximates the displacement field by nodal weighted shape functions. Often Hermitian polynomials are used for this approximation. They are formulated in natural coordinates, as depicted in Fig. 3. For a three node rod element the quadratic polynomials and their derivatives with respect to the natural coordinates are given by

$$H^1 = \tfrac{1}{2}\left(r^2 - r\right), \quad B^1 = r - \frac{1}{2}, \tag{8}$$

$$H^2 = 1 - r^2, \quad B^2 = -2r, \tag{9}$$

$$H^3 = \tfrac{1}{2}\left(r^2 + r\right), \quad B^3 = r + \frac{1}{2}. \tag{10}$$

The polynomials are assembled in the following matrices

$$\mathbf{H} = \begin{bmatrix} \mathbf{H}_m & \mathbf{0} \\ \mathbf{0} & \mathbf{H}_{el} \end{bmatrix} = \begin{bmatrix} H^1\,H^2\,H^3 & 0 & 0 & 0 \\ 0 & 0 & 0 & H^1\,H^2\,H^3 \end{bmatrix}, \tag{11}$$

$$\mathbf{B} = \begin{bmatrix} \mathbf{B}_m & \mathbf{0} \\ \mathbf{0} & -\mathbf{B}_{el} \end{bmatrix} = \begin{bmatrix} B^1\,B^2\,B^3 & 0 & 0 & 0 \\ 0 & 0 & 0 & -B^1\,-B^2\,-B^3 \end{bmatrix}, \tag{12}$$

where the index m stands for mechanical quantities and the index el refers to electrical quantities.

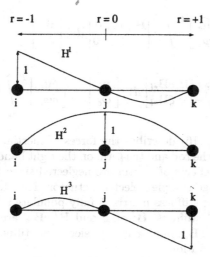

Figure 3. Hermitian polynomials

The continuous displacement field **u** and the continuous electric potential φ are approximated by the matrix of Eq. (11)

$$\mathbf{p} = \begin{bmatrix} \mathbf{u} \\ \varphi \end{bmatrix} = \begin{bmatrix} \mathbf{H}_m & \mathbf{0} \\ \mathbf{0} & \mathbf{H}_{el} \end{bmatrix} \begin{bmatrix} \mathbf{u}_e \\ \varphi_e \end{bmatrix} = \mathbf{H}\mathbf{p}_e. \tag{13}$$

The vector **p** in Eq. (13) contains the continuous displacement and potential field whereas $\mathbf{p}_e$ contains nodal values. The coordinates in the element state vector $\mathbf{p}_e$ are given by

$$\mathbf{p}_e = \begin{bmatrix} \mathbf{u}_e \\ \varphi_e \end{bmatrix} = \begin{bmatrix} \mathbf{u}^i \\ \mathbf{u}^j \\ \mathbf{u}^k \\ \varphi^i \\ \varphi^j \\ \varphi^k \end{bmatrix}. \tag{14}$$

The strain ε_m and the electric field $\mathbf{E}_{el}$ are approximated by the matrix of Eq. (12)

$$\begin{bmatrix} \varepsilon_m \\ \mathbf{E}_{el} \end{bmatrix} = \begin{bmatrix} \mathbf{B}_m & \mathbf{0} \\ \mathbf{0} & -\mathbf{B}_{el} \end{bmatrix} \begin{bmatrix} \mathbf{u}_e \\ \varphi_e \end{bmatrix}. \tag{15}$$

2.5. EQUATION OF MOTION

The equation of motion is obtained by implementing the finite element approximation in the coupled mechanical and electrical virtual work principle Eq. (7). This results in

$$\begin{bmatrix} \delta\mathbf{u}_e \\ \delta\varphi_e \end{bmatrix}^T \int_{V_e} \rho \begin{bmatrix} \mathbf{H}_m^T\mathbf{H}_m & 0 \\ 0 & 0 \end{bmatrix} dV \begin{bmatrix} \ddot{\mathbf{u}}_e \\ \ddot{\varphi}_e \end{bmatrix}$$

$$+ \begin{bmatrix} \delta\mathbf{u}_e \\ \delta\varphi_e \end{bmatrix}^T \int_{V_e} \begin{bmatrix} \mathbf{B}_m^T E\mathbf{B}_m & \mathbf{B}_m^T e\mathbf{B}_{el} \\ \mathbf{B}_{el}^T e\mathbf{B}_m & -\mathbf{B}_{el}^T \xi\mathbf{B}_{el} \end{bmatrix} dV \begin{bmatrix} \mathbf{u}_e \\ \varphi_e \end{bmatrix} = \begin{bmatrix} \delta\mathbf{u}_e \\ \delta\varphi_e \end{bmatrix}^T \int_{A_e} \begin{bmatrix} \mathbf{H}_m^T \\ \mathbf{H}_{el}^T \end{bmatrix} \begin{bmatrix} \mathbf{t} \\ q_s \end{bmatrix} dA.$$

(16)

The first term of Eq. (16) describes the forces of inertia, the second term is related to the stiffness of the rod and the term on the right hand side represents the external load. Electrical effects of inertia are neglected since these effects do not occur in the dynamic range of a piezoelectric actuator. Therefore the mass matrix has just one element. The stiffness matrix is fully populated. The term $\mathbf{B}_m^T E\mathbf{B}_m$ expresses the mechanical stiffness, $\mathbf{B}_m^T e\mathbf{B}_{el}$ and $\mathbf{B}_{el}^T e\mathbf{B}_m$ are electro-mechanical coupling terms and $-\mathbf{B}_{el}^T \xi\mathbf{B}_{el}$ represents the electrical stiffness. Eq. (16) can be rewritten in a compact notation

$$\delta\mathbf{p}_e^T (\mathbf{M}_e\ddot{\mathbf{p}}_e + \mathbf{K}_e\mathbf{p}_e - \mathbf{F}_e) = 0,$$

(17)

with the element mass matrix $\mathbf{M}_e$, the element stiffness matrix $\mathbf{K}_e$ and the element load vector $\mathbf{F}_e$. The load vector includes normal forces and electric charges at both ends of the rod. Volume forces, e.g. the gravity force, are not considered. Since the virtual state vector $\delta\mathbf{p}_e$ is arbitrary the term in parenthesis has to vanish and the equation of motion is deduced from Eq. (17)

$$\mathbf{M}_e\ddot{\mathbf{p}}_e + \mathbf{K}_e\mathbf{p}_e = \mathbf{F}_e.$$

(18)

By comparing Eq. (16) with Eq. (17) the physical meaning of the integrals becomes obvious. The first integral is the consistent mass matrix $\mathbf{M}_e$, the second is the stiffness matrix $\mathbf{K}_e$ and the third is the load vector $\mathbf{F}_e$. Using the shape functions (8,9,10) the matrices can be calculated as

$$\mathbf{M}_e = \frac{\rho A L}{30} \begin{bmatrix} 4 & 2 & -1 & 0 & 0 & 0 \\ 2 & 16 & 2 & 0 & 0 & 0 \\ -1 & 2 & 4 & 0 & 0 & 0 \\ 0 & 0 & 0 & 0 & 0 & 0 \\ 0 & 0 & 0 & 0 & 0 & 0 \\ 0 & 0 & 0 & 0 & 0 & 0 \end{bmatrix},$$

$$\mathbf{K}_e = \frac{A}{12L} \begin{bmatrix} 7E & -8E & 1E & 7e & -8e & 1e \\ -8E & 16E & -8E & -8e & 16e & -8e \\ 1E & -8E & 7E & 1e & -8e & 7e \\ 7e & -8e & 1e & -7\xi & 8\xi & -1\xi \\ -8e & 16e & -8e & 8\xi & -16\xi & 8\xi \\ 1e & -8e & 7e & -1\xi & 8\xi & -7\xi \end{bmatrix}, \quad \mathbf{F}_e = \begin{bmatrix} N^i \\ N^j \\ N^k \\ q^i \\ q^j \\ q^k \end{bmatrix}.$$

(19)

This FE formulation is used in the following chapter.

3. Active Vibration Isolation

A lot of technical systems are equipped with engines, pumps and other aggregates which generate vibrations. Due to mechanical connections the vibration propagates into the supporting structure. The concept of Active Vibration Isolation (AVI), which is described in [2], is used to cut the path of structure borne sound between the vibration source and the receiving structure. In Fig. 4 the vibration source is an electric motor. The passive isolation by resilient mounts is changed into an Active Vibration Isolation system by inserting a piezoelectric stack.

Figure 4. Electric engine and mounting

3.1. SDOF OSCILLATOR

For the presented investigations a simple test structure, a single-degrees-of-freedom (SDOF) oscillator, is used. The oscillator is shown in Fig. 5. Since the vibrating mass 5 is supported by the stack actuator 1 the dynamic reaction force on the foundation 6 can be controlled. This force is measured with the force transducer 4 whereas the acceleration of the mass 5 is sensed with an acceleration pick-up 3. The mass 5 is excited by using a second actuator 2.

On the left hand side of Fig. 6 the actuator 1 is depicted as a three node finite element. By using the derived FE formulation the following equation of motion is obtained

$$
\frac{\rho A L}{30}\begin{bmatrix} 4 & 2 & -1 \\ 2 & 16 & 2 \\ -1 & 2 & 4+\tilde{M} \end{bmatrix}\begin{bmatrix} 0 \\ \ddot{u}_2 \\ \ddot{u}_3 \end{bmatrix}+\frac{AE}{12L}\begin{bmatrix} 7 & -8 & 1 & 7d & -8d & 1d \\ -8 & 16 & -8 & -8d & 16d & -8d \\ 1 & -8 & 7 & 1d & -8d & 7d \end{bmatrix}\begin{bmatrix} 0 \\ \ddot{u}_2 \\ \ddot{u}_3 \\ \phi_1 \\ \phi_2 \\ \phi_3 \end{bmatrix}=\begin{bmatrix} F_1 \\ 0 \\ F_{exc} \end{bmatrix},
$$

$$(20)$$

260

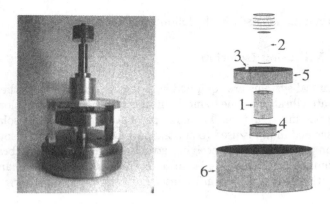

Figure 5. SDOF oscillator: Photo and components

where the part of the equation containing the electric charge is neglected. The piezoelectric field stress coefficient e is replaced by the piezoelectric field strain coefficient d ($[m/V]$ or $[C/N]$) and the lumped mass is $\tilde{M} = 30M/(\rho AL)$.

In order to calculate the required voltage level for the AVI application the FE model is interpreted in terms of a linear spring and an active componente (see right hand side of Fig. 6). The reaction force for the foundation equals the sum of the spring force F_s and the actuator force F_{act}. To cancel the reaction force the actuator force F_{act} has to be equal to the negative spring force $-F_s$.

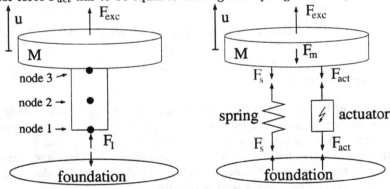

Figure 6. FE model and free body diagram

By neglecting the mass of the actuator the equation of motion is given as a special case of Eqn. (20)

$$M\ddot{u} + K_{act}u + K_{act}d_{33}U^*_{act} = \hat{F}_{exc}e^{i\Omega t}, \qquad (21)$$

where M is the mass of the oscillator, K_{act} is the stiffness of the actuator, d_{33} is the piezoelectric field strain coefficient, U^*_{act} is the 'virtual' voltage applied to the actuator, $\hat{F}_{exc}$ is the excitation force amplitude and Ω is the excitation frequency. The reaction force vanishes if the voltage is set to

$$\hat{U}^*_{act} = \frac{1}{\Omega^2 M d_{33}}\hat{F}_{exc} \rightarrow \hat{U}_{act} = \frac{1}{n_{PZT}\Omega^2 M d_{33}}\hat{F}_{exc}. \qquad (22)$$

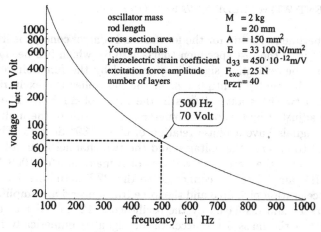

Figure 7. Benchmark example

In Fig. 7 the data for a benchmark example are listed. The graph of Fig. 7 shows the actuator voltage, which is needed for AVI, as a function of the frequency. The dashed lines mark the frequency which is used for the FE calculation and the experiment. The actuator voltage increases rapidly for low frequencies. Therefore a lower limit for the practical application of AVI exists. An upper frequency limit is given by the resonance frequency of the actuator.

3.2. FE SIMULATION

With the values of the benchmark example and the finite element represented by Eq. (20) and Fig. 6 left hand side a finite element simulation is performed. The time response of the force and the displacement are calculated and plotted in Fig. 8. Initially the actuator voltage is set to zero. Without control the reaction force is approximately equal to the excitation force. At $t = 0.05$ s the actuator voltage is set to the precalculated value (see Fig. 7). Therefore the reaction force drops from 25 N to 3 N and the associated displacement increases from $0.5\mu m$ to $1.2\mu m$.

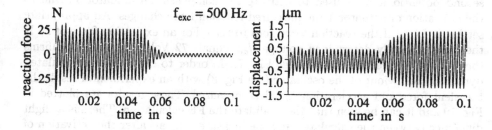

Figure 8. Results of FE-calculation

262

3.3. EXPERIMENT WITH FEEDFORWARD CONTROL

In Fig. 9 the experimental setup for the feedforward control concept is illustrated. In the center of Fig. 9 the test specimen (parts 1 to 6), which has been described above, is shown. The excitation signal is generated by the function generator 7 and then feed in the analog multiplier 8. In the multiplier the excitation signal is multiplied with two constant voltages in the range of ±1 Volt. The constant voltages can be adjusted by two potentiometers. As a result of the multiplication the two output signals have a phase relation of 0^0 or 180^0 degrees. The display module 9 is used to observe the voltages and the two channel high voltage (HV) amplifier 10 generates the necessary voltages of a maximum of 1000 Volt. The outputs of the HV amplifiers are connected to the PZT actuators 1 and 2. The signals of the acceleration pick up 3 and the force transducer 4 are amplified by the charge amplifiers 11 and displayed by the oscilloscope 12. In order to determine the displacement of the mass 5 the acceleration signal is numerically integrated with respect to time.

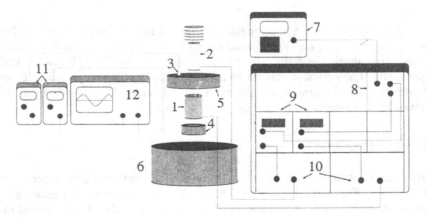

Figure 9. Experimental setup for feedforward control

The test procedure starts with the adjustment of the excitation force by tuning the potentiometer and observing the display of the oscilloscope. Then the second potentiometer is used to activate the compensation actuator 1. Due to the activation of actuator 1 the measured reaction force changes. An appropriate setting is found if the reaction force goes to zero. For an excitation force of 25 N the experimentally determined voltage U_{act} equals 72 Volt. This value confirms the calculation presented above (see Fig. 7). In order to compare the calculated dynamical behaviour of the oscillator (see Fig. 8) with an experimental result the transient force and acceleration signal is measured. These results are plotted in Fig. 10 and it can be seen that they validate the FE calculation. There is a slight difference between the calculated and the measured signals after the activation of the compensation during the transient periode. This difference is a result of the dynamics of the used HV amplifier which is not a part of the FE model.

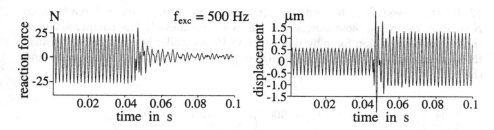

Figure 10. Results of the experiment with feedforward control

3.4. EXPERIMENT WITH FEEDBACK CONTROL

Since feedforward control requires a reference signal of the excitation its application is limited to technical systems where such a reference signal is available. If no reference signal is available the feedforward control concept has to be replaced by a feedback controller. The experimental setup depicted in Fig. 9 can be easily converted into a setup for feedback control. Therefore the function generator 7 is directly connected to the display module and the HV amplifier. The excitation signal is used to drive the excitation actuator 2 only. In contrast to the previous setup the measured signal of the reaction force, which is amplified within the charge amplifier 11, is feed in the analog multiplier 8. The potentiometer of the multiplier is used to adjust the proportional feedback factor and the output of the multiplier is connected with the display module and the second channel of the HV amplifier. With this setup a closed loop (force transducer → charge amplifier → multiplier → display module → HV amplifier → compensation actuator) is built.

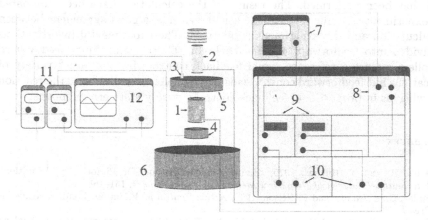

Figure 11. Experimental setup for feedback control

Because of the closed loop the feedback controller is not always stable. Therefore the feedback factor is limited. In this experiment the factor is evaluated by increasing the factor till the limit of stability is reached. The results are plotted in Fig.

12. It can be seen that the feedback controller reduces the reaction force from 25 N to 13 N but a complete cancelation can not be achieved. The compensation of the remaining reaction force by the use of an PI controller is not possible because the dynamics of the disturbance is higher than the dynamics of the system and the controller.

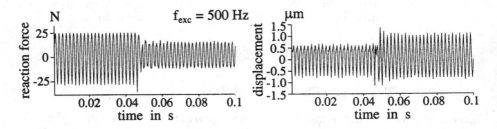

Figure 12. Results of the experiment with feedback control

4. Summary and Conclusions

A finite element formulation for piezoelectric rods has been derived. The formulation includes the electro mechanical coupling of the piezoelectric material but it neglects nonlinear and hysteretic effects which occur for high voltages and high stresses. To validate the FE formulation the concept of AVI has been presented and the calculation of a SDOF oscillator equipped with a piezoelectric stack actuator has been performed. The results of the calculation have been compared with experimental results and it has been shown that a good agreement between the calculation and the experiment is obtained. The experimental investigations are divided into feedforward and feedback control strategies. Since feedforward controllers require a reference signal feedback control allows a larger variety of applications but feedforward control leads to a complete cancelation of the reaction force whereas in proportional feedback control a steady state error remains.

References

1. Allik, H. and T. Hughes: 1970, 'Finite Element Method for Piezoelectric Vibration'. *International Journal of Numerical Methods in Engineering* **2**, 151–158.
2. Fuller, C., S. Elliott, and P. Nelson: 1996, *Active Control of Vibration*. London: Academic Press.
3. Gaudenzi, P. and K. Bathe: 1995, 'An Iterative Finite Element Procedure for the Analysis of Piezoelectric Continua'. *Journal of Intelligent Material Systems and Structures* **2**(6), 266–273.
4. Ghandi, K. and N. Hagood: 1996, 'Nonlinear Finite Element Modeling of Phase Transitions in Electro-Mechanically Coupled Material'. In: *Proceedings of SPIE*, Vol. 2715. pp. 121–140.
5. Naillon, M., R. Coursant, and F. Besnier: 1983, 'Analysis of Piezoelectric Structures by the Finite Element Method'. *ACTA Electronica* **4**(25), 341–361.

6. Shieh, R.: 1993, 'Finite Element Formulation for Dynamic Response Analysis of Multiaxial Active 3-D Piezoelectric Beam Element Structures'. In: *Collection of Technical Papers Proceedings of the AIAA/ASME Structures, Structural Dynamics and Materials Conference,* Vol. 6. pp. 3250–3258.

ACTIVE VIBRATION CONTROL OF A CAR BODY BASED ON EXPERIMENTALLY EVALUATED MODAL PARAMETERS

UWE STÖBENER* AND LOTHAR GAUL**
*Halcyonics GmbH
Tuchmacherweg 12, 37079 Göttingen, Germany
**Institute A of Mechanics, University of Stuttgart
Pfaffenwaldring 9, 70550 Stuttgart, Germany

Abstract. Active vibration control has been successfully tested for structures with simple geom-etry, such as beams and plates, by using modal controllers. Since the dynamical behaviour of a variety of mechanical structures can be expressed in terms of modal parameters, the application of modal control concepts can be extended to structures with more complex geometries. For such structures the evaluation of modal parameters from numerical calculations of local modes is complicated because the results strongly depend on proper boundary conditions of the truncated structure. Therefore the modal data are identified by an experimental modal analysis. The trans-formation of the experimentally evaluated mode shapes into a closed analytical formulation and the extraction of modal input and output factors for sensors and actuators connect experimental modal analysis and modal control theory. The implementation of the input and output factors into a modal state space formulation results in a modal filter for the point sensor array and a retransformation filter for the segmented actuator patches. In this study PVDF foil is used for sensors and actuators. The modal controller is implemented on a digital controller board and experimental tests with the floor panel and center panel of a car body are carried out to validate the proposed concept.

Key words: active vibration control, modal analysis, modal control, car body, PVDF sensors, PVDF actuators

Nomenclature

ω_r	natural frequency of the rth mode
ζ_r	viscous damping ratio of the rth mode
ϕ_r	mass normalized mode shape of the rth mode
q_r	modal coordinate of the rth mode
x, y	cartesian coordinates
F	mechanical load
f_r	modal mechanical load of the rth mode
$\{u\}, u$	input vector, actuator voltage

A. Preumont (ed.), Responsive Systems for Active Vibration Control, 267–284.

$[A]$	modal system matrix
$[B]$, b_r	modal input matrix, modal input factor
$[C]$, c_r	modal output matrix, modal output factor
$[G]$	gain matrix
$\{x\}$	state variable vector
$\{y\}$	sensor output vector
$\{R\}$	control vector
$\hat{G}_{FF}$, $\hat{G}_{XX}$	auto power spectrum
$\hat{G}_{FX}$	cross power spectrum
γ^2	coherence

1. Introduction

Structural vibration of a car body is caused by the engine, the wheels, the chassis and the airflow. The vibration propagates as structure borne sound from its source to the entire car body. Since car bodies are made of metal with low material damping the sound propagates with high efficiency. Especially at the roof, the doors and the floor panels vibration leads to sound radiation into the passenger compartment. The sound radiation can be reduced by covering the surface of the structure with foam mats or by active vibration control.

In the subject area of vibration control modal concepts have been successfully tested for the reduction of plate vibrations [11, 2, 20]. The determination of modal parameters for plates can be carried out by FE calculations or by analytical solutions of the plate equation. For structures with a complex geometrie, such as a car body, an analytical solution does not exist and even an FE calculation is complicated and expensive. As alternative an experimental modal analysis can be used to extract modal parameters, such as eigenfrequencies ω_r and mode shapes ϕ_r, from measured data. The mode shapes contains the necessary information for the sensor and actuator layout and placement. The problem of actuator and sensor placement is discussed in [12, 18, 13, 21]. After the sensor and actuator positions and dimensions are fixed, modal input and output matrices $[C]$, $[B]$ can be calculated. These matrices are derived in the form of a modal state space formulation which is a standard in modern control theory. Errors caused by slight non linearities of the structure or the curve fitting procedure of the modal analysis results in a discontinuity for the mode shape. Therefore a direct evaluation of the input and output matrices is not possible. The mode shapes have to be approximated by polynomials or cubic splines. Since the input and output matrices are used to create a real time modal filter the control law is formulated in modal coordinates q. The gain factors of the feedback loop correspond directly to the elements $-\omega_r^2$ and $-\zeta_r^2$ of the system matrix $[A]$ and can be determined by pole placement.

2. Test Specimen

The test specimen is a modern roadster car body, shown in Fig. 1. As mentioned above, the roof, the doors and the floor panel are the most important parts of the car body where sound radiation occurs into the passenger cell. Since this study is carried out for investigating the principles of active vibration control the realization of the mounting of the sensor and actuators should be as easy as possible. Therefore the floor panel has been selected for the experimental tests. The car body is supported by a heavy and very stiff steel frame. The high stiffness of the frame guarantees that for low frequencies the dynamics of the car body is not effected by the frame.

Figure 1.   Car body with supporting frame

3. Experimental Modal Analysis

Because of the rapid advancement of measurement and analysis techniques in the 1960s experimental modal analysis has become a powerful tool for the evaluation of modal parameters. Theory and the experimental application of modal analysis can be found in textbooks such as [4, 10, 15].

3.1. DEFINITION OF GEOMETRY

The photo in Fig. 2 shows an area of the floor panel and the center panel from inside of the car body at side of the co driver. This area is of special interest because maximum vibration amplitudes occur by the different exciting sources. In a first attempt this local influenced area was measured and modeled rather than considering the whole interiour of the car. The floor panel has a lenght of approximately 45 cm and a width of 20 cm. The dimensions of the center panel are 30 cm of lenght and 18 cm of height. The dark areas are the PVDF sheets which were bonded to the structure after the modal analysis has been carried out.

The geometry is approximated by a grid of 336 points. The grid points are marked on the car body. The resolution of the grid is 2x2 cm.

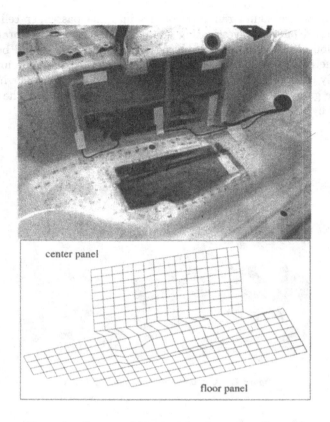

center panel

floor panel

Figure 2. Center and floor panel and corresponding grid

3.2. MEASUREMENT AND CURVE FITTING

Since experimental modal analysis requires the measurement of FRFs an appropriate setup has to be installed. This setup is illustrated in Fig. 3. The structure is excited by an electrodynamical shaker which is connected to the car body by a stinger. The stinger is fixed on the floor panel at the mounting of the gear box. The excitation force is sensed with a piezoelectric force transducer and the response is detected by piezoelectric acceleration pick-ups. Since the used FFT analyzer is equipped with only eight channels the measurement is repeated several times by changing the positions of the acceleration pick-ups.

The influence of noise and uncorrelated signals is evaluated with the coherence function

$$\gamma^2 = \frac{\left|\hat{G}_{FX}\right|^2}{\left(\hat{G}_{FF} \cdot \hat{G}_{XX}\right)}, \tag{1}$$

which can vary between 0 and 1. $\hat{G}_{FX}$ is the averaged cross power spectrum of the force and the acceleration signal and $\hat{G}_{FF}$, $\hat{G}_{XX}$ are the auto power spectrums. Good measurement results can be obtained for a coherence between 0.9 and 1. For the used setup these values have been achieved within a frequency range between 100 Hz and 600 Hz except for the anti resonances. Therefore this frequency range is selected for the acquisition of the FRFs by adjusting the sampling frequency of the FFT analyzer.

Since a continuous random noise signal is used for the excitation the measured data are windowed by a Hanning function, averaged and stored into files.

After the measurement is completed a curve fit process is used to obtain the modal parameters from the FRFs. The curve fitting is a complicated and sensible procedure. Therefore different methods were used, such as MDOF circle curve fitting, polynomial curve fitting and globals and residues. The last mentioned method was prefered because it leads to repeatable and satisfying results.

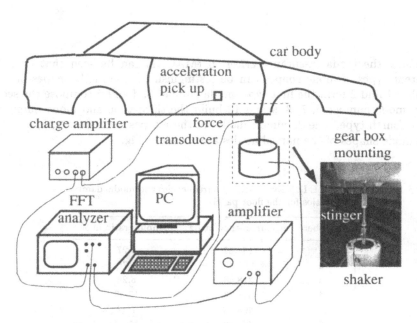

Figure 3. Experimental setup for the modal analysis

3.3. RESULTS OF THE EXPERIMENTAL MODAL ANALYSIS

The results of the experimental modal analysis are split into those for the center panel and the floor panel. In the frequency range from 195 Hz to 505 Hz eleven modes are evalutated for the center panel and eight modes are found for the floor panel. The eigenfrequencies and modal damping ratios are listed in Table I for the center panel and in Table II for the floor panel.

TABLE I. Evaluated eigenfrequencies and modal damping ratios for the center panel

number	eigenfrequency in Hz	damping ratio in %
1	196.56	1.10
2	220.26	7.98
3	281.65	1.00
4	286.77	2.24
5	310.27	7.35
6	411.50	1.51
7	436.29	1.77
8	457.02	1.45
9	464.21	1.34
10	477.34	2.58
11	500.31	1.25

Using the modal assurance criterion (MAC) it can be seen that only four different types of mode shapes can be distinguished. The mode shapes of mode number 1 and 2 form the first type, mode number 3, 4 and 5 compose the second type, mode number 6, 7, 8, 9 and 10 build the third type and mode number 11 is the fourth type. The dominant modes of these types are modes 1, 3, 8 and 11. The mode shapes of the named modes are shown in Fig. 4.

TABLE II. Evaluated eigenfrequencies and modal damping ratios for the floor panel

number	eigenfrequency in Hz	damping ratio in %
1	184.62	4.27
2	203.48	1.76
3	318.58	6.28
4	371.73	13.35
5	403.28	7.32
6	464.19	11.14
7	486.39	2.60
8	501.56	1.80

By neglecting the deformation at the border of the floor panel and focusing the interest to the center area the mode shapes of the floor panel can be classified as one type. All modes exhibit a deflection at the center area as it can be seen in Fig. 4.

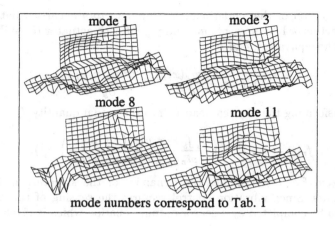

Figure 4. Evaluated mode shapes

4. Modal State Space Formulation

In this section a brief introduction to the modal state space formulation is presented. More informations about this topic can be found in [8].

For continuous systems the number of DOFs is infinity. This is a problem because control requires a reduction of the DOFs. The reduction can be achieved by modal truncation. Therefore the equation of motion is transformed in modal coordinates q_r

$$\ddot{q}_r + 2\zeta_r\omega_r\dot{q}_r + \omega_r^2 q_r = f_r, \tag{2}$$

where ζ_r is the modal damping ratio and f_r is the modal load. Eq. (2) characterizes the plate motion for a single mode normalized by the modal mass. The identity $\dot{q}_r = \dot{q}_r$ and $\ddot{q}_r = -2\zeta_r\omega_r\dot{q}_r - \omega_r^2 q_r$ are used to transform the second order DEq. (2) into a set of first order DEqs.

$$\left\{ \begin{array}{c} \dot{q}_r \\ \ddot{q}_r \end{array} \right\} = \left[\begin{array}{cc} 0 & 1 \\ -\omega_r^2 & -2\zeta_r\omega_r \end{array} \right] \left\{ \begin{array}{c} q_r \\ \dot{q}_r \end{array} \right\} + \left\{ \begin{array}{c} 0 \\ b_r \end{array} \right\} u \tag{3}$$

The modal load f_r has been replaced by the control input b_r and the driving actuator voltage u. It has to be noticed that the column vector $\{B\}$ becomes a matrix $[B]$ and the scalar u is replaced by a column vector $\{u\}$ if several control inputs are used. By collecting the modal coordinates in the state space vector $\{x\}$, the eigenfrequencies ω_r and the modal damping ratios ζ_r in the system matrix $[A]$, the modal input factor b_r of each actuator in the actuator input matix $[B]$ and the actuator voltages u in the input vector $\{u\}$ Eq. (3) can be written for a specified number of modes as

$$\{\dot{x}\} = [A]\{x\} + [B]\{u\}. \tag{4}$$

The calculation of the modal actuator input factors is carried out by determining the actuator line moments m_x and m_y for each actuator (described in [6]), calculating the equivalent load F

$$F(x,y) = \frac{\partial^2 m_x(x,y)}{\partial x^2} + \frac{\partial^2 m_y(x,y)}{\partial y^2}. \tag{5}$$

and transforming the physical load F into the modal quantity f_r

$$f_r(t) = b_r u(t) = \left(\frac{\int_a \int_b F(x,y)\phi_r(x,y)dydx}{\rho t_p \int_a \int_b \phi_r^2(x,y)dydx} \right) u(t). \tag{6}$$

The system Eq. (4) describes the dynamics of the structure including the actuator inputs. Since vibration control requires the sensing of the motion the model has to be completed by a sensor output equation which is given by

$$\{y\} = [C]\{x\}. \tag{7}$$

The output vector $\{y\}$ contains the sensor signals and $[C]$ is the $2mxn$ output matrix with m as the number of filtered modes and n as the number of sensors. Using the concept of point sensors allocated in an array the number of sensors has to be equal to the number of filtered modes ($m = n$). PVDF point sensors are defined as thin films with small size compared to the wavelenght of structural deformation. The sensor outputs are electrically coupled in order to sample an output equivalent to the output of a distributed sensor. The modal sensor output factors c_r are calculated by

$$c_r = -\frac{t_s}{\epsilon_{33}^S} \frac{(t_p + t_s)}{2A_s} \left(e_{31} \int_{x_1}^{x_2} \int_{y_1}^{y_2} \frac{\partial^2 \phi_r}{\partial x^2} dydx + e_{32} \int_{x_1}^{x_2} \int_{y_1}^{y_2} \frac{\partial^2 \phi_r}{\partial y^2} dydx \right), \tag{8}$$

where t_s is the sensor thickness, A_s is the sensor area, ϵ_{33}^S is the dielectric constant, e_{31}, e_{32} are the piezoelectric charge constants and x_1, x_2, y_1 and y_2 are the sensor coordinates.

5. Input and Output Matrices from Experimental Data

5.1. INTERPOLATION OF MEASURED MODE SHAPES

The elements of the input matrix $[B]$ are calculated by Eq. (6). Therefore the mode shape ψ_r is needed in an analytical closed form. The experimental results of the mode shapes are available in form of a table. Since small nonlinearities of the structure and curve fit errors lead to local deviations from the experimentally evaluated mode shapes a direct use of the table is not possible. The limitation of modal testing with respect to the linearity of structures is discussed in [5]. By interpolating the measured mode shapes with polynomials the local deviations are smoothed and a closed form is generated. The interpolating polynomial of degree

$N - 1$ through the N points $z_{x1} = f(x_1)$, $z_{x2} = f(x_2)$, ... , $z_{x3} = f(x_N)$ is given explicitly by Lagrange's interpolation formula,

$$\phi_{1_r}(x) = \frac{(x - x_2)(x - x_3) \cdots (x - x_N)}{(x_1 - x_2)(x_1 - x_3) \cdots (x_1 - x_N)} z_{x1} \tag{9}$$
$$+ \frac{(x - x_1)(x - x_3) \cdots (x - x_N)}{(x_2 - x_1)(x_2 - x_3) \cdots (x_2 - x_N)} z_{x2} + \cdots$$
$$+ \frac{(x - x_2)(x - x_3) \cdots (x - x_{N-1})}{(x_N - x_1)(x_N - x_2) \cdots (x_N - x_{N-1})} z_{xN}$$

There are N terms, each a polynomial of degree $N - 1$ and each constructed to be zero at all of the x_{xj} except one, at which it is constructed to be z_{xj}. The same interpolation is used in y-direction.

$$\phi_{2_r}(y) = \frac{(y - y_2)(y - y_3) \cdots (y - y_N)}{(y_1 - y_2)(y_1 - y_3) \cdots (y_1 - y_N)} z_{y1} \tag{10}$$
$$+ \frac{(y - y_1)(y - y_3) \cdots (y - y_N)}{(y_2 - y_1)(y_2 - y_3) \cdots (y_2 - y_N)} z_{y2} + \cdots$$
$$+ \frac{(y - y_2)(y - y_3) \cdots (y - y_{N-1})}{(y_N - y_1)(y_N - y_2) \cdots (y_N - y_{N-1})} z_{yN}$$

The displacement field in x- and y-direction is composed as the product of the polynomials $\phi_{1_r}(x)$ and $\phi_{2_r}(y)$

$$\phi_r(x, y) = \phi_{1_r}(x)\phi_{2_r}(y) \tag{11}$$

The sampling points x_j and y_j have to be selected from the measured data. This selection has to be made carefully. In practice the sampling points are chosen by regarding the mesh plots of the mode shapes for each mode (see Fig. 4). The lines of the mesh which show the maximum displacement and no local deviations are the reference lines. For each coordinate x and y one reference line has to be defined. The sampling points are picked from the reference lines and set in the Eqs. (9) and (10). The necessary number of sampling points depends on the complexity of the mode shape. Shapes with a simple deflection, e.g. one homogeneous curvature, require only three or four sampling points. In Fig. 5 the measured and interpolated mode shapes of the center panel are depicted.

5.2. DETERMINATION OF THE ELEMENTS OF THE INPUT MATRIX

The determination of the input matrix starts with the calculation of the integral in the numerator of Eq. (6). Since the mode shapes are interpolated by polynomials it would be possible to solve the integral analytically. The analytical solution requires the explicit calculation of the coefficients of the polynomials. The accuracy of the analytical solution depends on the quality of the interpolation and finally on the

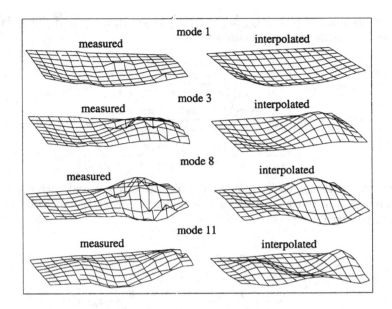

Figure 5. Measured and interpolated mode shapes of the center panel

quality of the measured mode shapes. Therefore the solution can be carried out by a discrete approximation method as well.

This solution uses the grid created for the experimental modal analysis. The contour of an actuator is approximated by choosing the appropriate grid elements and grid points. The points x_{a1}, x_{a2} and y_{a1}, y_{a2} are the grid points in x- and y-direction at the edges of the actuator and x_{a0} and x_{a3} are adjacent points to x_{a1}, x_{a2}. Each grid element constists of four points $p1$, $p2$, $p3$ and $p4$ and the lenght and the width of the grid is given by Δx and Δy.

For a mono orientated PVDF foil the actuator is replaced by the line moment m_x at the boundary of the PVDF sheet. The line moment is evaluated by the bending stiffness B, the slope of the strain distribution in z direction C_x and the Heaviside function which is used to determine the position of the actuator. Therefore the integral term in Eq. (6) follows as

$$I = \int\limits_a \int\limits_b F(x,y)\phi_r(x,y)dydx = BC_x \left[\frac{d\phi_{1_r}(x)}{dx}\right]_{x_{a1}}^{x_{a2}} \int\limits_{y_{a1}}^{y_{a2}} \phi_{2_r}(y)dy$$

$$\approx BC_x \left(\frac{\phi_{1_r}(x_{a3}) - \phi_{1_r}(x_{a2})}{\Delta x} - \frac{\phi_{1_r}(x_{a1}) - \phi_{1_r}(x_{a0})}{\Delta x}\right)$$

$$\cdot \sum_{j=1}^{L-1} \left(\frac{\phi_{2_r}(y_{j+1}) + \phi_{2_r}(y_j)}{2}\right) \Delta y \tag{12}$$

The second step is the determination of the so-called norm Λ which is identical to the integral in the denominator of Eq. (6). The approximation of the integral by a summation over the number M of grid elements leads to

$$\Lambda = \int_a \int_b \phi_r^2(x,y)dydx \approx \sum_{j=1}^{M} \left(\frac{\phi_r(p1_j) + \phi_r(p2_j) + \phi_r(p3_j) + \phi_r(p4_j)}{4} \right)^2 \Delta y \Delta x. \quad (13)$$

Finally the input factor is calculated by inserting the results of Eq. (12) and (13) into Eq. (6)

$$b_r = \frac{I}{\rho t_p \Lambda}. \quad (14)$$

5.3. DETERMINATION OF THE ELEMENTS OF THE OUTPUT MATRIX

For determinating the output matrix the integrals of Eq. (8) have to be solved. The solution can be found by calculating the coefficients of the polynomials of Eq. (9) and (10), taking the second derivatives of the polynomials with respect to x and y and calculating the integrals over the sensor area. An alternative solution can be created by transforming the integrals of Eq. (8)

$$I_1 = \int_{x_1}^{x_2} \int_{y_1}^{y_2} \frac{\partial^2 \phi_r(x,y)}{\partial x^2} dydx = \int_{y_1}^{y_2} \left[\frac{\partial \phi_r(x,y)}{\partial x} \right]_{x_1}^{x_2} dy, \quad (15)$$

$$I_2 = \int_{x_1}^{x_2} \int_{y_1}^{y_2} \frac{\partial^2 \phi_r(x,y)}{\partial y^2} dydx = \int_{x_1}^{x_2} \left[\frac{\partial \phi_r(x,y)}{\partial y} \right]_{y_1}^{y_2} dx, \quad (16)$$

approximating the derivatives by the gradients and solving the integrals by using the trapezoidal rule

$$I_1 \approx \left(\frac{\phi_r(x_3) - \phi_r(x_2)}{x_3 - x_2} - \frac{\phi_r(x_1) - \phi_r(x_0)}{x_1 - x_0} \right) (y_2 - y_1), \quad (17)$$

$$I_2 \approx \left(\frac{\phi_r(y_3) - \phi_r(y_2)}{y_3 - y_2} - \frac{\phi_r(y_1) - \phi_r(y_0)}{y_1 - y_0} \right) (x_2 - x_1). \quad (18)$$

This approximation can only be used if the dimensions of the point sensors are small related to the wavelenght of the mode shape. Finally the input factors are calculated by multiplying the integrals I_1 and I_2 with the constants described in Eq. (8)

$$c_r = -\frac{t_s}{\epsilon_{33}^S} \frac{(t_p + t_s)}{2A_s} (e_{31}I_1 + e_{32}I_2). \quad (19)$$

6. Modal Controller Concept

Modal controller concepts, as introduced in [16] for process control, [9] for control of structural dynamics and [14] for independent modal space control, have become more important since modern piezoelectric actuator and sensor technologies enable distributed control and sensing of structural motion. An overview of several control strategies is presented in [7] and more detailed aspects are discussed in [1, 17, 3]. Since modal control concepts require a modal truncation residual modes may excited by control inputs, which is known as the spillover problem. As mentioned in [19] robust control enables the reduction of spillover effects caused by neglecting high frequency modes.

In Fig. 6 the block diagram of the system and of the controller is shown. The system is represented by the system matrix $[A]$, the actuator input matrix $[B]$ and the sensor output matrix $[C_v]$. The index v indicates that the sensor signals are proportional to the velocity of the plate motion.

The matrix $[M]$ is a filter matrix which transforms the desired displacement w into appropriate actuator voltages $\{u_s\}$. Since the object of vibration control is to cancel the vibration amplitudes the set points of the displacement w and the actuator voltages $\{u_s\}$ are zero.

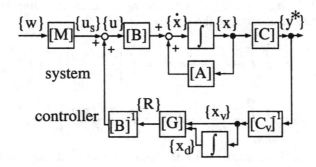

Figure 6. Block diagram of modal filtered feedback

The output vector $\{y^*\}$ contains the sensor signals of n PVDF point sensors which are located on the surface of the structure. Since the positions, the geometry and the material constants of the sensors are known, the output matrix can be calculated by using Eq. (19). Assuming that the output matrix $[C_v]$ is non-singular, the inverse matrix $[C_v]^{-1}$ exists. The inverse matrix $[C_v]^{-1}$ becomes the so-called mode analyzer by entering the output vector $\{y^*\}$ as input data. Therefore a system of linear equations is created (Eq. (20)) and the solution of this system is equal to the vector of modal velocities $\{x_v\}$. This type of filtering state space quantities by using an output vector is known as output feedback

$$
\left\{ \begin{array}{c} \dot{q}_1 \\ \dot{q}_2 \\ \vdots \\ \dot{q}_n \end{array} \right\} =
\left[\begin{array}{cccc} c_{11} & c_{12} & \cdots & c_{1n} \\ c_{21} & c_{22} & \cdots & c_{2n} \\ \vdots & \vdots & \ddots & \vdots \\ c_{n1} & c_{n2} & \cdots & c_{nn} \end{array} \right]^{-1}
\left\{ \begin{array}{c} y_1^* \\ y_2^* \\ \vdots \\ y_n^* \end{array} \right\},
\tag{20}
$$

$$\{x\} = \left\{ \begin{array}{c} 0 \\ \vdots \\ 0 \\ \dot{q}_1 \\ \vdots \\ \dot{q}_n \end{array} \right\} + \left\{ \begin{array}{c} \int \dot{q}_1 dt \\ \vdots \\ \int \dot{q}_n dt \\ 0 \\ \vdots \\ 0 \end{array} \right\} = \left\{ \begin{array}{c} q_1 \\ \vdots \\ q_n \\ \dot{q}_1 \\ \vdots \\ \dot{q}_n \end{array} \right\} . \tag{21}$$

Since the modal state space vector $\{x\}$ contains modal displacements and modal velocities the second part of the state space vector has to be evaluated by integrating the filtered modal velocities. The summation of the vector of modal velocities $\{x_v\}$ and the modal displacements $\{x_d\}$ results in the complete modal state space vector $\{x\}$ as formulated in Eq. (21).

The controller is defined by proportional feedback of the modal displacements and modal velocities. The gain matrix $[G]$ is composed of two diagonal matrices. The first diagonal matrix contains the gain factors to feedback the modal displacements and the second diagonal matrix is defined by the gain factors for the feedback of the modal velocities. Control theory has created several methods for the determination of the gain factors, e.g. the linear quadratic regulator (LQR). Since the gain factors have a direct influence on the poles of the closed loop system they can be chosen to change the modal damping ratios and the eigenfrequencies of the controlled system in a specified way. The pole requirement and the correlated calculation of gain factors is called pole placement. The multiplication of the gain matrix $[G]$ and the state space vector $\{x\}$ results in the control vector $\{R\}$.

$$\left\{ \begin{array}{c} r_1 \\ \vdots \\ r_n \end{array} \right\} = \left[\begin{array}{cccc} g_{q1} & & g_{\dot{q}1} & \\ & \ddots & & \ddots \\ & & g_{qn} & & g_{\dot{q}n} \end{array} \right] \left\{ \begin{array}{c} q_1 \\ \vdots \\ q_n \\ \dot{q}_1 \\ \vdots \\ \dot{q}_n \end{array} \right\} . \tag{22}$$

The elements of the control vector $\{R\}$ are related to the filtered modes. Therefore the controller works within the modal state space and a retransformation of the control vector $\{R\}$ into the physical space is required. The retransformation is carried out by the so-called mode synthesizer which is defined by the inverse modal actuator matrix $[B]^{-1}$. The multiplication of the inverse modal actuator matrix $[B]^{-1}$ and the control vector $\{R\}$ leads to the input vector $\{u\}$. The input vector $\{u\}$ has as many elements as actuators are available.

$$\left\{ \begin{array}{c} u_1 \\ u_2 \\ \vdots \\ u_m \end{array} \right\} = \left[\begin{array}{cccc} b_{11} & b_{12} & \cdots & b_{1n} \\ b_{21} & b_{22} & \cdots & b_{2n} \\ \vdots & \vdots & \ddots & \vdots \\ b_{m1} & b_{m2} & \cdots & b_{mn} \end{array} \right]^{-1} \left\{ \begin{array}{c} r_1 \\ r_2 \\ \vdots \\ r_n \end{array} \right\} . \tag{23}$$

7. Experimental Test

In Fig. 7 the experimental setup is depicted. The setup consists of two independent circuits. One circuit is used for the vibration control of the car body whereas the other circuit is used to monitore the resulting vibration.

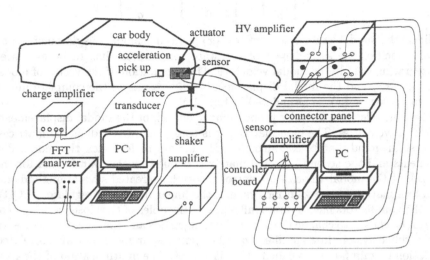

Figure 7. Experimental setup for the controlled structure

The main part of the controller circuit is a four channel digital signal processor (DSP). After the control algorithm is generated on the PC and downloaded to the DSP, the DSP works standalone with a sampling frequency of 20 kHz. The input channels of the DSP are connected via a sensor amplifier to four PVDF sensors. These sensors are placed on the back side of the center panel as depicted in Fig. 9. The sensor positions correspond to the locations of maximum displacement of mode 1, 3, 8 and 11. The sensor areas are 20x20 mm and the sensor thickness is 110 μm. The PVDF material is bi-orientated ($e_{31} = e_{32} = 4$ pC/N). Since the actuator on the back side of the center panel covers almost the complete area of the panel the sensors are bonded on the surface of the actuator. Nevertheless there are no collocated sensor/actuator pairs. The collocation of sensors and actuators is not required since the sensor signals are not fed back directly to the actuators but they are filtered to achieve modal displacements and modal velocities. In order to protect the sensor signals against the electro magnetic field of the actuator, the sensors are covered by an aluminum foil and connected with shielded cables to the signal amplifier. The amplification factor of the signal amplifier is set to 100 and it follows that the peaks of sensor signals are in the range of 3 to 5 volts. The controller board converts the signals into digital codes, calculates the modal velocities, integrates the modal velocities, collects the velocities and the displacements into the state vector, multiplies the state vector with the gain matrix, calculates the retransformation by using the inverse input matrix and converts the digital result into analog signals. These signals are amplified by four high voltage amplifiers. The maximum actuator voltage is limited to 230 volts. The outputs of the high voltage amplifiers are connected to the five PVDF actuators (Fig. 8 shows four actuators on the front side of the center panel and Fig. 9 shows one actuator on the back side of the center panel) via the connector panel. The actuator 1 and 3 on the left side of the center panel are connected to the same channel and are treated as one actuator. The actuator positions and shapes are adapted to the nodal lines of the modes 1, 3, 8 and 11.

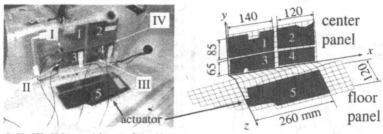

I, II, III, IV : positions of accelerometers

Figure 8. Front side of the center panel with PVDF actuators and pickups

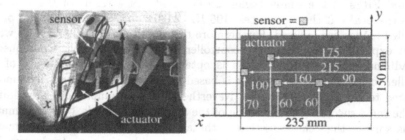

Figure 9. Back side of the center panel with PVDF actuators and sensors

As mentioned above the monitore circuit is independent from the control circuit. An electrodynamic shaker is used as vibration source and is integrated in the monitore circuit. The excitation signal is a continuous random noise within a frequency band of 150 to 600 Hz. The excitation force is sensed by a piezoelectric force transducer at the mounting of the gear box. The response of the structure is measured with four accelerometers at four different locations of the center panel (see Fig. 8). These locations are selected to achieve significant peaks in the FRFs at the eigenfrequencies of the four controlled modes. The FRFs are generated with the FFT analyzer and displayed on the monitore of the second PC.

Corresponding to the result of the experimental modal analysis (see Table I) mode numbers 1, 3, 8 and 11 are selected to be controlled. In Fig. 10 the four

282

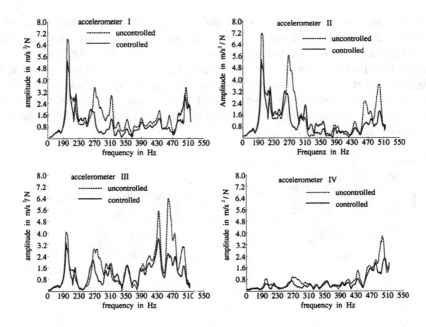

Figure 10. Measured FRF of the center panel and of the floor panel

measured FRFs of the center panel are depicted. As a result of the vibration control the peaks at the resonances (196 Hz, 281 Hz, 457 Hz and 500 Hz) of the controlled modes are reduced. Furthermore the amplitudes of other modes which are not explicitly included in the controller are reduced as well. Consequently a broadband reduction of vibration is obtained. The positive influence of the controller to the uncontrolled modes is based on the similar shape of these modes compared to the controlled modes. Nevertheless not all amplitudes are reduced with the same success. The reduction of the amplitude of mode 1 (approximately 25%) is smaller than the reduction of the amplitude of mode 3 (approximately 60%). The reason for the different reduction is given by the actuator layout. The more the layout is adapted to mode shape the better is the controllability of the mode. This means that errors for the experimentally evaluated mode shapes have a significant influence to the success of the control.

The control of the floor panel is not carried out by modal control but by a local proportional feed back controller. The PVDF sensor is located at the center of the floor panel. Therefore the sensor and the actuator (No. 5 in Fig. 8) are collocated. The feed back gain factor is adjusted with a potentiometer. The resulting FRF of the floor panel is shown in Fig. 11. A significant reduction of the amplitude can only be seen for mode 1 (196 Hz). For the following two peaks a slight reduction is obtained and for higher frequencies the amplitude increases. Obviously the local controller does not lead to a broadband vibration reduction. This result is reasonable since only for the first mode the sensor position corresponds to the location of maximum displacement. For higher modes the maximum displacement does not occur at the center of the floor panel and the sensor signal does not coincide with the modal displacements. Therefore the controller leads to spillover for the higher modes.

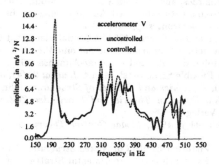

Figure 11. Measured FRF of the floor panel

8. Conclusions

The modal concept of modal analysis and modal control has been combined in order to create active vibration control. This combination is performed by using the results of an experimental modal analysis as input data of a modal state space formulation. Since this method is valid in the frame of modal theory it has great flexibility in applications to mechanical structures. The experimental test with a car body has shown that the modal controller works with high efficiency and that vibration can be reduced successfully over a wide frequency range. Nevertheless limitations of the proposed concept are given by the frequency band limitation of modal analysis. Deviations by curve fitting and truncation of residual modes reduce the accuracy and lead to suboptimal but for practical purposes efficient vibration reduction.

References

1. Beards, C.: 1995, *Engineering Vibration Analysis with Application to Control Systems.* Edward Arnold.
2. Callahan, J. and H. Baruh: 1995, 'Active Control of Flexible Structures Using Segmented Piezoelectric Actuators'. In: *Smart Structures and Materials*, Vol. 2443. pp. 630–642.
3. Clark, R., W. Saunders, and G. Gibbs: 1998, *Adaptive Structures.* Wiley Interscience.
4. Ewings, D.: 1984, *Modal Testing: Theory and Practice.* Research Studies Press LTD.
5. Ewins, D. and J. Sidhu: 1982, 'Modal Testing and the Linearity of Structures'. *Mecanique Materiaux Electricite* (389-390-391), 297–302.
6. Fuller, C., S. Elliott, and P. Nelson: 1996, *Active Control of Vibration.* London: Academic Press.
7. Gaul, L. and U. Stbener: 1999, 'Active Control of Structures'. In: J. Montalvão e Silva and N. Maia (eds.): *Modal Analysis & Testing*, Vol. 363 of *NATO SCIENCE SERIES: E Applied Sciences.* pp. 565–604.
8. Gawronski, W.: 1998, *Dynamics and Control of Structures.* Springer New York, Inc.

284

9.   Gould, L. and M. Murray-Lasso: 1966, 'On the Modal Control of Distributed Parameter Systems with Distributed Feedback'. In: *Transactions on Automatic Control*, Vol. 11. p. 79.
10. Heylen, W., S. Lammens, and P. Sas: 1997, *Modal Analysis Theory and Testing*. Katholieke Universiteit Leuven.
11. Hornung, T., H. Reimerdes, and J. Günnewig: 1998, 'Modal Analysis, Modelling and Control of Vibration Systems with Integrated Active Elements'. In: U. Gabbert (ed.): *Smart Mechanical Systems - Adaptronics*. pp. 163–176.
12. Lammering, R., J. Jia, and C. Rogers: 1994, 'Optimal Placement of Piezoelectric Actuators in Adaptive Truss Structures'. *Journal of Sound and Vibration* pp. 67–85.
13. Mackiewicz, A., J. Holnicki-Szulc, and F. Lopez-Almansa: 1996, 'Optimal Sensor Location in Active Control of Flexible Structures'. *AIAA Journal* **34**(4), 857–859.
14. Meirowitch, L.: 1990, *Dynamics and Control of Structures*. John Wiley.
15. Natke, H.: 1992, *Einführung in Theorie und Praxis der Zeitreihen- und Modalanalyse*. Wiesbaden: Vieweg Verlag.
16. Porter, B. and T. Crossley: 1972, *Modal Control Theory and Applications*. Taylor and Francis.
17. Preumont, A.: 1997, *Vibration Control of Active Structures*. Kluwer Academic Publisher.
18. Reynier, M. (ed.): 1998, 'Optimized Experimental Strategy for Updating Problems: The Sensor Placement'. NATO ADVANCED STUDY INSTITUTE.
19. Sivrioglu, S., Y. Kikushima, and N. Tanaka: 1999, 'An H_∞ Control Design Approach for Distributed Parameter Structures with Attached PVDF Sensors'. In: *Sixth International Congress on Sound and Vibration*. pp. 1669–1676.
20. Stbener, U. and L. Gaul: 1999, 'Active Control of Plate Vibration by Discrete PVDF Actuator and Sensor Segments'. In: N. Hagood and M. Atalla (eds.): *Ninth International Conference on Adaptive Structures and Technologies*. Cambridge, USA, pp. 349–358.
21. Stbener, U. and L. Gaul: 2000, 'Active Vibration and Noise Control for the Interiour of a Car Body by PVDF Actuator and Sensor Segments'. In: R. Ohayon and M. Bernadou (eds.): *10th International Conference on Adaptive Structures and Technologies*. Paris, France, pp. 457–464.

PVDF-TRANSDUCERS FOR STRUCTURAL HEALTH MONITORING

S. HURLEBAUS and L. GAUL
Institute A of Mechanics, University of Stuttgart,
Allmandring 5 b, D-70550 Stuttgart, Germany

Abstract. Structural health monitoring consists of both passive and active sensing monitoring. While passive sensing diagnostics serves for the identification of the location and force-time-history by external sources, active sensing diagnostics of structures serves for localization and determining the magnitude of a damage. This paper shows the application of polyvinylidene fluoride (PVDF) sensors and actuators for its use in structural health monitoring.

Key words: structural health monitoring, piezoelectric transducer, damage detection, impact identification, smart layer, PVDF

1. Introduction

To enhance performance and reduce weight, future structural systems will have to utilize both advanced composite materials and built-in sensors and actuators. Additional benefits, such as reduced operation and maintenance costs, could also be realized when these sensors and actuators are used for structural health monitoring (SHM) functions. Composite materials are the leading candidates for incorporating such sensors and actuators. Therefore diagnostic functions and techniques particularly suited for composite materials will have to be developed. Such technologies can be employed to monitor and control the processing of composite structures as well as monitor their in service health by detecting and evaluating damage in real time. Configurations of laminated composites with monitoring capabilities utilize embedded active material such as piezoelectrics. Piezoelectric materials have the ability to convert mechanical to electrical energy and vice versa. Thus piezoelectric materials can be utilized to provide both sensing and actuating capabilities which, along with fast response characteristics, is why these materials have been used extensively in the development of smart composite structural systems. A special candidate of piezoelectric material is polyvinylidene fluoride (PVDF), which shows a high piezoelectric effect due to elongation and following polarisation.

A lot of researchers [2, 3, 5] use PVDF as sensors, however, only few of them apply PVDF as actuators. Stöbener and Gaul [13, 14] were among the first who succesfully applied PVDF films on plates as actuators for active control of vibrations. However, they treated low frequencies up to 500 Hz only. This lecture

A. Preumont (ed.), Responsive Systems for Active Vibration Control, 285–296.
© *2002 Kluwer Academic Publishers. Printed in the Netherlands.*

demonstrates using PVDF material for SHM purposes in the higher frequency range as sensors and actuators.

The first section summarizes the basic constitutive equations of piezoelectrica. Furthermore, some background on the piezoelectric properties and benefits of PVDF are given. Then the application of PVDF material is shown by a passive and an active sensing diagnosis. First, the passive sensing monitoring is demonstrated by an impact force identification where the location and force time history of an impact is determined. Finally, the active sensing monitoring is presented by applying a smart layer consisting of an array of 10 x 10 PVDF elements which is used to determine an 'artificial' defect (location and size) in an aluminum plate. The result shows that the use of PVDF material is a promising tool in the field of SHM.

2. Piezoelectrica

Since the discovery of the direct piezoelectric effect in 1880 by the brothers Jacques and Pierre Curie, shown by some asymmetric crystals, the phenomena has raised a lot of interest. Quartz, some ceramics and polymers are widely used for mechanical and acoustic transducers taking advantage of the piezoelectric effect. This section summarizes the basic constitutive equations of piezoelectrica. Furthermore, some piezoelectric constitutive properties and benefits of PVDF are given.

2.1. CONSTITUTIVE EQUATIONS

The stresses and strains, electric field and electric displacements in a piezoelectric material can be fully described in the linear range by a pair of electromechanical equations. In a modified notation of IEEE Standard 176-1987 the following two pairs of equations are two such equivalent statements of the electromechanical relationships

$$\varepsilon_{ij} = s_{ijkl}^{\tilde{E}}\sigma_{kl} + d_{kij}\tilde{E}_k , \qquad (1)$$

$$\tilde{D}_j = d_{jkl}\sigma_{kl} + \tilde{\varepsilon}_{jk}^{\sigma}\tilde{E}_k , \qquad (2)$$

and

$$\sigma_{ij} = c_{ijkl}^{\tilde{E}}\varepsilon_{kl} - e_{kij}\tilde{E}_k , \qquad (3)$$

$$\tilde{D}_i = e_{ikl}\varepsilon_{kl} + \tilde{\varepsilon}_{ik}^{\varepsilon}\tilde{E}_k . \qquad (4)$$

ε_{ij} is the strain tensor, σ_{kl} is the stress tensor, $s_{ijkl}^{\tilde{E}}$ is the compliance tensor, d_{kij} and e_{ikl} are piezoelectric constants, $\tilde{E}_k$ is the electric field, $\tilde{D}_j$ is the electric displacement, and $\tilde{\varepsilon}_{ik}^{\sigma}$ is the permittivity. The superscripts $\tilde{E}$ and σ indicate that the values of the constants are obtained at constant electric field and constant electric stress, respectively. Equation (1) states that the strain in the piezoelectric material is proportional to both the applied stress (equivalent to the inverse of Hooke's law) and the applied electric field (the converse piezoelectric effect).

TABLE I. Properties of PVDF material [6]

Modulus of elasticity	E_{11}	1.6×10^9 N/m^2
Modulus of elasticity	E_{22}	1.6×10^9 N/m^2
Modulus of elasticity	E_{33}	1.6×10^9 N/m^2
Density	ρ	1780 kg/m^3
Piezoelectric coefficient	d_{33}	13×10^{-12} C/N
Piezoelectric coefficient	d_{31}	6×10^{-12} C/N
Piezoelectric coefficient	d_{32}	6×10^{-12} C/N
Relative permittivity	$\bar{\varepsilon}_{11}/\bar{\varepsilon}_0$	12
Relative permittivity	$\bar{\varepsilon}_{33}/\bar{\varepsilon}_0$	12
Dielectric constant	$\bar{\varepsilon}_0$	8.85×10^{-12} F/m

Equation (2) states that the electric displacement is proportional to both the applied stress (piezoelectric effect) and the applied piezoelectric field (dielectric effect). Equations (3) and (4) are physically equivalent to Equations (1) and (2) except that in Equation (4) the permittivity is measured at constant strain and in Equation (2) it is measured at constant stress.

2.2. PVDF

In 1969 Kawai [9] was the first one to discover a highly noticeable piezoelectric effect on PVDF. This material is the most studied and utilized piezoelectric polymer. Piezoelectricity can be obtained by orientating the molecular dipoles of polar polymers such as PVDF in the same direction by subjecting appropriate films to an intense electric field, this is called polarization. This polarization is attributable mainly to the spatial arrangement of the segments of the macromolecular chains, and the contribution of the injected charges to the piezoelectric effect is of secondary importance. PVDF is particularly suitable for the manufacture of such polarized films because of its molecular structure (polar material), its purity - which makes it possible to produce thin and regular films - and its ability to solidify in the crystalline form suitable for polarization.

The characteristic piezoelectric properties of a sample of a bioriented PVDF film are shown in Table I.

The benefits of PVDF-material over piezoelectric ceramics are the great flexibility, the high mechanical strength, the dimensional stability and the high and stable piezoelectric coefficients over time. Due to the great flexibility it even can be fixed on objects where the surface is curved and the contour of a film can be easily obtained using a cutter. Furthermore, the PVDF-film does not effect the original properties of the specimen.

3. Passive Sensing Diagnosis

Delamination of composite materials are failure modes which are often generated by impacts. Such delaminations are responsible for the change in stiffness of the structure that may result in safety reduction. It is not always possible to identify the structural damage because the structure is built in and is not accessible for visible inspection. Therefore, it is necessary to monitor the structure using the external load as parameter. If the determined load is above or below a distinct value which has to be obtained with the same material and geometry characteristics, it is possible to predict whether the structure is damaged or not.

In most cases, it is not possible to measure an impact load directly because the sensor has to be attached to the impactor. Thus, one has to place some sensors at selected points on the structure, measure the response of these sensors and has to identify the impact load.

It is the intent of this section to provide a brief explanation of the experimental setup and procedure. Furthermore, the application of the wavelet transform on strain-history-signals is described. A method of determining the location of the impact as well as the time lag and the group velocity is provided using an optimization technique. Finally, a procedure to obtain the force-time-history out of measured strain-history-signals is shown and the results are presented.

3.1. EXPERIMENTAL SETUP AND PROCEDURE

The experimental setup consists of a freely suspended plate which is impacted by a pendulum at the coordinates x_0 and y_0. The suspension of the plate is not critical for this investigation, because the considered time window is very short (3 ms). Therefore the dominant part of the signal consists of the initial flexural waves which are propagating from the location of the impact to the sensors. The side length of the square plate is 2 m and the thickness is $h = 10$ mm. The material of the plate is steel as well as the material of the tip of the pendulum (Young's modulus: $E = 210000$ N/mm^2, Poisson's ratio: $\nu = 0.3$, mass density: $\rho = 7860$ kg/m^3). Four piezoelectric films (PVDF) are used (thickness: 110μm, diameter 15 mm) and they are located in each corner of a square as shown in Figure 1.

The advantage of the piezoelectric film is that this sensor has the property of multidirectional sensitivity as opposed to strain gages and they can be used without any strain amplifiers. A laser Doppler vibrometer (LDV) together with the controller is used to compare the determined force-time-history with the 'actual' one. The oscilloscope, by which the signals are recorded, is triggered by one of the sensor signals and the data are sampled at 1μs. It is obvious that a measured strain history signal contains some undesired contributions from reflections. A simple way for smoothing digital data is to use moving averages of various amounts and points of application. Figure 2a shows an example of sensor 1 for smoothing a signal that incorporates an amount of reflections. Also shown in the figure is the removed part of the signal. The corresponding signals of the other sensors look similar. It is important to know that the strain is proportional to the output voltage of the piezoelectric films.

3.2. WAVELET TRANSFORM OF STRAIN HISTORY SIGNALS

The signal processing procedure can proceed once the experimental data are saved on a PC. It is important to window the signals before applying the Fourier transform. The signals are then transformed into the frequency domain using the Fast Fourier Transform (FFT) algorithm. Then the wavelet transform (WT) is applied using the Gabor wavelet. For further details on applying the WT and the WT itself the reader is refered to Gaul and Hurlebaus [5]. Figure 2b is the 3D plot of the magnitude of the wavelet level of sensor 1. The maximum of the plot indicates the arrival time of the flexural wave. Figure 2c shows the contour plot of the time-frequency-analysis using the WT of the signal of sensor 1.

3.3. DETERMINING OF THE IMPACT LOCATION, TIME LAG AND GROUP-VELOCITY

In this section, the impact location and the group velocity of flexural waves are determined from the time-frequency-analysis. As has been shown in [5] that the maximum of the magnitude of the wavelet transform denotes the arrival time of the wave with the group-velocity c_g. In order to get the arrival times the maximum of the magnitude of the WT has to be determined for each frequency scale. Simply the maximum of the wavelet transform denotes the arrival time of the incident wave. Figure 2d shows the location of the maxima in the time-frequency-plot. Now it is easily seen that the arrival time corresponding to small values of $\log_2(2\pi/\omega)$ is smaller than that of large values of $\log_2(2\pi/\omega)$. Once the arrival times are known, one has to specify the coordinates of the impact location, the time lag and the group velocity. The distances between sensor i and the impact location is l_i. The corresponding arrival times are given by b_i and the unknown time lag is b_0. The unknown coordinates of the impact are defined as x and y, respectively, while the side length of the squared configuration of the sensors is known ($\ell = 0.8$ m). The

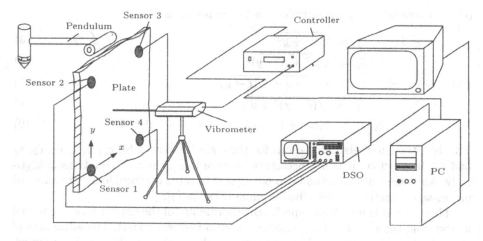

Figure 1. Experimental setup

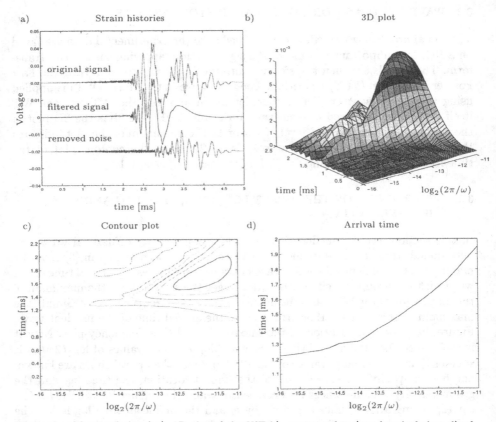

Figure 2. Measured signal a), 3D-plot of the WT b), contour plot c) and arrival time d) of sensor 1

relation between these magnitudes can be expressed as

$$\ell_1 = \sqrt{x^2 + y^2} \tag{5}$$

$$\ell_2 = \sqrt{x^2 + (\ell - y)^2} \tag{6}$$

$$\ell_3 = \sqrt{(\ell - x)^2 + (\ell - y)^2} \tag{7}$$

$$\ell_4 = \sqrt{(\ell - x)^2 + y^2} \tag{8}$$

$$b_i(\omega) = b_0(\omega) + \ell_i/c_g(\omega) \quad \text{for} \quad i = 1, 2, 3, 4. \tag{9}$$

This is a system of eight equations for the eight unknowns b_0, c_g, x, y, ℓ_1, ℓ_2, ℓ_3 and ℓ_4. In practice, it is not necessary to solve this system of equations analytically. An optimization method can be applied to solve this problem. The adopted optimization method is described in more detail in [8].

Therefore, it is possible to specify the coordinates of the impact location as well as the time lag and the group velocity for each frequency level. The advantage of the optimization method is that only some rough initial values have to be assumed. Thus, the exact material properties are not needed. By using the initial values for the optimization method one gets the coordinates of the impact location as well as the time lag and the group velocity of the dispersive waves. Figures 3a and 3b represent the values of x and y for different frequency scales. By comparing these values with the actual values of the impact location, all values of the coordinates of x and y which are obtained by the proposed approach are in the error range below 10%. However, averaging the identified values of x and y over the considered frequency scales leads to the coordinate 0.6092 m in x-direction and to the coordinate 0.4989 m in y-direction. Compared to the actual values (x_0=0.6 m, y_0=0.5 m) these are confident results. Moreover, Figure 3c and 3d shows the time lag and the group velocity for different frequency scales. The identified group velocity is compared with the analytically obtained group velocity. Again, these velocities correspond well.

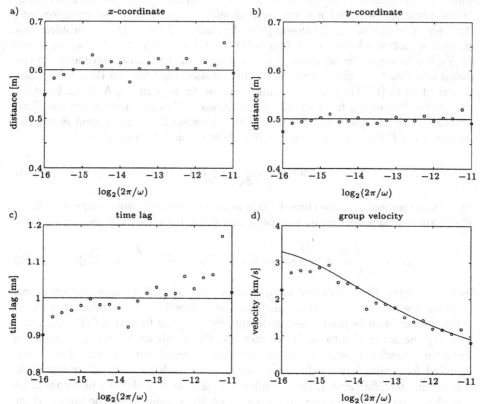

Figure 3. Identified (circles) and actual (full line) coordinates of the impact location a) and b), identified (circles) and actual (full line) time lag c) and theoretical (full line) and identified (circles) group velocity d)

Once the group velocity is known, it is easy to determine the wavenumber from

$$k(\omega) = \int_{\omega=0}^{\omega} \frac{1}{c_g(\tilde{\omega})} \, d\tilde{\omega}. \qquad (10)$$

Figure 4a shows the experimentally determined wavenumber and the wavenumber which is obtained by using the Mindlin Plate theory [12].

3.4. IDENTIFICATION OF THE FORCE TIME HISTORY

For identifying the force time history, the classical Kirchhoff plate theory is considered assuming that the generated wave field is circular and the field quantities are axisymmetric. The governing equation for the plate with thickness h then is

$$D\nabla^2\nabla^2 w + \rho h \frac{\partial^2 w}{\partial t^2} = q(t) \quad \text{with} \quad \nabla^2 = \frac{\partial^2}{\partial r^2} + \frac{1}{r}\frac{\partial}{\partial r}, \quad (11)$$

where $w(r,t)$ denotes the out of plane displacement, $q(t)$ is the load per area, D the plate stiffness and ρ is the mass density. Transforming the equation into the frequency domain and following the derivation of Doyle [4] the problem has an axisymmetric solution $\hat{w} = \mathbf{A}J_0 + \mathbf{B}Y_0 + \mathbf{C}K_0 + \mathbf{D}I_0$. The Bessel functions J_0, Y_0, K_0, I_0 are of the argument $z = kr$, where k is the wavenumber and r is the radial distance from the plate center. The notation used for the Bessel functions is according to [1]. There are four unknown constants, namely, $\mathbf{A}, \mathbf{B}, \mathbf{C}, \mathbf{D}$ which have to be determined from boundary conditions. Following the derivation of Gaul and Hurlebaus [6] who presented the relation between the displacement $\hat{w}$ and the impact force $\hat{P}$ the velocity at the impact location $\dot{w}(0,t)$ is obtained as

$$\dot{w}(0,t) = \frac{1}{8\sqrt{\rho h D}}P(t) \quad (12)$$

by transformation into the time domain and differentiating with respect to time. Furthermore, the radial strain is obtained by differentiating the displacement field

$$\hat{\epsilon}_{rr} = \frac{h}{2}\frac{\partial^2\hat{w}}{\partial r^2} = \frac{i\hat{P}h}{32D}[(J_0 - J_2) - i(Y_0 - Y_2) + \frac{i2}{\pi}(K_0 + K_2)]. \quad (13)$$

Now, by using the measured strain-history, the identified wavenumber k and knowing the distance between generation and detection of the flexural wave, it is possible to identify the force-time-history by applying Equation (13) and transforming the force $\hat{P}$ into the time domain. Figure 4b shows the reconstructed force-time-history as well as the one obtained using Equation (12). The reconstructed force-time-history does not completely vanish for times after the peak. This is due to reflections at the boundaries in the original signal which cannot be completely removed. However, the arrival time, the magnitude of the force and the shape of the force-time histories are in good agreement.

4. Active Sensing Diagnosis

In this section PVDF is used as self-sensing actuator in a smart layer for the ultrasonic inspection of specimens. The reason for developing such a smart layer is to use it as a tool for built-in structures [11]. Current nondestructive evaluation (NDE)

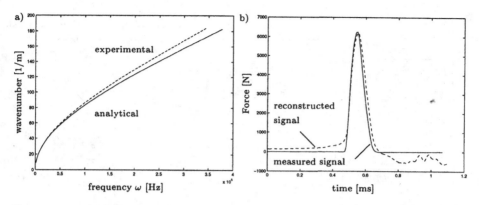

Figure 4. Analytical and experimentally determined wavenumber a), reconstructed force-time-history compared with the 'actual' one b)

methods are designed for testing the structures during the maintenance period when they are out of service. The present approach overcomes this shortcoming of conventional NDE methods by using the smart layer. Therefore, ultrasonic inspection no longer necessitates to put the considered structure in an immersion tank. It has become possible to monitor the inspected structure during operation.

First, the experimental setup of measuring propagating waves in structures using piezoelectric films is explained. Finally, a smart layer consisting of an array of 10 x 10 PVDF elements and signal processing techniques is used to determine an 'artificial' defect in an aluminum plate. The result shows that the proposed technique is a promising tool for identifying the location and size of damages at in-service structures. For further details see also Gaul and Hurlebaus [7].

4.1. EXPERIMENTAL SETUP

For generating of a short high voltage pulse in order to actuate the PVDF material a pulser-receiver is used (Figure 5). The pulser-receiver is driven in the pulse-echo mode, that means a single PVDF film is used as self-sensing actuator. In order to connect the PVDF films with the pulser-receiver a multiplexer as part of a VXI-measurement system is used. The pulse generator which is enclosed in the VXI-mainframe triggers the pulser-receiver in order to ensure that the pulser-receiver emits no high voltage when the multiplexer switches from one PVDF film to the other. The oscilloscope, again part of the VXI system, receives the signals of the pulser-receiver, namely, the sensor signal and the synchron signal. The VXI-system is connected to a PC via general purpose interface bus (GPIB). The PC is used for controlling the VXI-system, for storing the data and for further signal processing.

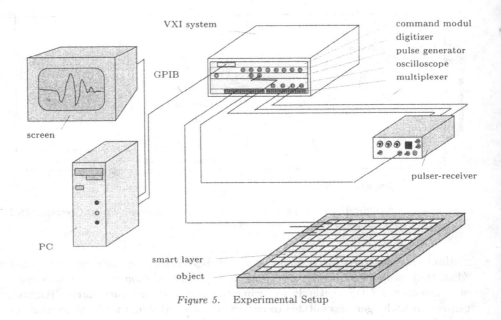

VXI system   command modul

GPIB

screen

pulser-receiver

PC

smart layer

object

Figure 5.   Experimental Setup

4.2. APPLICATION OF THE SMART LAYER

In this section the application of a smart layer is presented, which contains an embedded network of distributed piezoelectric polymers to identify an 'artificial' damage in an aluminum plate ($150 \times 150 \times 15$ mm^3). The artificial defect is created by milling out some bottom sections at the backside of the aluminum plate. The depth of the milled section is about 2 mm. The boundary of the milled section is shown in Figure 6 by the solid lines. On top of the aluminum plate, a smart layer is attached using couplant material. The smart layer is shown in Figure 6 with the conducting connections which are obtained by printed circuit techniques. The dark squared parts mark an array of 10 x 10 PVDF transducers. The side length of each PVDF element is 10 mm whereas the distance between two sensors are 2 mm. The transducers are connected with a multiplexer and the measurement are carried out by scanning each PVDF element. After measuring a time dependent signal is obtained. This one dimensional description gives the information at one distinct measurement point and is called A-scan. In order to visualize the information of a two dimensional scanned area the results can be shown in a C-scan. This scan provides a top view of the measured object out of the scanning area with the shown defects. Using the A-scan the size of the defect corresponds to the amplitude of the echo whereas the amplitude of the echo in the C-scan is shown with different colors [10].

    Figure 6 shows such a C-scan of the aluminium plate. It is obvious by inspection that the identified defect is much larger than the true one. However, this is a consequence of the amount and shape of PVDF transducers which are used. If one would use more PVDF transducers, the resolution of the identified defect would be finer and smoother. In general, the proposed method shows that it is possible to identify a defect using the developed smart layer. By considering the small amount of PVDF-elements used, it is obvious that the location and the size of the defect can be determined close to the actual one. This method can be used to monitor the so called 'hot spots' of a highly loaded structure where the functionality of this structure has to be always guaranteed.

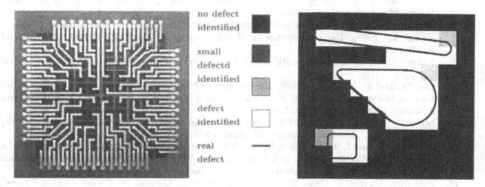

Figure 6. Smart Layer (left) and real and identified defect of the aluminum plate (right)

5. Conclusion

The paper presents results that were obtained by applying PVDF-transducers for SHM. These piezoelectric films were used for the localization and determination of the force-time-history of an foreign impact. Furthermore, prototypes of a smart layer which contains embedded PVDF films as sensors were fabricated and validation tests were performed. This smart layer has been used to measure the back wall echo of an aluminum plate that contains defects. The proposed layer shows good results by identifying the location as well as the size of the defect of the plate. However, for improving the resolution of the identification process more transducers have to be used. The advantage of the developed smart layer is that it can be embedded as built-in structure or can be fixed on the surface of the structure. The proposed technique is therefore applicable for existing structures.

It is shown that the PVDF material can be used for a possible application in SHM. The PVDF provides a progress for passive as well as active sensing diagnosis. Especially the smart layer is capable to detect delaminations in composite materials.

References

1. Abramowitz, M. and I. Stegun (eds.): 1965, *Handbook of Mathematical Functions*. New York: Dover Publications, Inc.
2. Day, R.: 1996, 'PVDF and Array Transducers'. *NDTnet* **1**(9), 1–10.

3. Deutsch, W., A. Cheng, and J. Achenbach: 1997, 'Self-Focusing of Rayleigh Waves and Lamb Waves with a Linear Phased Array'. *Research in Nondestructive Evaluation* 9(2), 81–95.
4. Doyle, J.: 1997, *Wave Propagation in Structures*. New York: Springer.
5. Gaul, L. and S. Hurlebaus: 1998, 'Identification of the Impact Location on a Plate Using Wavelets'. *Mechanical Systems and Signal Processing* 12(6), 783–795.
6. Gaul, L. and S. Hurlebaus: 1999, 'Determination of the Impact Force on a Plate by Piezoelectric Film Sensors'. *Archive of Applied Mechanics* 69(9/10), 691–701.
7. Gaul, L. and S. Hurlebaus: 2000, 'Application of PVDF-Films for Ultrasonic Testing'. In: J. A. Güemes (ed.): *European COST F3 Conference on System Identification & Structural Health Monitoring.* pp. 571–580.
8. Grace, A.: 1992, 'MATLAB Optimization Toolbox'. The MathWorks Inc., Natick, USA.
9. Kawai, H.: 1969, 'The piezoelectricity of poly(vinylidene fluoride)'. *Japanese Journal of Applied Physics* (8), 975–976.
10. Krautkrämer, J. and H. Krautkrämer: 1975, *Werkstoffprüfung mit Ultraschall.* Berlin: Springer Verlag.
11. Lin, M. and F.-K. Chang: 1998, 'Development of SMART Layer for Built-in Diagnostics for Composite Structures'. In: *The 13th Annual ASC Technical Conferences on Composite Material.* Baltimore, MD, pp. 1–8.
12. Mindlin, R.: 1951, 'Influence of Rotatory Inertia and Shear on Flexural Motions of Isotropic, Elastic Plates'. *Journal of Applied Mechanics* 18(1), 31–38.
13. Stöbener, U. and L. Gaul: 2000a, 'Active Vibration and Noise Control for the Interiour of a Car Body by PVDF Actuator and Sensor Arrays'. In: *Proceedings of the 10th International Conference on Adaptive Structures and Technologies.* pp. 457–464.
14. Stöbener, U. and L. Gaul: 2000b, 'Modal Vibration Control for PVDF Coated Plates'. *Journal of Intelligent Material Systems and Structures* 11(4), 283–293.

VIBRO-ISOLATION OF SENSITIVE EQUIPMENT

VYTAUTAS OSTAŠEVIČIUS
Kaunas University of Technology
Donelaičio 73, LT-3006 Kaunas, Lithuania

Abstract

At present a lot of high precision equipment is being developed and the systems of automatic control and regulation found a wide application. Macro and micro commutation elements that operate on the basis of vibro-impact principle constitute a big part among such equipment. The main indicators of vibro-impact systems – vibrational stability, speed of operation, reliability and longevity – are conditioned by both the continuous structure properties and the dynamic phenomena occurring in the course of operation.

Chatter in technological equipment is an objectionable manifestation of vibration in the machine, tool or workpiece; it affects surface finish, accuracy and adversely influences the life of tools.

Road micro profile irregularities affect the smoothness of automobile motion. Sensitivity of suspension parts to road profile is proved by frequency characteristics. Higher frequency excitation can by passed to car body not through a spring or a damper, but through parasitic connections. For the elements fastened on the automobile body, the conditions for vibrations excitation and recommendation for design are made more precise, at the same time keeping in mind that the main function of mentioned elements is not the reduction of noise. Laser holography method is chosen for investigation of the automobile element. For the analysis of holograms there was used the numerical hologram analysis method. The results of analysis of vibro-isolation of acoustic deflectors, exhaust tube and clutch coupling system are introduced.

1. Effects of Nonlinear Mechanics in Continuous Structures

The model consists of i=1,2…,m prismatic finite elements and j=1,2…,k motion limiters or supports (Fig. 1.1).

After proper selection of generalized displacements in the inertia system of coordinates, the dynamics of the model is described by the following matrix equation:

297

A. Preumont (ed.), Responsive Systems for Active Vibration Control, 297–324.
© 2002 *Kluwer Academic Publishers. Printed in the Netherlands.*

$$[M]\{\ddot{y}(t)\} + [C]\{\dot{y}(t)\} + [K]\{y(t)\} =$$

$$= \begin{cases} \{Q(t)\}, \text{if } \{\overline{\Delta}_j^i\} > \{y_i(t)\} \cap \{\Delta_j^i\} < \{y_i(t)\} \cup f_i(y_i, \dot{y}_i, t) \geq 0; \\ \{Q(t)\} + \{F(\dot{y}, \dot{y}, t), \text{if } \{\overline{\Delta}_j^i\} \leq \{y_i(t)\} \cup \{\Delta_j^i\} \geq \{y_i(t)\} \cap f_i(y_i, \dot{y}_i, t) < 0, \end{cases} \quad (1)$$

with initial conditions :

$$\{\dot{y}(0)\} = \dot{y}^0; \quad \{y(0)\} = \Delta_j^i, \quad (2)$$

where $\{F(\dot{y}, y, t)\}$ vector of impact interaction forces, the components of which $\{f_i(\dot{y}_i, y_i, t)\}$ express the reaction of elastic element making impacts against the support and has the following form:

$$\{f_i(\dot{y}_i, y_i, t)\} = \overline{K}_j^i \left[\left| \Delta_j^i \right| - \left| y_i(t) \right| \right] + \overline{C}_j^i \dot{y}_i(t), \quad (3)$$

where $\overline{K}_j^i, \overline{C}_j^i$ stiffness and viscous friction coefficients of the support, Δ_j^i distance from the i-th modal point of the elastic element to the j-th surface of the support.

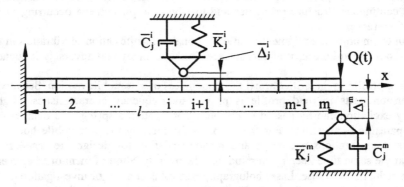

Figure 1.1. Finite element model of elastic mechanical system

Of a whole range of natural vibrations the first five modes were distinguished (I, II, III, IV, V) (Fig. 1.2) which in intersection with the axis-line formed nodal points marked by numbers that express the ratio between the distance x from the fixing place of the structure and its whole length l (x/l). The letters Y_{ii} and Φ_{ii} denote values of the maximum amplitudes of translation and rotation modes.

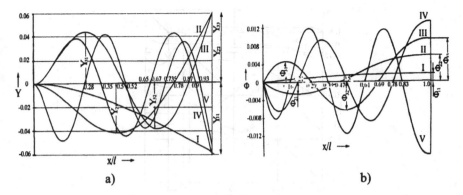

Figure 1.2. Natural traverse vibration modes of cantilever structure:
a) translation; b) rotation

As it can be seen from the diagram (Fig. 1.3), the smallest rebound amplitudes are typical for such structures in which support is located at point $x_0/l = 0.87$ or $x_0/l = 0.67$. A slight decrease in the rebound amplitude is observed at $x_0/l = 0.78$. The lower curve that asymptotically approaches the axis line corresponds to the deflection of free end of the structure at impact on the support.

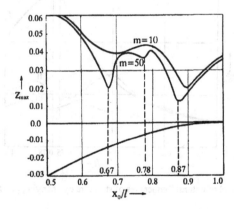

Figure 1.3. Cantilever structure rebound amplitude dependence on the position of support

Nodal points y_{ii} and ϕ_{ii} of the translation mode Y_{ii} and rotational vibration mode Φ_{ii} are marked by two indices. The first of which measures the number of vibration mode and the second – the sequence number or the nodal point from the structure fixing area. Having introduced the support, one nodal point is added to each mode as compared with the modes of free of support cantilever beam, marked $i=0$. In the diagram (Fig. 1.4) the diagonal line represents shifting of the support from the beam fixing area to the free end.

300

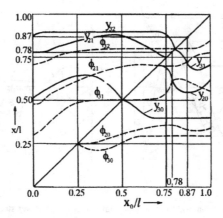

Figure 1.4. Control of nodal points position of structure vibration modes by changing position of support

The first natural frequency of the supported cantilever structure reaches maximum value when support is located at point $x_0/l=0.78$ and in this case the structure has highest vibro-stability (Fig. 1.5).

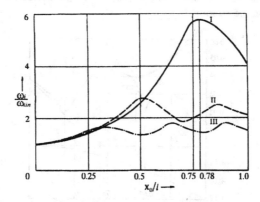

Figure 1.5. Control of natural frequencies of structure by changing position of support

If a support is at nodal point of the second vibration mode $x_0/l=0.78$, as its stiffness increases $\overline{K_4} > \overline{K_3} > \overline{K_2} > \overline{K_1}$ and reaches a certain value $\overline{K_3}$ or $\overline{K_4}$ it is possible noticeably increase the first natural frequency of the structure and to bring it closer to the second frequency (Fig. 1.6).

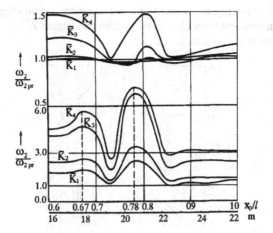

Figure 1.6. Control of natural frequencies of structure by changing stiffness and location of support

When the stiffness of support is $\overline{K}_3$ the first vibration mode completely disappears. This is confirmed by the appearance of nodal points y_{11} and φ_{11} in the first mode of transverse vibrations which is not characteristic of the investigated structure (Fig. 1.7). Stiffness $\overline{K}_3$ corresponds to such value in the presence of which the natural frequency of the support equals or exceeds the second frequency of transverse vibrations of the unsupported cantilever structure.

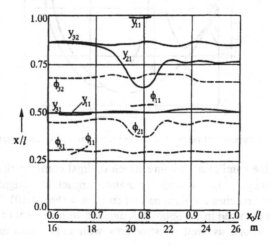

Figure 1.7. Control of nodal points position by changing location of support of $\overline{K}_3$ rigidity

The behavior of vibro-impact system in the case of harmonic excitation is best defined by the areas of existence of diverse vibration laws (fig. 1.8). For example law n=2+1 means that in the course of two excitation periods the cantilever makes one impact on the support and subsequently one impact in one excitation period (Fig. 1.9). As we can see from Fig. 1.9 the velocity can also increase after the impact $Z^{1+}(t)$. This is because of energy accumulation in the structure.

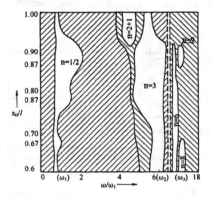

Figure 1.8. Dependence of existence of diverse law areas upon the position of the support

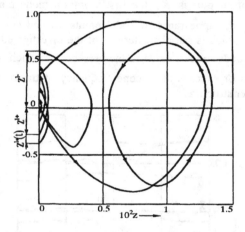

Figure 1.9. Phase trajectory of vibro-impact n=2+1 recurrent motion law of cantilever structure

As was expected the cantilever structure which optimal configuration III was obtained for natural frequency ω_3 has a settled vibration impact law (right-hand dashes) till excitation frequency reaches the third natural frequency (Fig. 1.10). Similar results also apply to structures obtained in the optimal design and with account of higher frequencies of transverse vibrations as well as structures with either longitudinal or torsional vibrations.

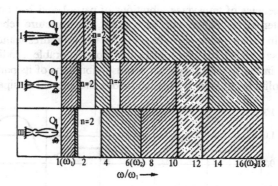

Figure 1.10. Diverse vibro-impact motion laws areas of optimal cantilever structures

The most stable in respect of vibrations in the presence of highest external kinematic excitation frequencies is a cantilever structure that has a support connected at distance $x_0/1=0.78$, i.e. at the nodal point of the second mode (Fig. 1.11).

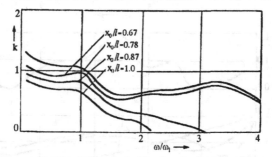

Figure 1.11. Vibrostability control by changing location of support

To explain the reasons of the decrease in the rebound amplitudes upon change of cross-sections area, the simulated cantilever structure must be analyzed as a structure consisting of two parts separated by the changed cross-section (Fig. 1.12). Vibro-impact vibrations of the part 2 evoke parametric vibration of the part 1. If an artificial pin-connection is installed at the place of the changed cross-section the part 1 will be equivalent to a pin-supported cantilever structure and the part 2 to a cantilever structure.

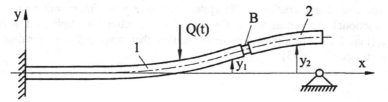

Figure 1.12. Changed cross-section elastic mechanical system

304

In case when frequencies of transverse vibrations of parts 1 and 2 are equal, $\omega_1 = \omega_2$ we have a case of internal resonance and, as a result, the structure rebound amplitudes should increase. In cases when $\omega_1 = 2\omega_2$ or $2\omega_2 = \omega_1$ (antiresonance) the rebound amplitudes decrease (Fig. 1.13). We obtain the most favorable conditions for a high energy dissipation in the structure material because vibrations of its parts are pre-phasic what increases amplitudes of higher vibration modes and internal dissipation of energy.

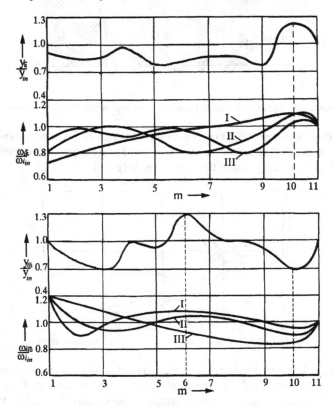

Figure 1.13. Control of vibration amplitudes and natural frequencies by changing different cross-section location

The aim of optimization is to select such geometrical parameters that would correspond to the technical characteristics of the system and give a minimum value to a certain quality functional. According to the simulation results vibration amplitudes of optimal structure (Fig. 1.14) are almost five times smaller than amplitudes of constant cross-section structure (Fig. 1.15).

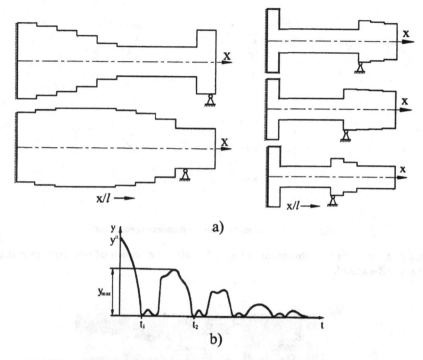

Figure 1.14. Vibro-stable configurations of a structure under kinematic excitation (a) and constant cross section cantilever motion trajectory in case of free damped impact vibration (b)

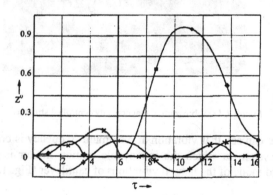

Figure 1.15. Vibration law of a cantilever structure under kinematic excitation: ■ - initial structure of constant cross-section; **x** - optimal structure

While investigating vibration-impact system it is necessary to verify the results experimentally and to reveal the phenomena related to the dynamics of it. Optical methods are most convenient (Fig. 1.16).

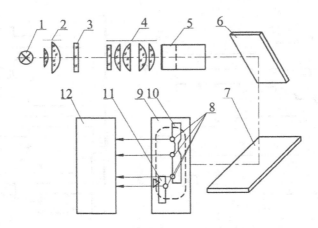

Figure 1.16. Principal diagram of photoelectric equipment

As we can see from the vibrogram in Fig. 1.17, after the impact of two cantilever beams they are self-excited.

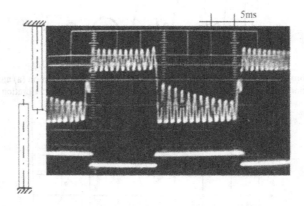

Figure 1.17. Vibrogram of magnetically controlled structure

From Fig. 1.18 we can see that vibration at maximum frequency is characteristic of the vibration-impact system in which the immovable motion limiter is positioned in the nodal point of third mode of the natural vibrations of cantilever (Fig. 1.19).

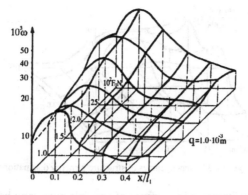

Figure 1.18. System frequency dependence on structural parameters

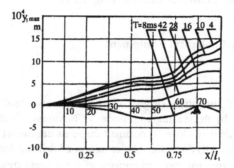

Figure 1.19. Vibration modes

As it can be seen from the curves in Fig. 1.20 frequency/amplitude characteristics are minimum when free vibration frequencies of two rebounding cantilevers differ by two times.

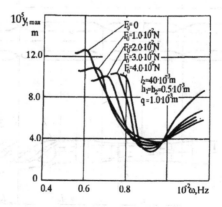

Figure 1.20. Amplitude-frequency characteristics of structure

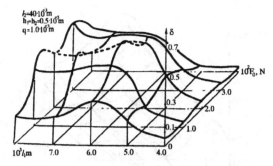

Figure 1.21. Dependencies of structure's energy dissipations.

Dissipation of energy increases in presence of multiple even ratio between free vibration frequencies of structural cantilever elements (Fig. 1.21).

2. Applications to Technological Equipment

2.1. DRILLING

The main purpose of invention (Fig. 2.1) is to increase tool's life and efficiency (productivity) of drill by decreasing amplitudes of torsional vibrations of cutting edges. Configuration of the drill is close to optimal shape of cantilever for the second given frequency of free torsional vibrations. Under action of drilling force such configuration is capable of exciting free torsional vibrations of the second mode Θ_2 as presented by dashed lines – lower amplitudes of cutting edges and higher frequencies than first mode Θ_1. It means that such drill acts like in process of vibrational cutting increasing the tool's life and efficiency.

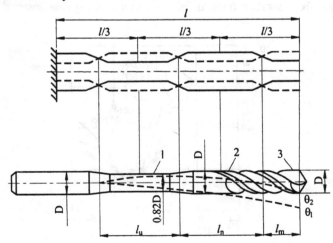

Figure 2.1. Variable cross-section twist drill (Certificate of invention No. 1357151)

2.2. INTERNAL BORING

Invention (Fig. 2.2) is attributed to mechanical treatment of metals by cutting and may be employed for internal boring, grinding and other hole treatment operations. The main purpose of the invention is amelioration of precision and quality of surface treatment owing to decrease in vibration amplitudes of insertion transmitted to cutting tool. Insertion consists of end part 1 for fixation in spindle of machine and cantilever part 2 of length L constant cross-section. In cantilever part 2 at distance 0.87L, which coincides with nodal point of the third mode of flexural vibrations of the cantilever, the cutting tool 3 is located. The third mode of such construction is dominant and is distinguished by lower amplitudes and intensive energy dissipations of vibrations as compared with the first mode. It means that amplitudes of vibrations of tool considerably decrease, precision and quality increase.

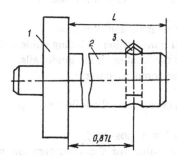

Figure 2.2. An insertion for cutting tool strengthening (Certificate of invention No. 1590206)

2.3. BALANCING PRINCIPLE OF WELDING FURNACE (FOR SPACE APPLICATION)

Welding furnace is inroduced in Fig. 2.3. It was necessary to decrease the influence of impacts to space structure during the welding process of furnace (ULB, Euro Space Agency project, 1995).

Figure 2.3. Welding furnace

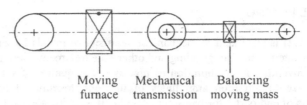

Moving   Mechanical   Balancing
furnace   transmission   moving mass

Figure 2.4. Reaction force compensator work principle

The reaction force due to furnace acceleration/deceleration can be compensated by the movement of an equivalent mass in opposite direction, in such a way that the center of mass remains fixed (Fig. 2.4).

The design of the reaction force compensator includes :

- choice of the compensation mass and gear ratio;
- choice of the mechanical transmission including guide rails;
- hoice of the number and the location of the guide rails;
- choice of the shape of the compensation mass.

These choices must be made considering the following requirements :

- a compensation mass as low as possible;
- a stroke as large as possible for the compensation mass within the available envelope of the breadboard;
- the path of the motion of the compensation mass as close as possible to the one of the furnace motion, to avoid any free inertia moments;
- a smooth and backlash free mechanical transmission.

3. Applications to transport systems

3.1 ACTIVE SUSPENSION

Making the vibration excitation conditions of the parts connected to the car body more precise, we investigated the micro profile records of different Lithuanian roads recorded with equipment used for IRI (International Roughness Index) measurement – laser surface roughness recorder.

Road micro profile irregularities affect the smoothness of automobile motion (Fig. 3.1). The half-wave length of irregularities is in the range of 100 – 10000 mm and their characteristics are defined by the road quality requirements. Such irregularities would excite the vibrations of frequency 0.19 – 360 Hz when driving at speeds of 70 to 130 km/h. In order to make the search for noise sources deeper one needs to keep in mind the texture of pavement which is defined by other road quality criteria. This category encompasses irregularities of wavelength smaller than 100 mm. Their parameters are related to the quality of contact between tires and road. This is usually not evaluated when analysing the operation of automobile suspension.

Some typical even and uneven road pavements and three road textures were separated after investigation of different road records (Fig. 3.1). The basic difference between these irregularities is in their type. The irregularities that fall into road texture category are of crest type and their dominant dimensions are similar. This corresponds to definition of noise-making excitation, but their disorder lets us assert that vertical motion of tire will not be excited.

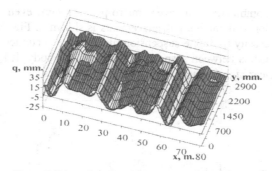

Figure 3.1. A section of road micro-profile 2900×9400 mm²

The height dispersion of irregularities of this type is correspondingly 0.28, 0.69 and 1.69. It was found that road texture is related with structure of pavement and road micro-profile is related with pavement laying technologies. That is why the characteristics of micro-profile irregularities are of different type. Even the roads that correspond with the highest standards have waviness (Fig 3.2.).

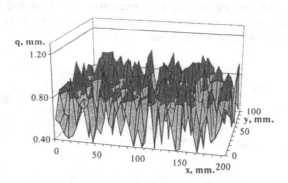

Figure 3.2. Road pavement texture 100×200 mm² area

The chosen records are taken imitating driving a car on different road pavements. Modeling was performed using a quarter-car model (Fig.3.3.) that corresponds to VW Golf parameters with soft and hard suspension.

312

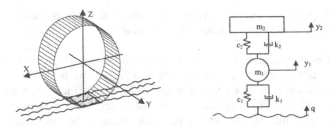

Figure 3.3. Quarter-car model

An influence of combination of even micro-profile with even texture towards automobile suspension and car body displacements is shown in Fig. 3.4. The obtained results show the necessity to define suspension model more accurately and to clear out construction peculiarities between suspension parts and car body. It is needed because one can see rather clear high frequency (33 – 61 Hz) harmonics.

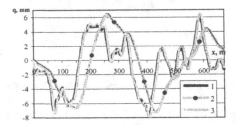

Figure 3.4. Kinematics excitation curves of a car with soft suspension: 1- road profile, 2- suspension displacements, 3- car body displacements

Sensitivity of suspension parts to road profile is proved by frequency characteristics in Fig.3.5. It also shows that higher frequency excitation can be passed to car body not through a spring or a damper but instead through parasitic connections like suspension lever. Also, it can be amplified by resonance effect.

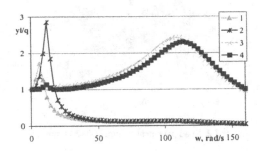

Figure 3.5. Frequency characteristics of car body (1,2) and suspension (3,4) displacements when suspension is of comfort type (1,3) and sport type (2,4)

When investigating simpler ways of excitation of cup-type parts, tire smoothened road profile, car body and suspension displacements, we performed harmonic analysis by determining members of the following series:

$$y(x) = \frac{a_0}{2} + \sum_{i=1}^{\infty} R_i \sin(i\omega + \varphi_i) \qquad (4)$$

Where $\omega = \pi\delta x$ V measurement interval of the road profile, m; V-car's velocity, m/s; R_i- amplitude; j_i-phase.

Some known regularities were got in analysis. There are a very big number of members in road micro-profile expression, much smaller number of members in suspension movement expression, narrow frequency interval for body displacement. However, at higher frequencies (starting with 12 Hz for hard suspension and 6 Hz for soft suspension) amplitudes of car body do not comply with amplitude ratios determined by frequency characteristics.

Amplitudes of tenths of millimeters are passed to car body as a background noise in a very wide frequency interval. In order to define border conditions we modeled a car going on a rough road. The height dispersion of irregularities is 13.7 mm and is similar to asphalt-concrete pavement. However, the irregularities are very different and one can distinguish short waves (Fig.3.6).

In this case, we get a similar result, that is, small frequency harmonics are higher than ones determined by frequency characteristics. That is why, when analysing resonance possibilities of various parts we have to evaluate the possibility of small amplitude, higher frequency vibrations going to car body. The best way of vibro-isolation of the car could be realized using active suspension (Fig. 3.7).

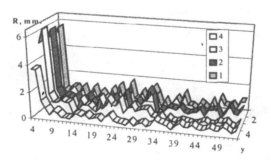

Figure 3.6. Road micro-profile (1), tire smoothened road profile (2), suspension (3) and car body (4) dependency of displacements on wavelength. Soft suspension automobile driving at 70 km/h on a rough road

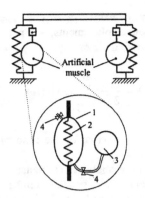

Figure 3.7. Active suspension of the car with artificial muscle

3.2 VIBRO-ISOLATION OF ACOUSTIC DEFLECTORS

We used a simplified model of a part for calculations of vibration modes (Fig.3.8). As one can see, the characteristic frequencies of vibrations of thermoplastic part are in between 100 – 400 Hz. Only indirect excitation is possible at these frequencies.

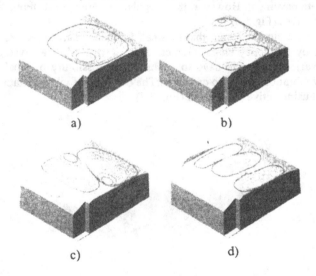

a) b) c) d)

Figure 3.8. Basic vibration modes of the simplified model. Natural frequencies are at a) 156,46 Hz, b) 310,01 Hz , c) 323,75 Hz and d) 434,29 Hz.

We analyzed how much vibration character can be changed when introducing additional elements like handles, or weakening sides with cut-outs (Fig.3.9.). As we expected, weakening of sides brings extra low frequency vibration modes (1-3) that cover sides with closest parts to bottom. Optimal choice of modes should be used. Unfortunately, when analyzing a real construction, the choices become small. In order to change vibration frequency, one may need to change rigidity of material. By evaluating the tendencies of change of elasticity module only one choice is left – increase in rigidity. This can be achieved by, for example filling parts, i.e. increasing natural frequency. Another problem is related with layout of supports because vibration modes are dependent on places of fastenings. In this case, changes are not easy either, because parts have to fulfill their primary functions.

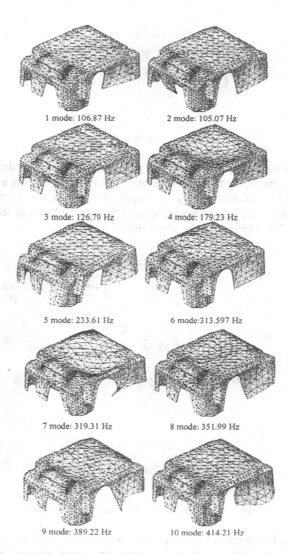

1 mode: 106.87 Hz   2 mode: 105.07 Hz

3 mode: 126.79 Hz   4 mode: 179.23 Hz

5 mode: 233.61 Hz   6 mode:313.597 Hz

7 mode: 319.31 Hz   8 mode: 351.99 Hz

9 mode: 389.22 Hz   10 mode: 414.21 Hz

Figure 3.9. Vibration modes of a detailed model. Natural frequencies are marked besides pictures

316

Experiment:
Form of a part and its properties are inevitably simplified when applying more complicated finite element models. This is a problem. We chose laser holography method for investigation of the part, because of previous experience with more complicated deformations of metallic parts at dynamic loading. The laser holography equipment is mounted on the special vibro-isolated basement. The test bench we used is shown in Fig. 3.10.

Figure 3.10. Test bench

The spots for laser holography are selected in the part in such way that they would be characteristic for investigated vibration mode (Fig.3.11). These spots are colored using the special paint. In addition gluing of stripes of stiffening or damping material is planned in order to change the characteristics of the part. In this way the stiffened parts or zones with modified properties are simulated.

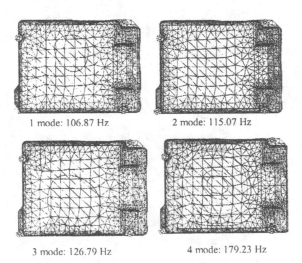

1 mode: 106.87 Hz   2 mode: 115.07 Hz

3 mode: 126.79 Hz   4 mode: 179.23 Hz

Figure 3.11. Control zones of first four vibration modes

For the analysis of holograms there was used the numerical hologram analysis method (Fig.3.12). Reflection intensity from hologram plate is determined by the following formula:

$$I \approx \left(\frac{a}{T} \int_0^T \exp\left(-i\left(\frac{2\pi}{\lambda}\left(\sin\theta + \int_0^t (\cos\theta_1 + \cos\theta_2)Rdt \right) \right) \right) dt \right)^2, \qquad (5)$$

where I – intensity; T – time of exposition; λ – length of the laser beam light, R – the vector of vibration, t – time; θ_1 – angle between the lightening and the vibration vectors, θ_2 – angle between the observation and the vibration vectors, θ – angle between the observation and normal vector of the hologram plane, a – the coefficient of the light reflection of the surface.

This method enables to compare the hologram of deformation obtained by numerical simulation using the method of finite elements with the one of a real surface. The carried out experiments affirmed the basic principles of numerical simulation and indicated the possibility for changing the dominating vibration modes by simple means.

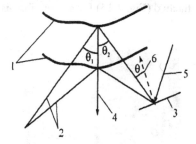

Figure 3.12. The scheme of experiment: 1-shell, 2-measurement laser beam, 3-hologram plate, 4-displacement vector, 5-reference laser beam, 6-normal of the hologram plate

The algorithm of holographic interferometry data processing is presented in Fig.3.13. Box 3 represents numerical identification of interference bands in the pre-processed digital image (Fig. 3.14). The output of this step is used as a direct input for the mathematical model processing algorithm (Box 5).

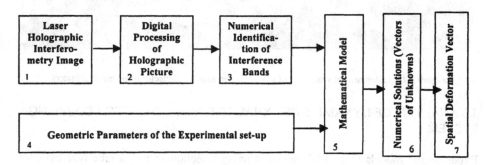

Figure 3.13. Diagram of the numerical hologram identification

318

As one of the most widely applied areas of laser holography analysis is vibration measurement, the analysis system is adopted for the identification of such kind of measurements.

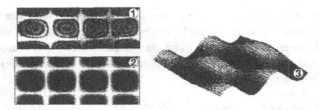

Figure 3.14. Experimental data processing: (1) - Laser holographic interferometry image of a vibrating plate; (2) - Digital image of interferogram; (3) - Deformation vector of the vibrating plate

The results of holography analysis gave the possibility to choose the points in which piezo-actuators should be attached (Fig. 3.15) for active vibro-isolation of acoustical deflectors (Fig. 3.16).

Figure 3.15. Acoustic deflector with active vibro-isolation

| f = 230 Hz | f = 310 Hz | f = 400 Hz |

Figure 3.16. Holograms of vibrations of vibro-isolated acoustic deflector using piezo actuators

3.3 ANALYSIS OF DYNAMICS OF EXHAUST TUBE AND CLUTCH COUPLING SYSTEM

Experimental models (Fig. 3.17) are built using a section of round tube type cylinder (exhaust tube) with fixing construction to a motionless foundation. The fixing construction is implemented in the form of rigid hinge type ring.

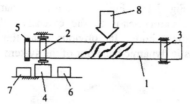

Figure 3.17. The scheme of the experimental setup: *1* - the analyzed cylinder; *2, 3* - fixing constructions; *4* - shaker table; *5* - piezo-electric ring; *6, 7* -driving generators; *8* - light from laser holographic system

The excitation of the cylinder is performed in two ways - either by the piezo-electric actuator fixed to the end of the cylinder, or via the basis which is excited by shaker.

Two different holograms forming methodologies were used:

1. Time-average method using He-Ne laser ($\lambda = 0.63$ μm). This method averages steady state dynamic processes and produces the statistical distribution of the deformations.

2. Time delay laser impulse method (double exposition) using impulse ruby laser ($\lambda = 0,690$ μm). This method is applied when the processes are transient or non-regular (chaotic), when averaging in time may demolish the whole picture of deformations.

The optical measurement was performed using time-average holographic interference method. As example the hologram of the cylindrical body is presented in case when the angle between the incident light vector and the axis of the cylinder is different (Fig. 3.18). Such method of investigation enables to define the main characteristics of the analyzed object and its response to harmonic excitation at defined fixing conditions. Anyway, if the response to the harmonic excitation turns to be more complicated due to the non-linearity of the system, the discussed analysis method turns to be not applicable. Thus the time-delay laser impulse method (double exposition) is applied for the vibration processes which enables to register transient and non-regular processes of vibration.

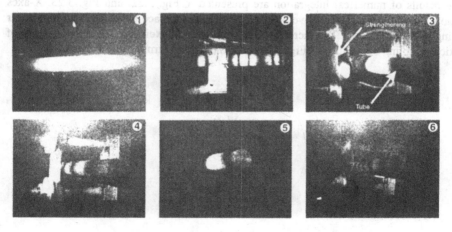

Figure 3.18. Hologram of the exhaust tube - harmonic reaction at $p = 1.25$ kHz.
The incident light vector angle: 1) 75°, 2) 90°, 3) 50°, 4) 55°, 5) 45°, 6) 40°.

320

It can be clearly seen from Fig. 3.19 - Fig. 3.21 that the variation of the amplitude of forced vibrations (of the actuator) leads to the extensive change of the shape of deformations of the cylindrical body. Its obvious, that the increase of excitation amplitudes generated stochastic dynamic behavior of the system.

Figure 3.19. Hologram of the exhaust tube nearby of the fixing joint at a_0=0.5 mm, p = 1.25 kHz

Figure 3.20. Hologram of the exhaust tube nearby of the fixing joint at a_0=0.55 mm, p = 1.25 kHz

Figure 3.21. Hologram of the exhaust tube nearby of the fixing joint at a_0=0.6 mm, p = 1.25 kHz

3.3.1 *Numerical Analysis*

The results of numerical integration are presented in Fig. 3.22. and Fig. 3.23. X-axes denote displacement, Y-axis - velocity of the cylinder surface at the point of non-linear fixing in the direction z. The increase of the amplitude of excitation leads to a cascade of period doubling bifurcations, (Fig. 3.24) until the motion turns to be stochastic.

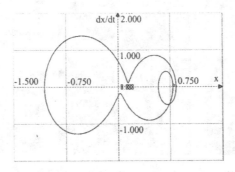

Figure 3.22. The motion of the cylinder surface point at a_0=0.5 mm, p = 1.25 kHz

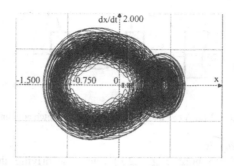

Figure 3.23. The motion of the cylinder surface point at a_0=0.56 mm, p = 1.25 kHz

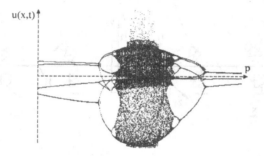

Figure 3.24. The cascade of period doubling bifurcations at increasing amplitudes of excitation force, p varies between 0.4 and 0.7, u ranges between - 0,1 and 1,0 μm

The analysis of excited transmission system is complex problem in the sense of multiple dynamic processes taking place in the system. The synthesis of a well operating system requires not only perfect motion energy transfer, but also the exploitation of some useful phenomena as stabilization of steady state rotary motion. Such analysis would be impossible without common application of mathematical, numerical and laser holography interference methods.

3.3.2 *Mathematical Model of the clutch coupling*

The system model is presented in Fig. 3.25, here mass 1 is loaded by external moment M_1 and is coupled with mass A via linear elasticity c. Mass A rotates with constant angular velocity ω. Mass 1 is coupled with mass 2 via inertial coupling 1 - 2. Mass 2 is loaded by external moment M_2. I_1 and I_2 denote constant inertia moments of masses 1 and 2.

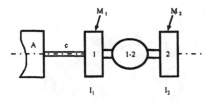

Figure 3.25. A two mass system with inertia clutch coupling

3.3.3 *Experimental analysis*

The experimental model is built (Fig. 3.26.) where the driving shaft 1 and the driven shaft 2 are connected by half clutches i ($i = 1,2$).

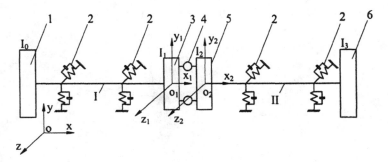

Figure 3.26. Dynamic model of the rotor system: *1* – engine rotor; *2* – elastic supports; *3* – first half clutch; *4* – half-clutch elastic connecting couplings (composite elastic elements); *5* – second half clutch; *6* – rotor of the driven mechanism; *I* – driving shaft; *II* – driven shaft; I_i ($i = 0,1,2,3$) –inertia moments: I_0, I_3 – inertia moments of the engine and the driven mechanism; I_1, I_2 – inertia moments of the first and second half-clutches

The scheme of the analyzed clutch is presented in Fig. 3.27.

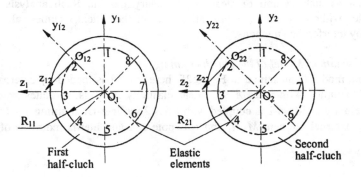

Figure 3.27. Scheme of the geometry of the clutch

The output vibrations on the body of the driven mechanism were measured by piezoelectric accelerometer. The results of the measurements are presented in Fig. 3.28. It is important to note that Fig. 3.28 presents not frequency spectrum but total vibration levels at different frequencies of the shaker table.

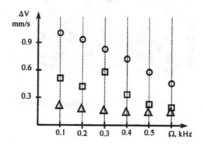

Figure 3.28. Increase of total vibration level versus excitation frequency, here O denotes the engine's vibration level increase; □ denotes the output mechanism's vibration level increase; Δ denotes the theoretical output vibration level increase

The analysed half-clutch was mounted onto the measurement table. The fixing construction was analogous to the mounting on the driving shaft. The table was excited by a shaker thus enabling the registration of deformation patterns on the surface of the half-clutch.

Fig. 3.29 shows the hologram of the half-clutch at excitation frequency 0,3 kHz. Obviously, the half-clutch deformations are in resonance. Fig. 3.30 show the same half-clutch at the same excitation viewed at different angles what proves that vibrations are taking place both in normal and tangent directions.

Fig. 3.29-3.30 prove that the analyzed half-clutch turns to be itself a source of vibrations in case the harmful engine vibration is taking place at 0.3 kHz. Naturally, some changes in the design of the half-clutch are necessary to make the operation of the whole system to be more smooth.

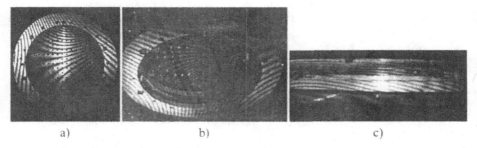

a)   b)   c)

Figure 3.29. Hologram of the half-clutch: a) viewing angle 90° b) viewing angle 45° c) viewing angle 10°

324

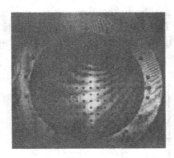

Figure 3.30. Hologram of the modified half-clutch

References

1. Preumont, A. (1997) *Vibration Control of Active Structures. An introduction,* Cluwer Academic Publishers, pp. 258.
2. De Man, P., Ostaševičius, V. (1995) *Reaction Force Compensator Report,* Mechanical Engineering and Robotics Department, ULB, Bruxelles.
3. Ostaševičius, V. and all, *Twist Drill With Variable Cross Section,* Sertificat of invention SU Nr: 1357151.
4. Ostaševičius, V. and all, *An Insertion for Strenghening the Cutting tool,* Sertificat of invention SU Nr: 1590206.
5. Barauskas, R., Ostaševičius, V. (1999) *Analysis and Optimization of Elastic Vibro-Impact Systems,* "Technologija", Kaunas, pp. 213.
6. Ostaševičius, V., Sapragonas, J. (1999), *Analysis of Dynamic Characteristics of Acoustic Deflectors* 1, Proc. 1999 Noise And Vibration Conference, Michigan, Vol. 3, pp. 1449-1455.
7. Ostaševičius, V., Sapragonas, J., and Dagys, D. (2001) *Analysis of Dynamic Characteristics of Acoustic Deflectors* 2, Proc. Automotive And Transportation Technology congress and Exhibition, SAE Barcelona, Spain October 1-3, 2001, Vol. 1, pp. 215-221.
8. Ostaševičius, V., Gaidys, R. (1988) *Optimal Design of Vibration-Impact Systems,* J. Computers in Industry, Vol. 3, pp. 36-42.

AN APPROACH TO SMART STRUCTURE DESIGN USING MEMS TECHNOLOGY

P. MINOTTI
Institut des Microtechniques de Franche Comté
Laboratoire de Mécanique Appliquée R. Chaléat
Université de Franche Comté
24 rue de l'Epitaphe, 25030 Besançon Cedex, France

1. Introduction

The arrival of integrated circuits with very good performance/price ratios and relatively low-cost microprocessors and memories has had a profound influence on our way of life. In particular, electronic circuits were introduced on a large scale in the measurement and control field, leading to very sophisticated systems and opening new perspectives on active control of structures. However, most of current control systems involve sensors and actuators that were conceived a few decades ago. Accordingly, it becomes necessary to investigate new machining technologies that will allow sensors and actuators to be produced with performance/price ratios which would approach that of modern electronic circuits [1].

Increasing demand for new electromechanical devices combined with electronic circuits initiated worldwide research and development programs on microsystems technology (MST). Most of investigations focused on silicon micromechanics so that, with reference to the microelectronic model, the opinion was that modern electromechanical devices would have to be monolithically integrated with integrated circuits (ICs) in one single chip [1]. Thus, silicon has become familiar as the material from which electronic components and modern control systems are made [2]. So, even if conventional precision mechanics also made important contributions to microsystems technology, it should not be overlooked that basic concepts of MST and MEMS (e.g. the modern Anglo-Saxon acronym for microelectromechanical systems), originate actually in microelectronics.

A decisive element in the MEMS technology is photolithography by which the patterns designed on the computer are transferred to the silicon monocrystal wafer. According to optical imaging limitations, only two-dimensional patterns can be transferred on the work piece. Consequently, shaping limitations using photolithography techniques lead to a major drawback of the MEMS technology because we are accustomed to think, design and manufacture in the three-dimensional world. However, photolithography allows the machining of devices much smaller than conventional tool-machined components because pattern details are limited by the wave length of light only. In addition, photolithography allows extremely high accuracy of reproduction as well as batch manufacturing potentialities, according to very high flows of information due to parallelism of optical imaging. Therefore, photolithography techniques allowing thousands of devices to be produced in parallel on a silicon wafer, stand out from most of other tool manufacturing

A. Preumont (ed.), Responsive Systems for Active Vibration Control, 325–377.

techniques. Accordingly, despite the rise in total expenditure of manufacture, a drastic reduction in costs of a single component has been possible, therefore opening new design perspectives in the field of micrometer size systems.

2. MEMS Origins

2.1. THE REVOLUTION OF ELECTRONICS INDUSTRY

For almost 50 years after the turn of the 20th century, the electronic industry had been dominated by vacuum tube technology. But vacuum tubes had inherent limitations such as fragility, unreliability, power consumption and heat production. So when 50 years ago the transistor was invented by W. Shockley, this marked the onset of a technological change which had a permanent influence on the lives of all of us and will continue to do so. Nevertheless, even if the contact point transistor developed in 1947 by Bell Telephone Laboratories still remains the major invention of the 20th century (see Fig. 1), it wasn't until that time that the vacuum tube problem was solved.

Compared with vacuum tubes, transistors were minuscule, more reliable, produced less heat and consumed less power, leading to design ever more complex electronic circuits containing thousands of discrete components. But the problem was that these components still had to be interconnected to form electronic circuits, and hand-soldering thousands of components to thousands of bits of wires was expensive and time consuming. Thus, the challenge was to find cost-effective reliable ways of producing these components and interconnecting them.

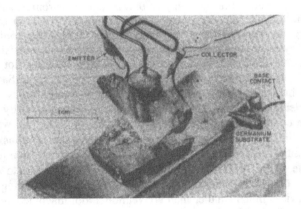

Figure 1. The "Contact Point Transistor" by W. Shokley (Nobel Prize in Physics – 1948).

In 1958, Jack Kilby (TEXAS INSTRUMENTS) conceived the first electronic circuit in which all the components , both passive and active, were fabricated in a single piece of semiconductor material 7/16-by-1/16-inches in size. As shown in Figure 2, Kilby's microchip was a rough device made from a slice of germanium with only a transistor and a few other components interconnected through external gold wires. But when Kilby pressed the switch, the first IC worked, showing an unending sine curve undulated across the oscilloscope screen. Even if, in a first time, industry reacted skeptically, the monolithic integrated circuit shown in Figure 2 was about to revolutionize the electronics industry and

Jack Kilby will awarded the Nobel Prize in Physics 42 years after his invention.

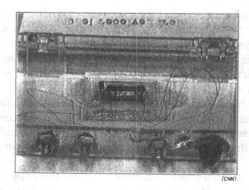

Figure 2. The first Integrated Circuit by J. Kilby (1959). After Texas Instrument, http://www.ti.com.

In 1959, J. Hoerni improved former IC's external connections by using flat metallic circuit lines deposited on a silicon oxide insulated layer. A few years later, R. Noyce (who co-founded INTEL with G. Moore in 1968) developed the first industrial IC-fabrication process that will create a few years later the modern computer industry, transforming former room-size machines into today's array of mainframes and personal computers.

Since early sixties, the growth in memory chip performance has obeyed the Moore's Law (e.g. each new chip integrates roughly twice as much transistors as its predecessor every 18 months). Accordingly, number of transistors on a chip has increased more than 3,200 times, from 2,300 (on the *"4004 Microprocessor"* that was the first Intel's processor commercialized in 1971) to 7.5 million on the *Penthium II* processor developed from 1997.

Today, Moore's Law still holds true as we enter the 21st century. The new *Pentium IV* processor generation recently debuted with 42 million transistor and circuit lines of 0.18 microns (see Fig. 3), with initial speed of 1.5 gigaHertz, while Intel's first *4004 microprocessor* ran at 108 kilohertz 30 years ago. If automobile speed had increased similarly over the same period, we would be able to drive from San Francisco to New York in about 13 seconds !

Figure 3. The processor "Pentium IV" . After Intel, http://www.intel.com.

2.2. SILICON MICROMECHANICS USING IC-TECHNOLOGY

From early sixties, Richard Feynmann has been certainly the main pioneer in the field of silicon micromechanics. On December 26, 1959, he gave a talk at the annual meeting of the American Physical Society at the California Institute of Technology, which has been transcript afterwards into the famous paper *"There's Plenty of Room at the Bottom"*[3]. His descriptions of the micro world were intriguing and more than 40 years later, the content of his talk still addresses many current research issues.

Another speech from Feynman, which is a sequel to the famous 1960 talk, was delivered in 1983 at the Jet Propulsion Laboratory (JPL). The second Feynman's talk was also remarkable in many ways and led to a manuscript entitled *"Infinitesimal Machinery"*, today available from the Journal of Microelectromechanical Systems [4]. As cited in [4] by Stephen D. Senturia, *"Feynman anticipated the sacrificial-layer method of making silicon micromotors, the use of electrostatic actuation , and the importance of friction and contact sticking in such devices. He explored the persistent problem of finding meaningful applications for tiny machines, touching a range of topics along the way. And he looked at the future of computation using a register made of atoms, and quantum-mechanical transitions for computation operations"*.

Feynman's anticipation gave rise to a worldwide motivation for microsystems and fundamentally contributed to the development of the modern MEMS technology derived from IC-fabrication processes. Thus, after tiny valves, nozzles and pressure sensors, new generations of sophisticated mechanical systems such as micrometer size actuators were progressively developed from late1980's, using IC-planar technologies combined with the sacrificial-layer technique.

3. Top-Down" and "Bottom-Up" Design Approaches in Microsystems

3.1. TOOL- MAKING AND MASK-MAKING PHILOSOPHIES

Tool-making for mechanical machining began perhaps a million years ago when humans learned to walk erect and had hands free to grasp objects. Since that time, humans have been cutting metal to make the myriad tools, machines and other devices our civilization demands [5].

On the other hand, the "IC philosophy" has been recently successfully applied to a wide variety of non-IC products through modern mask-making approaches. Thus, the design of a microsystem can be potentially investigated through both tool-making and mask-making technologies that are two fundamentally distinct ways :

- the *"top-down"* approach, which makes use of historical tool-machining technique so as to downsize conventional mechanisms through miniaturization of existing devices,

- the *"bottom-up"* approach, which makes use of revolutionary mask-machining techniques in such a way to design futurist structures combining many elementary components duplicated within a parallel process flow (see Fig. 4).

So when designing a microsystem, engineers have now the choice between traditional precision engineering and modern micromachining techniques and it is not yet clear which approach gives the most appropriate answer.

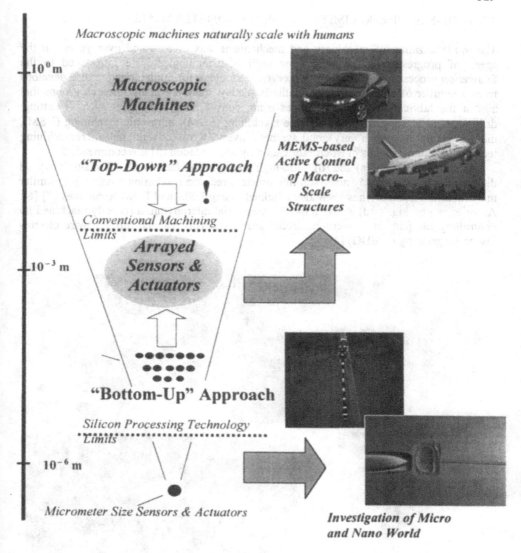

Figure 4. Designing electromechanical systems through "*Top-Down*" and "*Bottom-Up*" approaches.

3.2. TOOL-MACHINING LIMITS ON THE MICROMETER SCALE

The miniaturization of machinery and mechanisms was investigated over years but the speed of progress was relatively slow until IC-technologies were introduced to the fabrication process of mechanisms. However, technology for miniaturization also develops from a number of other fabrication methods. Today, there are many research groups that aim at the fabrication of small devices using conventional techniques such as cutting, drilling, sand blasting, electro-discharge machining (EDM), ultrasonic machining (USM), mold injection etc. For a very good synthetic overview of conventional micromachining techniques, Fundamentals of Microfabrication by Marc Madou [5] is recommended.

As shown in Figs. 5(a) and (b), ultra-high precision machines with sharp single-crystal diamond tools have already made sub-micrometer precision machining possible [6]. Similar manufacturing accuracy has been also obtained using USM and EDM techniques [7] [8]. As an example, Fig. 5 (d) shows a three-dimensional micropyramid recently machined by controlling the path of a micro-electrode (see Fig. 5(c)) realized through wire electro-discharge grinding (WEDG) [8].

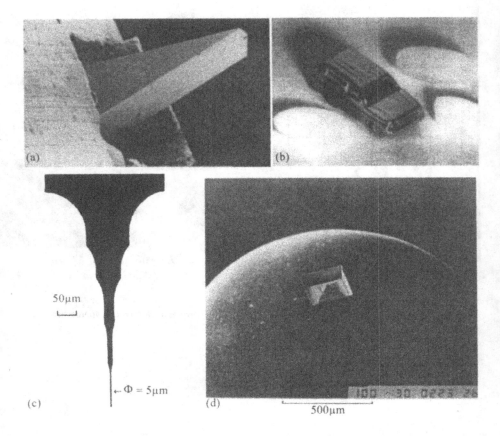

Figure 5. Ultra-high precision tool-machining : (a) single crystal diamond tool, (b) tool-machined rice tip-sized car (after [6]), (c) micro-tool realized through WEDG and d) three-dimensional micropyramid respectively machined by controlling the path of a WEDG electrode. After [8].

Finest details that can be machined using conventional tool-machining technology are today in the range of 10 to 100 microns. Thus, even if such a manufacturing accuracy remains one to two orders larger than what photolithography makes possible (the smallest details in current ICs are 0.18 micron in size), conventional machining must be seen as a set of tools to solve practical problems on micrometer scale devices. Even more: conventional techniques allows the machining of components which are not restricted to flat substrates, therefore avoiding shaping limitations inherent to the photolithography technique.

The main limitation of traditional machining approaches is that using conventional machining technology, the components of a system must be made piece by piece(e.g. serial machining approach). So, even if conventional precision mechanics satisfies most of current miniaturization constraints, it should not be overlooked that cost-effective parallel integration of microsystems cannot be achieved using serial machining approach.

3.3. POTENTIALITIES OF MASK-MACHINING TECHNIQUES

Technologies involving optical imaging through the photolithography technique naturally overcome the major limitation of serial machining approaches on the micrometer scale. In photolithography, light impinges a transparent plate with opaque regions which is called a mask (see Fig. 6). Below the mask a photosensitive material is placed, which can be developed after illumination. The illuminated regions are dissolved (or the non-illuminated ones), so that the underlying material becomes accessible to further processing. Thus, due to intrinsic parallelism of optical imaging, thousands of components can be machined within a common process flow, therefore satisfying most of economic constraints related to miniaturization of systems.

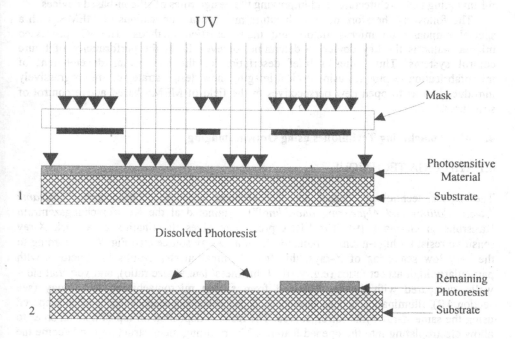

Figure 6. Basic photolithography process

3.4. MINIATURIZATION AND SYSTEMS

Microthings are necessarily systems because of the small details in micrometer size systems. Because micrometer size objects are too small to be assembled at reasonable costs, microsystem components have to be designed during the design of a whole system (e.g. there is no design phase of the system separate from the design phases of components). Accordingly, the design of a microsystem is an integral process. Even more, the fabrication of the microsystem is an integral process which is being provided in particular by silicon miromachining.

When designing a microsystem, the fabrication process of the system must be designed too. Because of the strong correlation between a micromachining process and the resulting mechanical architecture, both system design and process should be integrated in a single design process. Thus, microsystem design approaches differ from macrosystem design rules because conventional assembling procedures using available components (such as motors and wheels in car industry...) is impossible with microthings [1].

In contradiction with IC-design and processing were there are today strict and clear design rules and well established processes, MEMS designers must have a large number of skills and should be able to design both the system and the fabrication process. In addition to these constraints, MEMS designers have to work in groups of engineers of different disciplines, because microsystems usually have complex functions which have roots in various physical domains.

During the last decades, electrical engineers contributed to the development of the MEMS technology and nobody doubts the feasibility of microsystems using IC-fabrication techniques. However, the involvement of mechanical engineers has become indispensable for inventing new architectures and improving the design rules of silicon-based devices.

The following therefore deals with future mechanical applications of MEMS, with a special emphasis on microactuators and their actuation methods. The IC- processed microactuator is the key device to determine, or even limit, the performance of future control systems. Thus, after a brief description of the spectacular development of microfabrication techniques using optical imaging, new design strategies are progressively introduced so as to open new perspectives in the field of MEMS-based active control of structures.

4. Micromachining Techniques using Optical Imaging

4.1. THE LIGA TECHNIQUE

The LIGA technology (based on the German acronym "*Lithography, Galvanik (electroplating) and Abformung (moulding)*"), originated at the Kernforschungszentrum Karlsruhe in Germany [9]. The LIGA process involves illumination of a thick X-ray sensitive resist by high-intensity radiation from an X-ray source (see Fig. 7). According to the very low scattering of X-rays, this first illumination step results in structures with particularly high aspect ratios (e.g. vertical thickness/ lateral size ratio), and very flat side-walls, compared with devices obtained from silicon micromachining technology (see section 4.2). Illumination is done through an X-ray absorbing mask which is often prepared using the same X-ray synchrotron source. The resist is deposited on top of a metal so as to allow electroplating into the opened features. The resulting metal structure can become the final part, or can be used as an injection mold for parts made out of a variety of plastics.

Polymers microcomponents fabricated this way are on the market now. However, the development of microsystems using LIGA process is very expensive, according to the illumination step cost and to the numerous masks that must be prepared before getting a stabilized process flow.

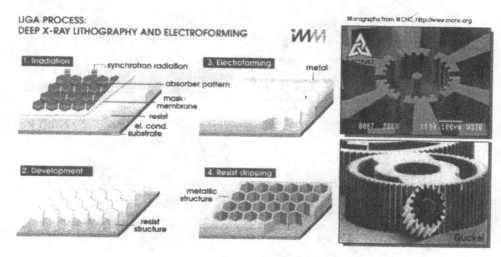

Figure 7. The LIGA process. After [9].

4.2. THE SILICON MICROMACHINING TECHNOLOGY

The arsenal of technologies for silicon micromachining includes photolithography, thin film deposition, doping, etching by wet chemical and plasmas and waferbonding. These technologies have been developed from early sixties in order to allow self-assembled microsystems, such as pressure sensors, to be inexpensively manufactured. Silicon micromachining starts from wafers used by the electronics industry. As shown in Figure 8, wafers are sliced out of a single pure crystal of silicon. Bulk-micromachining and surface-micromachining are the two main techniques that rely on making whole mechanical systems in place, etching out unwanted material, and mass assembly of thousands of devices with a single operation [10].In the following, we give a brief account of notions strictly necessary to understand architectural limits (and corresponding design constraints) that deal with silicon micromachining. More information on silicon micromachining can be found in recent literature [11] [12] [13].

Figure 8. Preparation of silicon wafers. Wafers are sliced out of a single pure crystal of silicon and then polished before being used for further processing.

4.2.1. *Bulk-Micromachining of Single Crystal Silicon*

In most manufacturing, the metric is defined by the tooling used to machine the part. In bulk-micromachining, the metric is also contained in the part being made 14]. Under the correct conditions, chemicals etch the different crystallographic planes within the single crystal silicon at different rates. As shown in Figure 9, these anisotropic etches usually remove material from the faces <111> planes much more slowly than other planes. This differential etching allows the fabrication of an amazingly wide range of parts.

During manufacture of a part, all the wafer is coated with a protective layer, except for carefully designed windows. When the wafer is placed in the anisotropic etch, the exposed silicon is etched away until a stop plane is reached. Isotropic etching can also be achieved so as to leave rounded cavities, by using different chemicals and processing conditions. More complicated shapes can be made by bonding different wafers and other materials, such as glass plates, together.

A breakthrough was achieved in early 1980's when bulk-micromachined pressure sensors were introduced in the market. Conventional metal membranes were replaced by monocrystalline silicon membranes, which suffer much less from creep, fatigue and hysteresis. Furthermore, the combination of small size and the high elastic modulus and low density of silicon allowed the development of sensors having a very high resonance frequency. Such technical advantages combined with excellent performance/price ratios logically led to the exponential growth of the MEMS sensor market [10].

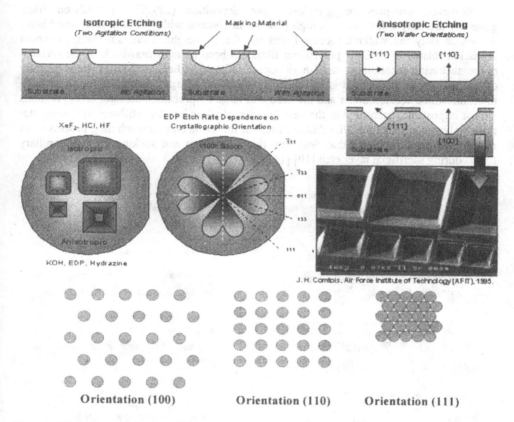

Figure 9. Anisotropic and isotropic etching of single crystal silicon. Chemicals etch the different crystallographic planes at different rates, leading to a wide range of parts. After L. Ristic, ed., Sensor Technology and Devices, ArtechHouse, 1994).

4.2.2. *Surface-Micromachining of Polycrystalline Silicon*

Surface-micromachining is a self-micro assembly technique that allows parallel fabrication of free standing and freely moving microstructures such as microactuators. The process idea is given in Figure 10. Layers are successively deposited and patterned on a single crystal silicon wafer acting as a substrate. Although many combination of materials is possible with surface-micromachining, this technology has developed to a high standard for the combination of silicon dioxide as the sacrificial material, and polycrystalline silicon (polysilicon) as the main structural material.

Mechanical properties of polycrystalline silicon are excellent. Polysilicon strength is as high as 2 – 2.5 GPa, and is thus stronger than steel which as usually a strength of 200 MPa – 1 GPa. In addition, polysilicon films are extremely flexible (e.g. the maximum strain before fracture is ~ 1%) , and do not readily fatigue. Even more important : polysilicon is currently used as the primary material comprising the gate electrode of transistors and thus, appears to be directly compatible with modern IC fabrication processes. Accordingly, polysilicon surface micromachining is being aggressively pursued and applied by many in the MEMS community [10].

336

However, low-pressure chemical vapor deposition (LPCVD) polysilicon films generally are only a few microns high (low z), in contrast with wet bulk micromachining for which only the wafer thickness limits the feature height. Thus, structural thickness limitation related to LPCVD polysilicon films has been a major drawback of the surface-micromachining technology, until high aspect ratio polysilicon structures were recently developed. As an example, it is still difficult to fashion a large proof mass for an surface-micromachined accelerometer and the design of large IC-processed actuator arrays remains almost impossible, according to the very low out-of-plane bending stiffness of micrometer thick polysilicon structures. In addition, users of the surface-micromachining process must master two basic problems that deal with thin film stress and sticking through capillary forces during sacrificial layer etch [10] [14].

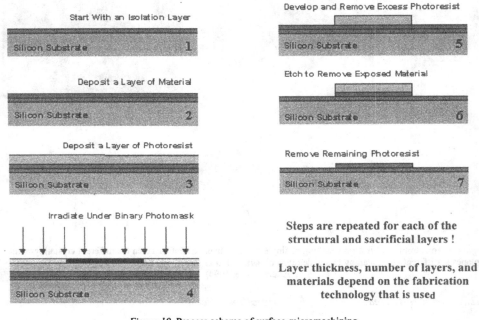

Figure 10. Process scheme of surface-micromachining.

5. Processing Issues for the Monolithic Integration of Microelectronics with MEMS Sensors and Actuators

5.1. MULTILAYER SURFACE-MICROMACHINING PROCESSES

5.1.1. *Three-Layer Polysilicon Surface-Micromachining Process*

Silicon micromachining requires significant investments for specific equipment and clean rooms are not yet available in most of universities and research institutions involved in advanced MEMS development. Thus, commercial programs providing surface-micromachining process facilities through MEMS foundries have been recently developed in Europe and United States. The multi-user MEMS processes (MUMPs) from *CRONOS Integrated Microsystems* is one of these programs. The MUMPs program provides governmental and academic institutions with cost-effective surface - micromachining fabrication.

Figure 11 gives an example of an IC-processed electrostatic micromotor using *CRONOS'* facility. The standard three-layer polysilicon surface micromachining process shown in Fig. 11 has been derived from work performed at the Berkeley Sensors and Actuators Center (BSAC) at the University of California in late 80's [15] [16]. Polysilicon is used as the structural material, silicon oxide is used as the sacrificial layer, and silicon nitride is used as electrical insulation between the polysilicon and the substrate. The MUMPs process is designed to be as general as possible, and to be capable of supporting numerous different designs on a single silicon wafer, therefore allowing many customers to share the same fabrication run. Since the process was not optimized with respect to one specific device, the thickness of both structural and sacrificial layers were chosen to suit most potential users. Nevertheless, according to the micrometer thickness of polysilicon films, the external diameter of IC-processed micromotors using *CRONOS'* facility remains restricted on the sub-millimeter scale.

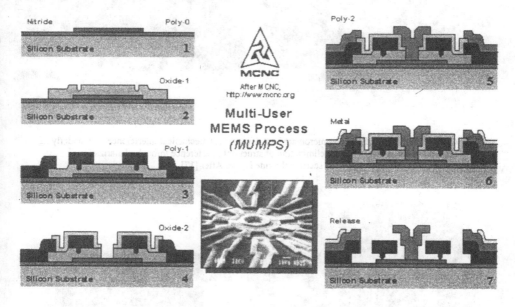

Figure 11. Process scheme of an IC-compatible electrostatic micromotor using the MUMPs process. The device integrates three polysilicon structural layers . After *MUMPs Design Handbook*, http://www.mcnc.org.

5.1.2. *Five-level Process using Chemical-Mechanical Planarization of Sacrificial Layers*

The complexity of surface-micromachined devices strongly depends on the number of mechanical layers available in the fabrication process. With one or two mechanical layers, one can create basic mechanisms such as respectively a comb-drive actuator and a gear constrained to rotate on a hub. But an entirely new range of design possibilities would be feasible by adding additional mechanical layers.

The challenge associated with adding extra layers of polysilicon to a surface micromachining process has been mainly related to residual film stress and device

338

topography [17]. MEMS foundries experienced significant difficulties to reproducibly reduce polysilicon stress to an acceptable level. Device topography, which results in structural interference that severely constrains the range of potential designs, has been a major drawback of the surface micromachining technology too. As an example, surface-micromachined structures such as shown in Figure 12(a) systematically exhibited mechanical interference with underlying polysilicon layers. However, Sandia National Laboratories recently developed a proprietary process by which it maintains stress levels to typically less than 5 MPa [17]. In addition, Sandia has implemented a new planarization method based on chemical-mechanical polishing (CMP) [18]. As shown in Figure 12(b), this planarization method results in the top layer of polysilicon being planar, enabling far greater design flexibility than in a non-planar process (see Figs. 13 and 14) [17].

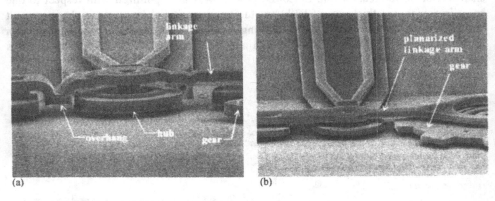

(a)   (b)

Figure 12. Topography of surface-micromachined devices : (a) mechanical interference with underlying polysilicon layers and (b) elimination of structural interference through planarization of sacrificial oxide layers. After [17].

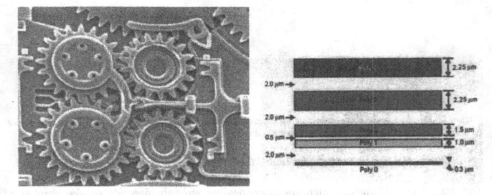

Figure 13. The standard SUMMIT V Technology developed at SNL [19]. After Sandia National Laboratories, http://mems.sandia.gov/micromachine.

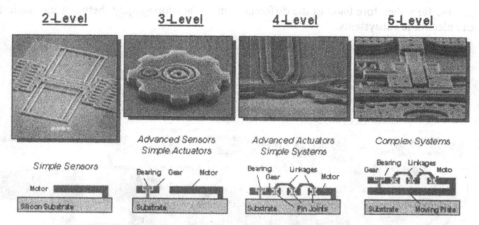

Figure 14. Device complexity as a function of the number of mechanical layers available in a surface micromachining process. After [20].

The *SUMMIT V* process, which is today the most sophisticated surface micromachining process commercialized by Sandia National Laboratories (SNL), uses up to 14 individual masks in the process (about the same number as many CMOS IC processes), therefore opening a truly enormous design flexibility. Much taller surface-micromachined devices will thus be made (up to 12 microns high), enabling greater stiffness and mechanical robustness. The additional structural height will also be used to achieve larger forces from actuators. In particular, increased structural thickness will probably allow macroscopic scale MEMS arrays to be machined on the near future, opening new perspectives in the field of MEMS-based active control of macro-scale structures.

5.2. TOWARDS CMOS/MEMS MONOLITHIC INTEGRATION

The creation of microsystems that sense, think, act and communicate, naturally requires electronic circuitry coupled with mechanical components. Accordingly, a great deal of interest has developed in manufacturing processes that will allow the monolithic integration of MEMS with driving, control, and signal processing electronics [21]. This integration promises to improve the performance of sophisticated control systems by combining the micromechanical devices with electronic subsystems in a same manufacturing and packaging process called CMOS/MEMS technology.

Previous attempts to integrate MEMS and CMOS have tried to process the MEMS devices after the CMOS circuits, using refractory metals that can withstand the high temperature annealing cycles required to relieve stress in structural polysilicon. Still other approaches have tried to interleave the necessary MEMS processing steps with the CMOS steps, leading to a rigid and constrained process flow [22]. One of the most promising approach toward integrating surface-micromachined components with the conventional CMOS circuits has been probably developed at SNL (see Fig. 15). Because MEMS devices are generally thicker than conventional CMOS circuits (e.g. respectively 4 to 10 microns against 1 to 2 microns), the MEMS components are fabricated in a shallow trench below the wafer surface prior to CMOS processing. The SNL integrated micromechanical/ CMOS process flow therefore handles the differences in surface topography between mechanical and electronic subsystems.

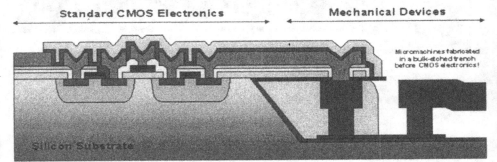

Figure 15. Integrated MEMS/CMOS Technology developed at SNL for the monolithic integration of CMOS control circuits and MEMS sensors and actuators on the same silicon chip. After [21].

5.3. FULLY-MONOLITHIC INTEGRATION LIMITS

Despite spectacular developments described in section 5.2, fully integrated microsystems technology is actually not realistic because the combination of so many differing processes would be associated with serious problems of yield. Therefore, former fully monolithic integration concept has been recently substituted with a more realistic way of looking at things [1]. Actually, the hybridization concept will probably succeed as a technological middle course with the components machined in different processes joined to a system on a substrate.

The main motivation for MEMS microassembly is the growing recognition of the limitations of silicon lithography and etching as the universal platform for silicon-based microsystems. Monolithic integration of electronics and micromechanics inevitably compromise both subsystems. Modern CMOS processes use 8 inches diameter wafers and are shifting to 12 inches presently, while MEMS are still built on 4 inches or 6 inches substrates. Therefore, the fully monolithic integration of CMOS and MEMS on the same substrate would need the use of premium CMOS process equipment on smaller wafers, leading to higher production costs and lower-performance electronics [23].

The concept of parallel microassembly technologies shown in Figure 16 will thus be the key for future MEMS industrialization. Parallel assembly involves the simultaneous precise organization of arrays of microcomponents, which can be achieved through direct microstructure transfer between aligned wafers (e.g. *"deterministic parallel microassembly"*), or by massive manipulation of MEMS using various self-assembly concepts (e.g. *"stochastic microassembly"*) [23].

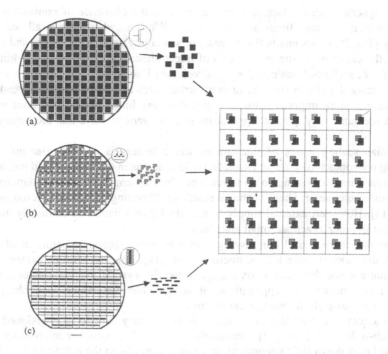

Figure 16. Concept of parallel microassembly concept involving massive hybridization of various MEMS components such as CMOS electronic circuits, mechanical microstructures and photonic devices. After Cohn and Böhringer [23].

Parallel micromanipulators involving actuator arrays and programmable vector fields constitute one of the most promising approach towards massive MEMS hybridization. In particular, recent investigations on controlled self-microassembly via electromechanical modulation of surface properties will be discussed later, as a prospective research in the emerging field of MEMS-based active control.

6. Some Aspects Regarding Scaling of Microsystems

It is interesting to consider what the problems are, when designing smaller machines. Firstly, with parts stressed to the same degree, the forces go as the area we are reducing, so that weight and inertia are of relatively no importance. The strength of materials, in other words, becomes very much greater in proportion. On the other hand, the metals commonly used to build macro-scale machines have a grain structure, and this would be very annoying at small scale because the material cannot still be considered to be homogeneous. Therefore, plastics, glass and things of this amorphous nature are very much more homogeneous , and so we would have to preferably make micrometer size machines out of such materials [3].

There are thus a few matters one needs to keep in mind when miniature technical systems are designed. A few things, in particular, are quite obvious. Linear dimensions shrink slower than surfaces and surfaces shrink slower than volumes. So, friction associated with surfaces becomes relatively more important than masses. But also stiffness, electric forces, magnetic forces, adhesive forces as well as the character of convection, surface finish and characteristics times associated with diffusion and resonance all scale in their own way [10]. It is thus instructive to analyze scaling by comparing forces and effects by studying the exponent of the length scale of linear dimensions l. Then, linear dimensions l scale as $l \sim l^1$, surfaces S scale as $S \sim l^2$ and volumes V scale as $V \sim l^3$. So when a system is reduced isomorphically in size, the ratio of surface area to volume increases, rendering the surface forces more important. Because shrinking the linear dimensions does not shrink forces in the same way, physical properties scale in various ways that are summarized in Table I.

The size scaling issues of micrometer size actuators is of particular interest when designing micromachines. As an example, rapid heat loss in the micro world would prevent the gasoline from exploding, so an internal combustion engine would be inappropriate on the sub-millimeter scale. Other chemical reactions, liberating energy when cold would thus be used [3]. However, external supply of electrical power would be obviously much more convenient in order to actuate small machines.

Of course, there is still controversy over whether magnetic actuators or electrostatic actuators are more suitable for the micro domain [14]. The magnetic properties on a very small scale are not the same as on a large scale. So, even if magnetic actuators are still dominant for macro-scale applications, it is not clear yet which technology will be dominant for future electromechanical systems.

Microactuators cannot simply be scaled down, so they have to be redesigned carefully on the sub-millimeter scale [24]. One scaling law that is helpful for microactuators is the increased break down field strength of very small gaps due to the Paschen effect [1]. Thus, for small sizes, the obtainable electrostatic forces can be stronger than electromagnetic forces [25]. In addition, since electrostatic devices can be constructed using only conductors and insulators, they can be made compatible with silicon micromachining techniques [26]. Nevertheless, recent improvements in material processing now allow deposition of thin-film magnetic materials, so integrated micro-magnetic systems are now possible. The suitability of an actuator therefore depends greatly on the application and besides scaling, many other factors need to be considered when deciding upon a certain type of actuation (see Table II).

TABLE I. Scaling of various physical properties. After Elwenspoek and Wiegerink [10].

Linear Dimension	l	$\sim l^1$
Surface	A	$\sim l^2$
Volume	V	$\sim l^3$
Electric Energy	W_{el}	$\sim l^3$
Electric Force	F_{el}	$\sim l^2$
Magnetic Energy	W_{mag}	$\sim l^5$
Magnetic Force	F_{mag}	$\sim l^4$
Structure Deflection under Own Weight	ξ	$\sim l^2$
Stable Length	L_{cr}	$\sim l^{4/3}$
Diffusion Time	τ	$\sim l^2$
Drift Velocity	υ	$\sim l^2$
Transient Time	τ	$\sim l^2$
Electric Resistance	R_{el}	$\sim l^{-1}$
Hydraulic Resistance	R_{hy}	$\sim l^{-3}$
Reynolds Number	R_e	$\sim l^2$
Resonant Frequency	f	$\sim l^{-1}$

TABLE II. Comparing electrostatic and magnetic actuation

Attributes of Electrostatic Actuators (I^2)	Attributes of Electromagnetic Actuators (I^4)
- Thin insulated layers exhibit breakdown voltage as high as 2MV/ cm.	- Despite scaling effects, the achievable absolute forces are still very large
- Corresponding energy density: $7 \cdot 10^5 J/m^3$	- In conventional motors using iron, the magnetic induction is restricted to 1.5 T
- Contracting pressure: 1.3 MPa	- Corresponding energy density: $9 \cdot 10^5$ J/m^3 *(more than twice the achievable electrostatic energy in a 1 µm gap)*
- 100 V driving voltage sufficient to generate the strong field mentioned	
- Simple actuation (pair of electrodes)	- Thin ferromagnetic films yielded 2T fields
- Voltage switching far easier and faster than current switching	- Unfortunately, 10 to 15 T superconducting magnets fall outside the micromachining domain
- Low energy loss through Joules heating	
- Low weight and power consumption	- Friction easier to avoid in magnetic actuators
- IC - compatible	- Less sensitive to dust and humidity

Besides scaling, many other factors need to be considered when deciding upon a certain type of actuation.

7. IC-Processed Actuators

7.1. INTRODUCTION

While electrical devices such as sensors and electronic circuits are today well established technically, the study of sophisticated mechanical devices such as micromechanisms and microactuators began only fifteen years ago when the first MEMS workshop was started. Since then, researchers have achieved remarkable progresses that are briefly summarized in the following.

The first IC-processed electrostatic micromotors with diameters of 60 to 120 microns were developed in 1987 (section 7.2.1). Rotational speed was relatively low (e.g. on the order of 500 rpm) because of the parasitic friction between the stator and the rotor. Later improvement by Mehregany enabled the rotational speeds up to 15,000 rpm and continuous operation fore more than a week (section 7.2.2). Even for the improved micromotors, friction has been identified to be a major problem, leading subsequently to replace the sliding contact at the center with rolling contact of an eccentric rotor that rotates without slipping at the contact with the shaft (section 7.2.3). Wobble micromotors exhibited two advantages compared with former electrostatic actuators, e.g. reduction of friction and higher torque at low speed.

Another way to avoid friction has been investigated from early 1990's by using elastic suspensions coupled with the moving part (e.g. the rotor). The most popular one was the comb-drive actuator which is supported by cantilever beams and actuated by interdigitating

344

polysilicon comb-like structures (section 7.2.4). Nevertheless, structural height limitations of surface-micromachined comb-drive actuators resulted in unusable electrostatic forces (e.g. typically less than 10 μN) that cannot address most of prospected industrial applications [27].

7.2. DRIVING LIMITATIONS OF EARLIER ELECTROSTATIC ACTUATORS

7.2.1. *Side-Drive Electrostatic Actuators*

Very first micromotors that have been produced using integrated circuit processing were studied from late 1980's at the Berkeley Sensor and Actuator Center (BSAC) [15] [16]. Side-drive electrostatic actuators of the type shown in Figure 17 were formed from 1.0 to 1.5 micrometer thick polycrystalline silicon films deposited on a phosphosilicate glass (PSG) sacrificial layer. Taking into account the restricted micrometer thickness of the polysilicon films and the resulting parasitic structural flexibility, the diameter of the former electrostatic actuators has been limited to a few tens micrometers (typically 60 to 120 microns). Accordingly, the shaping limitations did not allow significant electrostatic energy to be stored within former electric field driven actuators, leading to unusable loading characteristics. Thus, with a potential V applied across the driving electrodes of a side-drive actuator, the stored energy within the stator/ rotor interface is given by :

$$U(\vartheta) = -\frac{1}{2}C(\vartheta)V^2 = -\frac{n\varepsilon_r\varepsilon_0 S(\vartheta)V^2}{2d} = -\frac{n\varepsilon_r\varepsilon_0 hR\,\vartheta V^2}{2d} \tag{1}$$

where :
- C is the capacitance across the driving electrodes,
- n is the number poles of the actuator,
- ε_0 is the free-space permittivity ($8.85 \; 10^{-12}$ F/m),
- ε_r is the effective dielectric constant between the capacitor plates,
- h is the thickness of the polysilicon film (e.g. altitude of the driving electrodes),
- R is the radius of the stator/ rotor interface
- d is the gap between driving electrodes

The torque (Γ) exerted by the electric field onto the polysilicon rotor can be expressed in terms of the stored energy (U), by taking the derivative with respect to the rotor angular position (ϑ). Neglecting fringing fields, the electrostatic torque can be approximated by :

$$\Gamma = -\frac{\partial U}{\partial \vartheta} = \frac{n\varepsilon_r\varepsilon_0 hRV^2}{2d} = \frac{1}{2}n\varepsilon_r\varepsilon_0 hRdE^2 \tag{2}$$

with :

$$E = \frac{V}{d}, \tag{3}$$

is the electrostatic field across the capacitor plates.

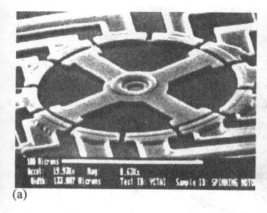

(a)   (b)

Figure 17. (a) First IC-processed side-drive micromotor. After [15]. High-driving voltage ranging from 300 to 400 Volts was needed to initiate stepwise rotations of the polysilicon rotor, (b) SEM micrograph showing the stator/rotor interface of a surface-micromachined side-drive micromotor. After LAAS CNRS-France.

Equation (2) suggests that if the driving electrodes are slightly displaced with respect to each other, an electrostatic torque (Γ) is developed to realign the driving electrodes. Neglecting again fringing effects, the torque remains theoretically constant when the driving electrodes have a substantial overlap. On the other hand, when the driving electrodes do not overlap, or overlap completely, the fringing effects are predominant and equation (2) becomes a rough approximation.

In addition to shaping limitations such as restricted altitude (h) of the driving electrodes, parasitic friction appeared to be a major problem. The coefficient of friction between two smooth, small surfaces tends to be large, and even worse, quite variable [14]. Thus, according to predominant parasitic friction forces on the micrometer scale, the maximal rotation speed of the side-drive micromotor shown in Fig. 17 was restricted to 500 rpm (against 100,000 rpm theoretically expected), while the driving torque was estimated to be on the order of 10^{-6} µNm, leading to a maximal output mechanical power of 10^{-11} Watts that is actually unusable to perform a concrete task.

7.2.2. *Top-Drive Electrostatic Actuators*

Many other varieties of IC-processed microactuators have been developed so as to overcome the driving limitations of the first side-drive electrostatic actuator. The design of top-drive electrostatic actuators such as shown in Fig. 18 was the first alternate strategy to increase the torque of electric field driven actuators, through higher capacitance between the driving electrodes [28].

Taking into account the stator/rotor overlap shown in Fig. 18, the capacitance of the top-drive actuator is given by :

$$C(\vartheta) = n\varepsilon_r\varepsilon_0 \frac{S(\vartheta)}{d} = \frac{n\varepsilon_r\varepsilon_0\vartheta}{d}(R_1 - R_0)\left(\frac{R_1 + R_0}{2}\right)$$

(4)

where R_0 and R_1 are respectively the inner and the outer radius of the stator/ rotor interface.

Figure 18. Former top-drive electrostatic actuator. After Mehregany et al. [28].

The electrostatic torque developed by the actuator shown in Fig. 18 is thus given by :

$$\Gamma = \frac{1}{2}V^2\frac{\partial C(\vartheta)}{\partial\vartheta} = \frac{n\varepsilon_r\varepsilon_0 V^2}{4d}\left(R_1^2 - R_0^2\right) = \frac{n\varepsilon_r\varepsilon_0 RV^2}{2d}\left(R_1 - R_0\right)$$

$$= \frac{1}{2}n\varepsilon_r\varepsilon_0\left(R_1 - R_0\right)RdE^2 \tag{5}$$

with :

$$R = \left(R_1 + R_0\right)/2 \tag{6}$$

Compared with former side-drive actuators (see equation (2)), the electrostatic torque of a top-drive actuator has been consequently increased as a function of the following geometric ratio :

$$G = \frac{R_1 - R_0}{h} \tag{7}$$

which strongly depended on the stator/ rotor overlap.

Unfortunately, the overlap of the stator and rotor poles was quite restricted because of the parasitic deflection of the micrometer thick polysilicon stator roof built over the rotor disk. Therefore, this first alternate strategy to overcome the driving limitations of former side-drive actuators did not allow a significant amplification of the electrostatic torque (e.g. $G \leq 20$). Top-drive actuators also exhibited out-of-plane unbalanced forces due to the problems of static-charge, so they have not been suitable to perform concrete external tasks such as manipulation of objects.

7.2.3. Harmonic-Drive Actuators

The idea of the harmonic-drive motor was developed from early 1990's so as to significantly increase the driving torque of IC-processed electrostatic actuators by using an offset rotor in the motor (see Fig. 19) [28][29][30]. An harmonic motor (also called a wobble motor) has two paths of different lengths that roll upon each other. A circular stator hole containing a slightly smaller disk rotor is an example of this harmonic motion. As the rotor roles on the inside of the stator hole without slipping, it also rotates slightly. Thus, more torque at lower speed can be obtained. The integrated speed reduction which results in the harmonic motion shown in Fig. 19, is given by :

$$N = \frac{\omega_R}{\omega_S} = \frac{R-r}{r} = \frac{\delta}{r} \tag{8}$$

where :

- ω_s is the frequency at which the stator poles are cycled,
- ω_r is the rotor output frequency (e.g. the rotation speed),
- R and r are respectively the inner stator radius and the outer rotor radius,
- $\delta = R - r$ is the maximal gap between the driving electrodes.

Accordingly, the integrated torque amplification of the wobble motor is given by :

$$G = \frac{1}{N} = \frac{r}{\delta} \tag{9}$$

With typical control of gap sizes and wobble motor diameters, integrated gear ratios up to several hundreds have been obtained., leading to a significant increase of the driving torque. Furthermore, wobble motors, which are less friction dependent than side-drive and top-drive actuators, exhibited higher efficiency. Finally, more electrostatic energy has been generated with reasonable voltages because the driving electrodes come in very close contact. In particular, high field strengths on the order of a few hundreds MV/m have been achieved with thin film insulating materials such as silicon nitride deposited on the stator electrodes. However, it has not been possible to overcome the problem of coupling the rotor to external loads. Thus, despite obvious advantages compared with former IC-processed actuators, silicon-based harmonic-drive actuators have not yet been shown to be capable of transmitting significant force (torque) to the external world.

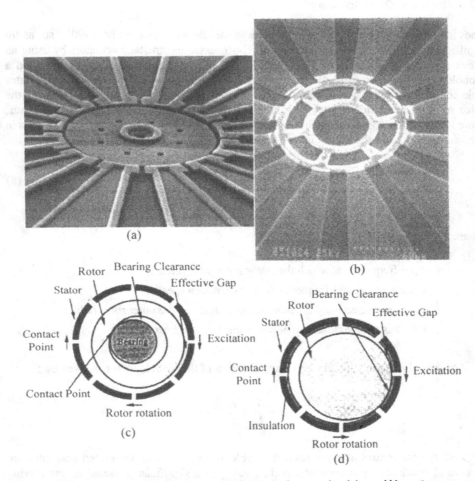

Figure 19. Harmonic-drive electrostatic actuators (a) IC-processed actuator involving wobble motion around the shaft, (b) rolling motion of a polysilicon rotor inside an insulated stator, (c) and (d) corresponding driving mechanisms . After Mehregany et al. [28].

7.2.4. *Comb-Drive Actuators*

Comb-drive actuators has been the most popular approach to overcome parasitic friction on micrometer scale mechanical devices. An example of such an actuator, in which interdigiting teeth contract under a voltage difference, is shown in Fig. 20. In order to prohibit friction, cantilevered beams are used to provide an elastic suspension for the moving comb. Depending on the flexibility of the suspension, large motions can be obtained. However, the maximum force these actuators can deliver is quite small, therefore one needs many of them in parallel to achieve significant electrostatic forces. Early comb-drive actuators have been basically fabricated by polysilicon surface micromachining techniques using a basic one-mask fabrication process [31]. Such actuator consists of two interdigitated finger structures, were one comb is fixed and the other is connected to a compliant suspension. The capacitance between two engaged comb finger arrays such as shown in Fig. 20 can be expressed as :

$$C = \frac{2n\varepsilon_0 h(y - y_0)}{d} \qquad (10)$$

where n is the number of fingers, ε_0 is the dielectric constant in air, h is the thickness of the comb fingers, y_0 is the initial comb finger overlap, y is the comb displacement and d is the gap spacing between the fingers.

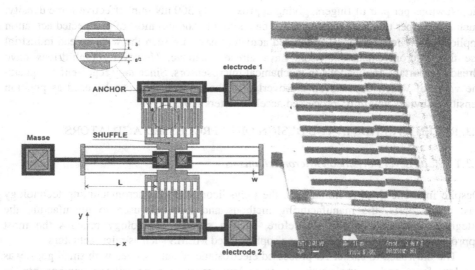

Figure 20. Surface-micromachined comb-drive actuator involving tangential electric field interaction between interdigitated arrays of polysilicon fingers. The movable comb shown on the SEM micrograph is connected to a folded-flexure elastic suspension.

The lateral electrostatic force exerted in the y-direction is thus given by :

$$F_y = \frac{1}{2}\frac{\partial C}{\partial y}V^2 = \frac{n\varepsilon_0 h}{d}V^2 \qquad (11)$$

This electrostatic force is acting on the spring to which the movable comb is connecting, resulting in a deflection :

$$y = \frac{n\varepsilon_0 h}{k_y d}V^2 \qquad (12)$$

where k_y is the linear spring constant of the elastic suspension.

The deflection of the movable comb structure strongly depends on contrary forces from the suspension spring which thus needs to be very compliant in the y-direction, while being very stiff in the orthogonal directions. The folded-flexure design shown in Fig. 20 has been the most popular design because such mechanical architecture strongly reduces the development of parasitic axial forces and exhibits a much larger linear deflection range compared with that of other designs (e.g. clamped-clamped or crab-leg flexure designs).

Micromechanical comb-drive actuators involving tangential electric field interactions can be used for positioning of small masses such as mirrors or tips in read/write heads in data storage devices. However, the maximum driving force, even with clever designs, remains severely limited because of the restricted micrometer thickness of surface-micromachined structural polysilicon layers. As an indication, comb-drive actuators such as shown in Fig. 20 (3.5 μm thick) develop a maximum force in the order of a few tens picoNewtons per pair of fingers, giving approximately 300 $\mu N/mm^2$ effective force density. Such a force per unit area remains far to be enough to address most of prospected actuation applications. Thus, earlier IC-processed actuators have not seen the wide-spread industrial use that micromechanical sensors have already achieve. However, comb-drives have already important applications in mechanical microsensors. Since they represent a capacity the value of which depends on the overlap of the fingers, they can be used as position sensitive devices within IC-processed accelerometers [10].

7.3. RECENT TRENDS IN THE DESIGN OF IC-PROCESSED ACTUATORS

7.3.1. *Multi-Level Polysilicon Micromechanisms*

Despite inherent shaping limitations, the polysilicon surface micromachining technology has, as its basis, the manufacturing methods and tool sets used to manufacture the integrated electronic circuit. Therefore, such a specific technology remains the most appropriate method for manufacturing sophisticated MEMS such as microactuators.

The drawback to earlier IC-processed electrostatic actuators, even with small gaps, was the low force and torque obtainable from low-aspect ratio polysilicon components. In addition, integrated force amplification mechanisms such as multi-level polysilicon gear-speed-reductions were prohibited for a long time, until Chemical Mechanical Polishing (CMP) was introduced to surface micromachining.

Historically, the primary obstacles to multi-level polysilicon fabrication were related to the severe wafer topography generated by the repetition of film depositions and etching. However, CMP applied to multi-level surface micromachining has largely removed these issues and opened significant avenues for device complexity [18].

The primary devices to benefit from multi-level SMT are micromechanical actuators. Thus, the SUMMIT V Technology (e.g. 5 polysilicon structural levels) recently developed at *Sandia National Laboratories*, has aided in producing higher-force actuators through monolithic integration of complex mechanisms such as polysilicon gear-boxes shown in Fig. 21.

However, as a MEMS fabrication technology, low aspect ratio polysilicon surface-micromachined structures tend to be most sensitive to stiction. This is mostly due to the surface-to-volume ratio of polysilicon components and the scaling behavior of various surface effects on the micrometer scale. Accordingly, experience with geared polysilicon mechanisms pointed out that the device surface phenomena of stiction, friction and wear present the greatest impediment to further industrialization of IC-processed actuators [32]. Consequently, the mechanism shown in Fig. 21 actually works as a speed-reduction rather than a torque multiplication unit. Therefore, other processing approaches dealing with high-aspect ratio polysilicon structures have been recently investigated so as to overcome frictional losses in polysilicon gear-speed-reduction units.

Figure 21. Scanning electron micrograph of a five-level polysilicon gear-speed reduction unit. After [32].

7.3.2. High-Aspect Ratio Electric Field Driven Actuators

The height-to-width aspect ratio of an electrostatic actuator is another extremely important design parameter for at least two reasons. First, for a fixed capacitive gap, the actuator output force (or output torque) is linearly proportional to the structural height and therefore increases linearly with aspect ratio. Second, it is important that any uncontrollable cross-axis resonance be well above the servo bandwidth [33].

Modern methods for producing high-aspect ratio structures have thus been recently developed so as to meet the design requirements for efficient electric field driven actuators.

Silicon on Insulator (SOI) Technology. The history of SOI development started in the 1960's, when the SOI structure was created so as to develop the first microprocessor using silicon-on-sapphire (SOS). Since that time, various methods of forming the SOI structure on a Silicon wafer were proposed. In 1998, IBM announced a breakthrough in manufacturing a high-performance, low power CPU using the SOI technology. The SOI CPU delivers 30% faster performance and two-thirds lower power than a bulk-silicon CPU. This achievement marked a turning point in the 30-year history of SOI research and development.

The SOI wafer is drawing attention throughout the microelectronics industry as a high-potential next-generation wafer. The conventional silicon wafers found in existing semiconductor devices usually have a thickness of several hundred microns, while the electrically active domain of a wafer is limited to its surface (e.g. within a thickness of a few microns). Thus, the excess material that is used as a substrate cause both a rise in power consumption and a fall in operating speed of the device. The SOI wafer incorporates an insulating oxide layer between a thin active domain (usually a few micron thick), and its much thicker substrate. The substrate is then isolated and can thus no longer deteriorate the speed or efficiency of the active layer.

Today, the SOI wafers can also be ordered according to the MEMS designer's specifications. In such a case, the SOI layer (e.g. the single crystal silicon layer on top of the insulated layer) is usually much thicker (usually a few tens microns) than that used in the semiconductor industry. Thus, SOI wafers ordered by MEMS designers consist of device wafer, buried oxide, and handle wafer as shown in Fig. 22. Device wafer (e.g. SOI wafer) has very low resistivity to remove the further doping process and is polished to 10-75 microns for high-aspect ratio structural elements. Buried oxide is typically 2 micron thick so that it is sacrifially etched at a reasonable time and the device wafer does not stick to the substrate wafer in the release process. Substrate wafer has the same low resistivity as the device wafer to eliminate the parasitic capacitance by grounding the MEMS structures.

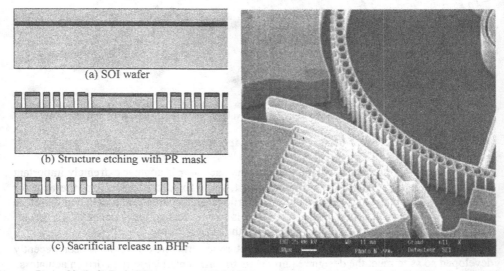

Figure 22. (Left) Basic SOI process, and (right) high-aspect ratio mechanism made via the SOI technology. The structural height of the single crystal silicon comb actuator is 50 microns. After LMARC/IMFC-Besançon (France) in cooperation with EPFL-Lausanne (CH).

As in Fig. 22, the SOI process begins with structure patterning of photo-resist (PR) which has the enough resistance to the inductively-coupled-plasma (ICP)/reactive ion etching (RIE) process. After structure etching and PR stripping, buried oxide is sacrificially etched in BHF solution and the high-aspect-ratio MEMS structure is successfully released without any sticking.

The recent introduction of the SOI technology in MEMS manufacturing methods significantly improved the mechanical performances of IC-processed sensors and actuators. Since former surface-micromachined actuators involved a polycrystalline silicon film of only about 2 micron thick in their structure, they were subjected to sticking, cross-axis parasitic resonance, low capacitance and unusable driving forces. So, high-aspect-ratio single crystal silicon structures today available from SOI technology open new perspectives in the field of IC-processed actuators.

High-Aspect Ratio Surface-Micromachining Technology. Another promising approach, the "HexSil" fabrication process, has been developed in late 1990's so as to develop high-aspect-ratio MEMS structures using IC-compatible surface-micromachining of polycrystalline silicon [33]. The HexSil process allows the fabrication of high-aspect-ratio polysilicon structures that are molded into deep trenches, previously etched through reactive ion etching (RIE), in a reusable silicon mold wafer. The depth of the trenches etched determines the height of the fabricated polysilicon structure.

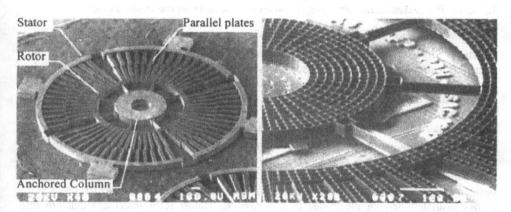

Figure 23. Scanning electron micrographs of a 100 μm-high/ 2.6 mm-diameter electrostatic actuator made via the HexSil process from a 3 μm thick LPCVD polysilicon film. After [33].

High-aspect ratio electric field driven actuators such as shown in Fig. 23 have been recently developed using molds wafers which were etched to a depth of 100 microns. The actuator consists of a fixed outer ring, or stator, and a mobile inner ring (e.g. rotor), which is connected to an anchored central column via narrow polysilicon structures. Actuation is accomplished via capacitive parallel plates, which are attached to the rotor and the stator in opposing pairs. A voltage applied across the driving electrodes results in distributed normal electrostatic forces which provide mechanical oscillations of the central rotor. Electrostatic torque is produced by $2n$ capacitive plate pairs, half of which rotate the actuator clockwise, the other half counterclockwise.

Applying a voltage v_1 to one half the structure and v_2 to the other half creates a torque :

$$\Gamma(v_1, v_2, \theta) = \frac{1}{2} nR\varepsilon_0 S \left[\left(\frac{v_1}{x_n - R\theta} \right)^2 - \left(\frac{v_2}{x_n + R\theta} \right)^2 \right] \tag{13}$$

where R is defined to be the centroid of the plate to the center of rotation of the rotor, S is the area of each plate, ε_0 is the permittivity of air, ϑ is the rotor oscillation (assumed to be small), and x_n is the nominal capacitive gap with zero rotation [33].

Taking advantage of the considerable increase of the actuator's bending stiffness which drops as the third power of the polysilicon structure height, the diameter of the actuator shown in Fig. 23 has been significantly increased, compared with former low-aspect ratio electrostatic actuators described in section 7.2.

Thus, millimeter scale actuators fabricated using the HexSil process are suitable for concrete industrial applications such as two-stage servo systems for magnetic disk drive. In particular, such high-aspect ratio electrostatic actuators have been shown to be capable of positioning read/write transducers of a 30% slider (e.g. a picoslider that contains a magnetic head), over a +/- 1μm range, with a predicted bandwidth of 2 kHz [33].

7.3.3. IC-Processed Actuators Using Contact Interactions on the Wafer Level

The height-to-width aspect ratio has been for a long time considered to be the main design parameter of IC-processed electrostatic actuators, leading to various high-aspect ratio microfabrication methods such as discussed above. However, one can imagine other design rules for highly-efficient electrostatic actuators using low-aspect ratio surface-micromachined polysilicon structures. As an example, new micrometer thick IC-processed electrostatic actuators recently pointed out unusual mechanical performances by using "*contact interactions*", instead of conventional "*electric field interactions*" [34].

Electrostatic actuators using contact interactions (EACI) are partly similar to traveling wave and standing wave piezoelectric motors, which make use of frictional contact interactions between a vibrating stator and a rigid rotor. Thus, as shown in Fig. 24, EACI basically operate using a friction-based mechanical energy transduction which takes place across insulated driving electrodes (e.g. across the stator/ rotor interface) that are into intimate contact.

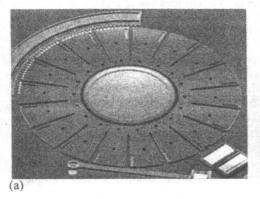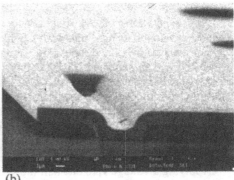

(a)   (b)

Figure 24. (a) φ 500 microns IC-processed electrostatic actuator using contact interactions on the wafer level, (b) SEM micrograph showing an elementary actuation cell. After [34].

Figure 24 shows SEM micrographs of a micrometer thick annular micromotor involving distributed stator/ rotor contact interactions. The rotor is a flexible polysilicon disk plate on which radial grooves have been etched so as to shape elementary actuation cells. Each actuation cell is actuated by applying a driving voltage on the underground electrode located near the lower rotor surface, all along the rim of the actuator. Micrometer height contact plots that are embedded into the actuation cells provide contact asymmetries which are fundamentally needed to actuate the stepping motion of the rotor. By applying a square voltage across the driving electrodes (e.g. across the stator/ rotor interface), the electrostatic pressure periodically twists the annular array of actuation cells. The corresponding rotation of the distributed contact plots therefore provide nanometer stepping motion that can be easily controlled through open loop frequency control.

The annular IC-processed micromotor shown in Fig. 24 obeys the Coulomb friction Law and thus develop an high-driving torque that linearly depends on the electrostatic pressure exerted onto the flexible polysilicon rotor. Because of the non-linear mechanical behavior which results from progressive clamping effects, the electrostatic pressure applied to the rotor can reach several tens MPa, against typically 10^{-2} MPa using conventional parallel-plate capacitors (see Fig. 25) [35].

Accordingly, micrometer size annular micromotors supplied with electrical square pulses having +/- 100 Volts peak amplitude develop a nominal driving torque on the order of 1.5 µNm [36], while the maximum output torque can be as high as 4 µNm (see Fig.29) [37]. Such an unusual driving torque is roughly 1,000 to 10,000 times the output torque usually developed by conventional electrostatic actuators having a similar size, and addressed with a similar voltage.

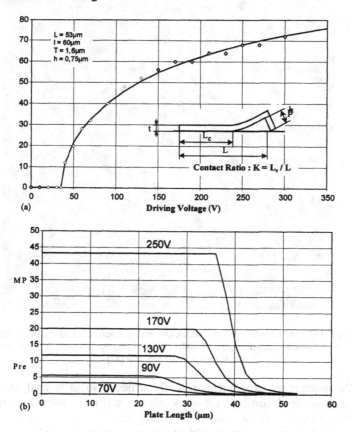

Figure 25. Numerical simulation of the stator/ rotor interface of an electrostatic actuation cell using contact interactions : (a) contact rate as a function of the driving voltage, (b) electrostatic pressure acting onto the flexible actuator plate as a function of the applied driving voltage. After [35].

356

Torque measurements have been achieved through electrostatic probing of polysilicon actuators on the wafer level. Because polysilicon is known to be a brittle material, actuators had thus been combined with purely elastic polysilicon torque sensors machined in a common process flow [36]. Monolithic test structures such as shown in Fig. 26 led to ideal boundary conditions for getting micrometer size actuator loading characteristics. Elastic torque sensors involving polysilicon beam arrays embedded into the silicon substrate (wafer), were monolithically connected to the polysilicon rotor. Accordingly, external mechanical loading that was applied from the elastic torque sensor deflection has been directly measured as a function of the angular stiffness of the polysilicon beam array. More recently, output mechanical power limits have been investigated using an high-frequency CMOS camera combined with an image analysis software (see Fig. 27) [37].

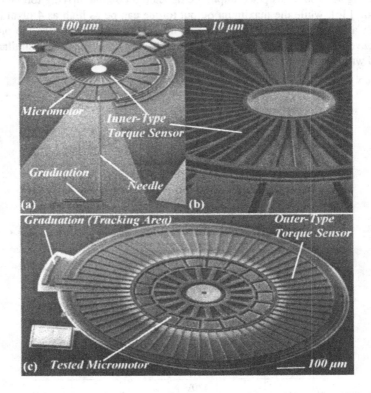

Figure 26. (a) and (b): Micrometer size test structure combining annular electrostatic micromotors with inner-type and outer-type elastic torque sensors, (c) magnified SEM micrograph showing an inner-type polysilicon torque sensor. After [36] and [37].

Figure 28 summarizes experiments recently achieved so as to analyze the dynamic behavior of an annular electrostatic actuators using stator/ rotor contact interactions. The driving frequency supplied to the actuator has been gradually incremented from 20 kHz, up to 100 kHz, while the amplitude of the driving voltage was maintained to a maximal level of 200 Volts. The rotation speed of the tested actuators decreases as a function of the loading torque (applied from the elastic torque sensor), which progressively brakes the rotor until stopping.

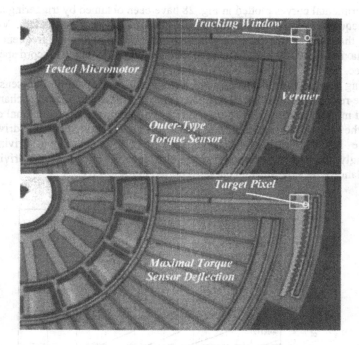

Figure 27. Video frames showing a test structure during electrostatic probing of the actuator : (a) before actuation and (b) configuration during testing. After [37].

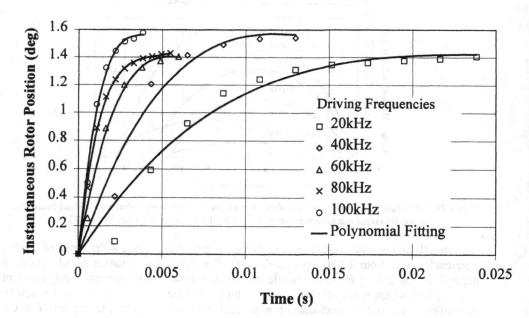

Figure 28. Dynamic behavior of a tested actuator for various driving frequencies in the range 20 to 100 kHz. Experiments are performed by tracking a target pixel in the area delimited with the white window shown in Fig. 27. After [37].

358

Experimental curves plotted in Fig. 28 have been obtained by triggering acquisitions of images sequences with electric signals supplied to the tested actuators. Very short test duration, that were on the order of 4 milliseconds when the driving frequency is increased up to a maximum of 100 kHz, have been recorded by switching the record speed up to 1828 images/sec.

Taking into account the non-linear angular stiffness of the torque sensor, Figure 29 shows the resulting torque/ speed characteristics as well as the output mechanical power of the tested micromotors as a function of the driving frequency. The maximal driving torque remains the same whatever the driving frequency that is supplied onto the driving electrode, while the actuator rotation speed linearly depends on the input driving frequency. Accordingly, the output mechanical power also linearly depends on the driving frequency, in accordance with experimental curves reported in Fig. 29 (b).

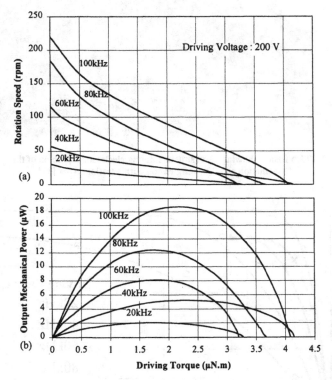

Figure 29. (a) Torque/ speed characteristics of electrostatic micromotors using stator/ rotor contact interactions, and (b) output mechanical power as a function of the driving frequency. After [37].

Table III summarizes experiments that have been performed in a wider range of driving frequencies (e.g. from 1 kHz up to 100 kHz). The free rotor's rotation speed obviously depends on the driving frequency, while the driving torque remains approximately constant (e.g. 4 µNm, which is roughly ten times the torque needed to move the tips of a watch !). Accordingly, the output mechanical power reaches 18 µWatts. Thus, taking into account the overall mass of the rotor (which is on the order of $2 \ 10^{-10}$ kg), the maximal output power per mass unit of the tested actuator is approximately 100 Watts/gr.

Such an unusual mechanical performance, which is roughly 100 to 1000 times the output power/mass ratio of conventional macroscopic size electromagnetic actuators, opens new actuation perspectives in the field of MEMS-based active control of structures. Nevertheless, recent experiments exacerbated surface related phenomena such as stiction, friction and wear which still limit the operating lifetime of electrostatic actuators using contact interactions, prohibiting numerous industrial applications on the near term. As an example, experiments performed on the first generation annular prototype shown in Fig. 25 recently exhibited an operating lifetime of the order of 3×10^{7} elementary stepping motions, meaning that prototypes operated roughly 50 minutes with an average driving frequency of 10 kHz. Considering the diameter of the actuator (ϕ 500 microns) as well as the nanometer range of the elementary stepping motion (which is roughly 30 nanometers), the overall traveled distance was of the order of one meter, which, on the other hand, corresponds to 1500 rotor revolutions [38].

TABLE III. Loading characteristics of ϕ 500 microns electrostatic micromotors using stator/ rotor contact interactions on the wafer level. After [37].

Driving Frequency (kHz)	Free Rotation Speed (rpm)	Locking Torque (μN.m)	Maximal Mechanical Power (μW)
1	1.4	3.6	0.15
2	2.1	4.1	0.4
3	3.3	3.6	0.57
4	5.1	4.1	0.77
5	6.4	3.9	0.82
6	7.8	3.7	0.9
7	9.5	3.9	1.02
8	11.7	4	1.1
9	13.2	3.5	1.15
10	15	3.4	1.2
20	30.7	3.3	2.1
30	46.7	3.5	3.8
40	57.1	4.1	5.0
50	85	3.8	6.3
60	115	3.2	7.9
70	140	3.3	9.4
80	184	3.7	12.1
90	200	3.9	15
100	220	4.1	18.2

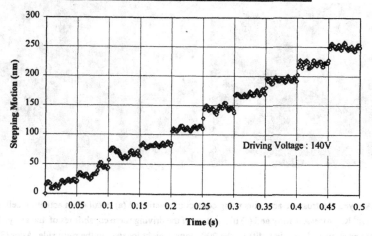

Figure 30. (a) Experimental plots showing the nanometer stepping motion of an electrostatic actuator using contact interactions. Measurements are performed with a high-frequency CMOS camera combined with an image processing software. After LMARC/IMFC-Besançon (France).

7.3.4. *Arrayed Electrostatic Actuators Using Contact Interactions*

Despite limitations discussed above, one can imagine devices such as disposable active microcatheters and micro-antenna spreading systems that may suit with the short lifetime of electrostatic actuators using contact interactions.

Recent works consequently focused on disposable arrays of high-efficiency actuation cells involving contact interactions [39]. Figure 31(a) gives the details of an elementary actuation cell . The overall surface of the actuator is smaller than the hair's cross section (typically 50 x 50 microns). Thus, using the silicon processing technology, actuation cells can be easily duplicated within arrays so as to produce significant forces through the superposition of many elementary driving forces.

Recent investigations pointed out unusual driving force densities that were as high as 400,000 μN/mm^2, considering millimeter scale arrays of the type shown in Fig. 31(b). Considering such a considerable driving force, the 6 mm^2 IC-processed actuator array would be theoretically able to lift-up the 200 grams weight shown in Fig. 31(c) ! Nevertheless, according to the stress gradient within the frame (see Fig. 32(a)), *serial connection* between the actuator array and the gravitational load would lead to irreversible damages such as shown in Fig. 32 (b).

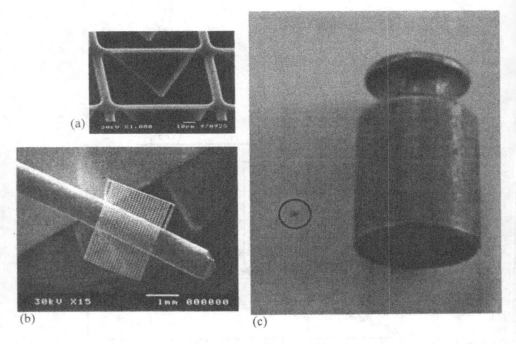

Figure 31. Arrayed electrostatic actuators using contact interactions : (a) detail of an actuation cell, (b) SEM picture of a 6 mm^2 IC-processed array and (c) illustration of the driving force capabilities of the array. The circled actuator array would be able to lift-up the 200 grams weight located on the right side. After [39].

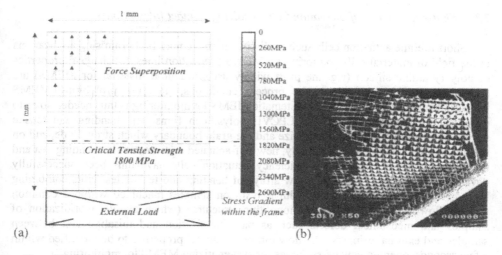

Figure 32. Limit of the *serial connection* principle :(a) visualization of the tensile stress gradient within the polysilicon frame using serial connection between the actuator array and the external load (b) video picture showing the loss of integrity during actuation of the array. After [39].

Thus, instead of conventional serial connection, *parallel connections* must be directly established between each of the distributed actuators and the external load [39]. As an example, Fig. 33 shows a millimeter size tubular electrostatic which involves parallel connections between the arrayed IC-processed actuators, which act as a flexible polysilicon stator, and the load (e.g. the cylindrical rotor). Each of the elementary actuation cells comes in direct contact with the external load whatever its location in the polysilicon sheet, thus keeping the mechanical integrity of the array during actuation of the tubular rotor [40].

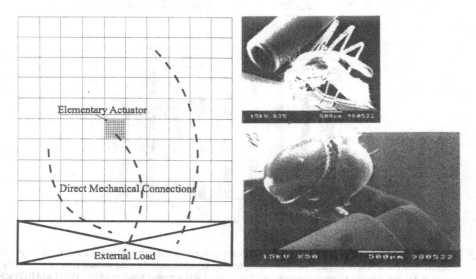

Figure 33. Millimeter scale tubular electrostatic micromotor involving *parallel connection* between the rotor and the IC-processed microactuators distributed within the flexible polysilicon stator. After [40].

362

7.3.5. *First Applications of Actuation Cells Involving Contact Interactions*

Short lifetime actuation cells such as discussed above may find promising applications in the field of material self-monitoring on MEMS production lines. Mechanical properties of polycrystalline silicon (e.g. the most widely used structural material for MEMS) are known to strongly depend on the process conditions, therefore prohibiting MEMS industrialization on the near term. Systematic MEMS lot monitoring is thus needed so as to expertise mechanical properties of LPCVD polysilicon films (e.g. bending and tensile strength), as a function of the grain size and the grain boundary which strongly depend on the processing conditions. Accordingly, new self-actuated micrometer size bending test and tensile test units involving high-efficiency actuation cells have thus been successfully developed [41]. Figs. 34 and 35 show arrays of bending and tensile test units combining high-driving force actuation cells with polysilicon samples machined within a common process flow. The resulting ideal monolithic structures (which avoids hybridization of samples and mechanical defects such as parasitic friction and misalignment between samples and external grippers...), allow intrinsic material properties to be identified within a few seconds, opening new perspectives for systematizing MEMS lot monitoring.

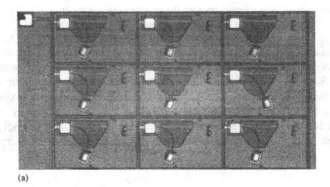

(a)

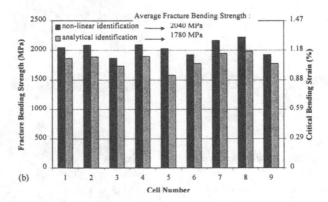

(b)

Figure 34. IC-processed array of 9 LPCVD polysilicon bending test units: (a) video frame showing maximal elastic deflection of self-actuated samples during testing, (b) histogram showing critical strain and bending strength measured through electrostatic probing of an array of self-actuated bending units. After LMARC/IMFC-Besançon (France).

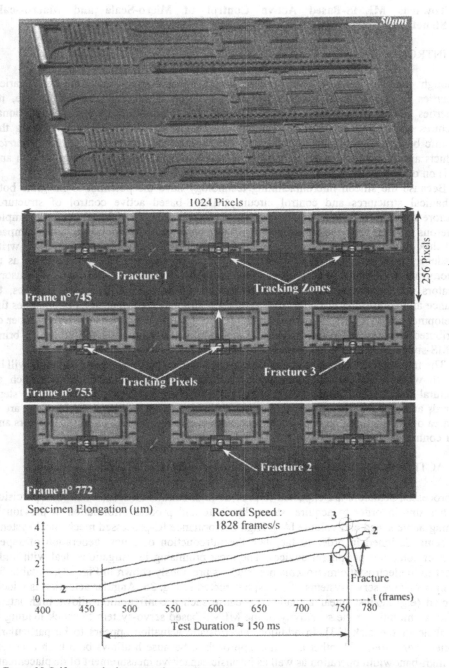

Figure 35. Self-actuated micrometer size tensile test units: (a) SEM of an array of 3 test units having an overall surface on the order of 0.075 mm², (b) video frames showing real time tracking of tensile specimen elongation through an high-frequency CMOS camera coupled with a specific image processing software. The overall test duration is 150 ms and the cross-section of the specimens is 1,000 times smaller than that of hair. After LMARC/IMFC-Besançon (France).

8. Towards MEMS-Based Active Control of Micro-Scale and Macro-scale Structures

8.1. INTRODUCTION

Although today's structure and product designer have many materials with varied properties to choose from, once a particular material is selected and cut to size, its properties remain fixed. As a consequence, most of individual parts of conventional structures are unable to adapt to changing conditions. On the other hand, making the dynamic behavior of matter programmable has the potential to enable *"control- centric"* products and processes that actively adapt to changing conditions, achieving strength and precision operation through intelligent adjustment of their dynamic behavior [42].

Because the silicon micromachining technology have the potential to integrate both mechanical structures and control circuits, MEMS-based active control of structures therefore appears to be one of the most promising challenges of MST. As an example, numerous projects are being developed in order to enhance active control of super-compact rigid disk drives, using auxiliary IC-processed actuators directly connected to read/ write transducers (see section 8.2). Considerable efforts also began a few years ago so as to demonstrate techniques that will enable distributed MEMS-based arrays of sensors, actuators and computational elements embedded within materials and on surfaces, to enhance and control the behavior of sophisticated structures. Current research includes the development of software and architectures for coordinating the actions of large number of distributed devices. New manufacturing methods are also investigated so as to bring MEMS-style batch fabrication to bear on macro-scale objects.

The futurist ideal system is one in which large numbers of sensors and actuators will be able to work relatively independently to achieve global performance criteria such as structural stability and mechanical modulation of surface properties. As one of the steps towards autonomous distributed systems, the following gives the current state of the art in the area of distributed IC-processed devices, with a special emphasis on microactuators and their control strategies.

8.2. ACTIVE CONTROL OF READ/ WRITE MAGNETIC HEAD

IC-processed actuators are expected to provide high-bandwidth servo-controlled precision mechanisms in order to acquire, store, distribute and process information at previously unimaginable speed and volume [43]. High-performance IC-processed mechanical systems are being designed in order to facilitate the introduction of a new generation of super-compact computer peripheral devices. The most promising investigations deal with high-aspect ratio electrostatic microactuators (such as previously shown in Fig. 23), suitable for use in two-stage servo-systems of the type depicted in Fig. 36. Angular actuation is widely selected (e.g. against linear actuation), because such a configuration allows high lateral stiffness, minimizing the sensitivity of the MEMS-based servo-system to shock loading in the plane of the disk [33]. In addition, electrostatic actuation appears to be particularly efficient compared with other actuation approaches, because it allows both high accuracy and high-bandwidth operation as well as intrinsic capacitive measurement of displacement.

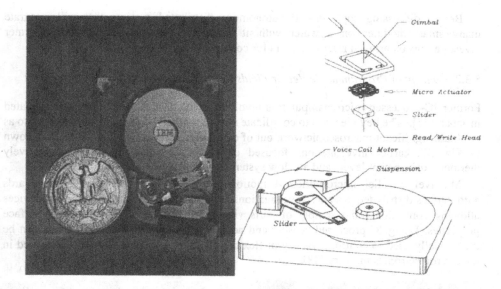

Figure 36. (a) Tiny disk drive after IBM, http://www.storage.ibm.com. (b) Servo positioning mechanism scheme of a magnetic disk drive including an IC-processed actuator between the slider and the gimbal. After [33].

8.3. PARALLEL MANIPULATION USING MECHANICAL MODULATION OF SURFACE PROPERTIES

8.3.1. *Concept of Distributed Microactuators*

Most of prospective applications of IC-processed actuators presently deal with optics, transportation and aerospace, robotics chemical analysis systems, biotechnologies, medical engineering and microscopy using scanned micro probes [44]. These applications have a common feature in that only very light objects such as mirrors, magnetic heads, valves, cells and microprobes are manipulated and that little physical interaction with the external environment is necessary. The main reason is that most of present micrometer size IC-processed actuators are still primitive and large forces cannot be transmitted to the external world.

Although the small size of IC-processed components is a very distinctive feature of the MEMS technology, it has other, may be even more attractive features. Thus, a decisive advantage offered by the silicon micromachining technology is the possibility of implementing sensors and actuators within interactive arrays. Cooperative work of many actuation cells can perform a large task, even one single IC-processed actuator can only produce unusable force or perform insignificant motion. Therefore, in addition to *miniaturization*, the concept of *multiplicity* appears to be another key to successful microsystems.

The third key to successful microsystems deals with the *integration of microelectronics* which is also essential for distributed microactuators to cooperate with each other and to perform a given macroscopic task.

366

Before discussing sophisticated autonomous distributed systems that will integrate many smart modules on a wafer without assembly, the following analyses former investigations on arrayed microactuators for conveyance systems.

8.3.2. *Concept of Programmable Vector Fields using IC- Processed Micromanipulators*

Former IC-processed micromanipulators involving actuator arrays have been investigated in Japan [45]. The key idea was to coordinate simple motions of many microactuators so as to obtain significant macroscopic work out of distributed IC-processed actuators. As shown in Fig. 37, earlier investigations focused on conveyance systems using respectively thermally driven cantilevers and air flow systems [46] [47].

Massively-parallel microfabricated motion pixels has been the first step towards autonomous distributed systems. Investigations are now moving to new automation devices allowing controlled self-assembly of parts via electromechanical modulation of surface properties. Using IC-processed sensors and actuators, programmable vector fields can be theoretically used to control a variety of flexible planar part feeders that will be needed in the microelectronics industry [48].

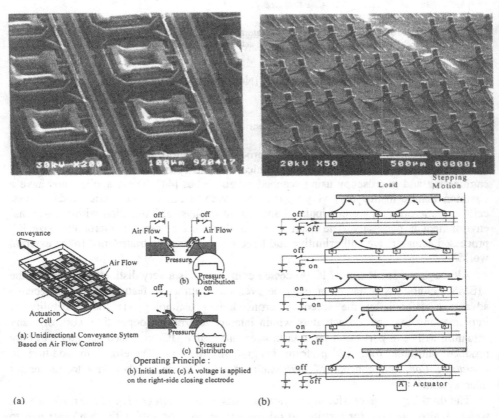

Figure 37. (a) Air-flow conveyance system involving distributed nozzles and (b) Ciliary motion system using thermal actuation. After [46] [47].

Figure 38 gives an illustration of the principle of tactile shape recognition using contact detection of various manipulated parts [49]. However, sophisticated systems combining many IC-processed sensors and actuators are yet challenging. Thus, recent attention focused on sensor-less strategies such as illustrated in Fig. 39 [50].

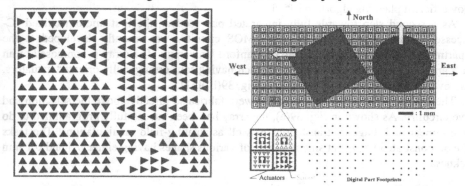

Figure 38. (a) Concept of multidirectional motive surfaces using modern array technologies and (b) principle of tactile shape recognition using contact detection of manipulated parts. After [49].

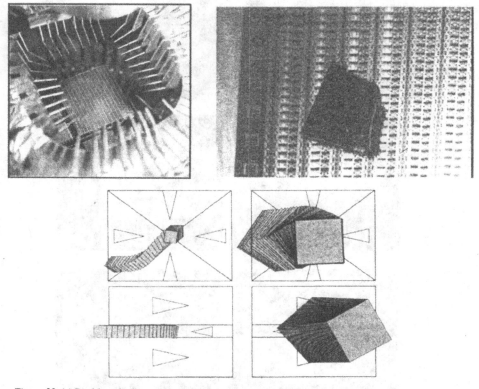

Figure 39. (a) Partitioned micromanipulator involving distributed IC-processed electrostatic oscillators [49].
(b) Sensor-less sorting strategy using force vector fields : parts of different sizes are first centered and subsequently separated depending on their size. After Böhringer et al. [50].

When a part is placed on top of an actuator array, the programmed vector field induces a force and moment upon it. Then, by chaining together sequences of local vector fields, the part is conveyed until the desired final state is reached. Assuming the part may come to rest in a dynamic equilibrium state over time, the resulting strategy requires no sensing and enjoy efficient planning algorithms [50].

As a second step towards fully integrated part feeders, the first micromachined IC-processed actuator array with on-chip CMOS circuitry was recently developed at the Department of Electrical Engineering of Stanford University. The array is composed of an 8 x 8 array of elementary motion pixels, each having four orthogonally oriented actuators, in an overall die size of 9.4 x 9.4 mm (see Fig. 39(b))[51].

The polyimide-based ciliary actuators were fabricated directly above the selection and drive circuitry. As shown in Fig. 39(a), the array has been successfully programmed to do simple linear and diagonal translations as well as rotating-field manipulations. The tasks were demonstrated using flat silicon pieces of various shapes having either 0.5 and 0.1 mm thickness.

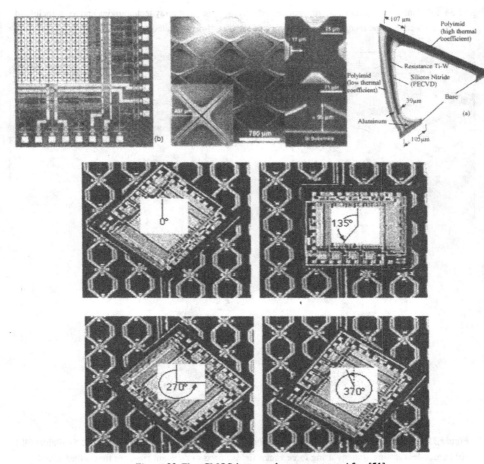

Figure 39. First CMOS-integrated actuator array. After [51].

8.4. TOWARDS MEMS-BASED CONTROL OF ACOUSTICAL IMPEDANCE

There appear to be a number of applications in acoustics that could benefit from MEMS-based electromechanical transducers which efficiently radiate acoustic waves into an adjacent medium, and comes in the form of a flexible sheet, that can be bounded to the surface of macro-scale structures [52] [53]. In the case were the medium is a gas, such as air, examples of potential applications include active noise control, acoustic flow control and specialty audio systems such as CMOS-MEMS membranes for audio-frequency acoustic radiation [54].

As a first step towards MEMS-based active control of noise, Fig. 40 shows an experimental platform which integrates a vibrating plate isolated with an array of surface-micromachined acoustic transducers. The transducers have been designed so that the radiated pressure in far field would be significantly decreased at frequencies on the order of a few kHz. The vibrating plate is made of a 500 micrometer thick silicon wafer that is clamped on a ϕ 50mm circle. The obtained circular plate is then sustained in transverse vibration through input external forces that are supplied with a driving frequency set at the first out-of-plane resonance frequency of the clamped silicon plate. Thus, the vibrating plate works as a single crystal silicon resonant membrane which radiates significant acoustic pressure in the audible range.

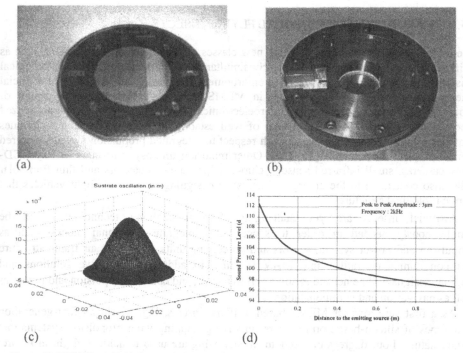

Figure 40. MEMS-based active control of noise using surface-micromachined array of polysilicon acoustic transducers. (a) and (b) Photographs showing respectively the silicon wafer before and after clamping into the metallic frame. The array of 8000 acoustic transducers is located at the center of the wafer. (c) and (d) Acoustic radiation of the uncontrolled silicon membrane which is expected to be killed through open-loop control of MEMS-based transducers. After LMARC/ IMFC-Besançon (France).

As shown in Figs. 40 (c) and (d), with maximal peak-to-peak oscillations of 3 microns at the center of the membrane, the radiated acoustic pressure level is on the order of 100 dB at 2 kHz driving frequency which is closed to the resonance frequency of the silicon membrane.

The MEMS-based acoustic transducer array that is distributed over the emitting surface of the silicon membrane combines 8000 parallel plate polysilicon capacitors. Quasi-static control of the out-of-plane motion of the elementary acoustic transducers can be achieved from dc up to 15 kHz, with an ac amplitude response on the order of a few microns under low driving voltage (e.g. roughly 20 Volts). So, even if out-of-plane oscillations are still quite restricted, parasitic acoustic pressure levels on the order of 100 dB at 2kHz are expected to be actively killed, by controlling the local motion of each MEMS-based acoustic transducer. In practice, the driving annular electrodes deposited over the silicon wafer (e.g. the emitted surface), are supplied in such a way that the absolute displacement of acoustic transducers becomes nearly negligible, whatever the local displacement δ of the vibrating silicon substrate.

Industrial applications involving MEMS-based acoustic transducers are mainly expected in aeronautics and car industry so as to significantly reduce parasitic acoustic radiation of aircraft reactors and internal combustion engines. Similar developments are also expected in the field of MEMS-based flow sensors and actuators for drag reduction in next generation aircraft and spacecraft.

8.5. TOWARDS MICROMECHANICAL FLYING INSECTS

The high level of interest in developing new classes of micro air vehicles (MAV), such as artificial insects, is the result of the nearly simultaneous emergence of their technological feasibility and an array of new needs in environment. Technological feasibility of artificial insects follows from recent advances in MEMS technology that allow the design of processes and systems combining microelectronics components with comparably-sized mechanical elements [55]. The arsenal of well established micro-fabrication techniques provides a high degree of optimism with respect to integrated propulsion systems inspired from biological mechanisms [56] [57]. Other maturing microsystems such as tiny CCD-array cameras, small infra-red sensors, chip-sized substance detectors and thin film solar cells, also contribute to the emerging interest in designing very small flight vehicles that mimic biological structures.

Artificial insects are yet challenging. Several enabling technologies will be simultaneously needed in order to successfully integrate functional blocks such as propulsion and power system, sensors and processing units, as well as air frame structure and communication units into MAV systems. In addition, innovative solutions and original designs will be needed to solve aerodynamics and control, propulsion and power as well as navigation and communication [55].

As a first step towards MEMS-based artificial insects, Fig. 41 shows first-generation prototypes of silicon-based ornithopters involving flapping wing propulsion systems that mimic nature. Four degrees of freedom in each wing are used to achieve flight in nature: flapping, lagging, feathering and spanning [58]. However, not all flying animals implement all of these motions. Thus, most insects do not use the spanning technique (e.g. successive expanding and contracting of the wingspan). In addition, insects with low wing beat frequencies on the order of a few tens Hz, generally have restricted lagging capabilities (e.g. forward and backward wing motion parallel to the flying body). Thus, flapping flight is possible with only to degrees of freedom that respectively satisfy flapping and feathering.

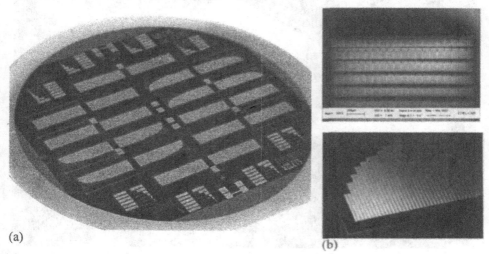

Figure 41. Silicon-based ornithopters. (a) four inches SOI wafer including various wings covered with distributed microactuators, (b) SEM micrograph and 3D video image showing arrayed actuators distributed over the wing's surface. After LMARC/ IMFC-Besançon (France).

The MEMS-based wings shown in Fig. 41 have been designed so as to allow both flapping (e.g. angular movement about an axis in the direction of flight), and feathering (e.g. angular movement about an axis in the center of the wing which tilts the wing to change its angle of attack). So when they are coordinated, these two degrees of freedom are expected to provide lift not only on the down stroke, but also on the up stroke, as it is the case for flying insects.

Figure. 42 shows a SEM micrograph of a wing that has been recently released from the silicon wafer. The wing is going to be actuated through the 20,000 elementary actuation cells that are distributed over the upper surface of the centimeter scale single crystal silicon structure. According to the micrometer thickness of both the single crystal silicon wing (e.g. 5 to 10 microns), and the distributed polycrystalline silicon actuators, the overall mass of the flapping wing system is restricted to a few milligrams. The design specifications of the proposed MEMS-based flapping system are summarized in Table IV.

Wingspan lateral dimensions	~ 40 x 10 millimeters
Wingspan structural thickness	~ 5 to 10 microns
Distributed actuation cells per wing	~ 20,000
Overall mass of the flapping wing propulsion system	~ 1.2 milligram
Wingbeat frequency	~ 30 to 50 Hz
Flapping amplitude range	~ +/- 50°
Expected on-board energy density (Lithium Battery)	~ 360 Joules/ gram

TABLE IV. Design specifications of the MEMS-based flapping wing system shown in Fig. 41.

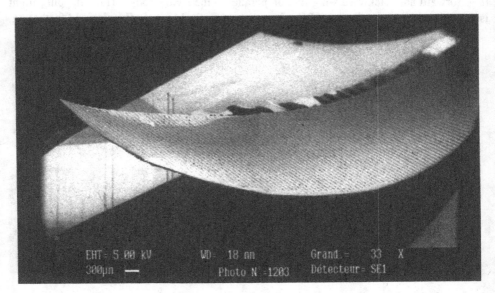

Figure 42. First released prototype of a self-actuated flapping wing involving 20,000 distributed actuation cells. After LMARC/ IMFC-Besançon (France).

9. Conclusion

Historically, conventional ultra-precision tool-machining historically made significant contributions to microsystems technology (MST), but the speed of progress was relatively slow until IC-technologies were introduced to the fabrication process of mechanics. Also, MEMS technology, which basically makes use of the tools of microelectronics, gave rise to a growing motivation for MST. According to remarkable progresses briefly summarized in this contribution, MEMS technology has recently exhibited the potential to integrate silicon-based mechanical structures, CMOS circuits as well as fluidic and photonic devices with very small dimensions on a common substrate. It is thus clear that MST and MEMS will have a profound influence in many areas dealing, in particular, with mechanical engineering.

Remarkable progresses that have been achieved from early 1990's in microsystem technology (MST), will certainly result in a growing involvement of mechanical engineers in the field of MEMS-based active control of micro and macro-scale structures. So far, electrical engineers devoted themselves in MEMS processes and materials, but the involvement of mechanical engineers becomes obviously essential for inventing new actuation mechanisms and improving design tools towards MEMS-based smart structures. It will be important, and even decisive for the success of MST, that in design of mechanical structures, the means offered by the MEMS technology will be adequately applied. Miniaturization of systems using linear reduction of macroscopic scale structures would not adequately take into account both scaling effects and potentials of the MEMS technology. Additionally, it would not be reasonable to use MEMS technology to produce conventional components on a reduced scale, because taking advantage of batchwise manufacture only, would not be wholly satisfactory.

Mechanical engineers will thus have to learn MST and MEMS before investigating completely new design strategies that will optimize the advantages offered by MEMS technology. MEMS-based active control of macro-scale structures using bottom-up design strategies will be clearly of particular interest (e.g. mainly because humans are living on the macroscopic scale), and will probably decide whether or not MST and MEMS can gain ground on the market over conventional technologies. The proposed approach to smart structure design using MEMS technology is dedicated to various readers. To students and those who are not familiar with MEMS, it is expected to help them to better understand the meanings of miniaturization and integrated systems. To experts in mechanical engineering, it finally provides an insight on how both MEMS-based structures and fabrication processes should be simultaneously designed to successfully integrate functional blocks into MEMS-based responsive systems for active control of micro and macro-scale structures.

374

10. References

1. Fukuda, T. and Menz, W. (2002) *Micro Mechanical Systems : Principles and Technology*, Handbook of Sensors and Actuators 6, Elsevier.
2. Angell, J.B., Terry, S.C. and Barth, P.W. (1983) Silicon micromechanical devices, *Scientific American Journal* **248**, 44-55.
3. Feynman, R.P. (1959) There's plenty of room at the bottom, American Physical Society Meeting, Pasadena CA. (USA), reprinted in *Journal of Microelectromechanical Systems* **1**, 60-66.
4. Feynman, R.P. (1983) Infinitesimal machinery, Jet Propulsion Laboratory Meeting, Pasadena CA. (USA), reprinted in *Journal of Microelectromechanical Systems* **2**, 4-14.
5. Madou, M. (1997) *Fundamentals of Microfabrication*, CRC Press.
6. Teshiguhara, A., Kawahara, N., Ohtsuka, Y. and Hattori, T., (1995) Performance of a 7mm microfabricated car, *Journal of Microelectromechanical Systems* **4**, 76-80.
7. Egashira, K., Masuzawa, T., Fujino M. and Sun, X.Q. (1997) Application of USM to micromachining by on-the-machine tool fabrication, *International Journal of Electrical Machining* **2**, 31-36.
8. Masuzawa, T. and Tönshoff, H.K., Three-dimensional micromachining by machine tools, *Annals of CIRP* **46**, 621-628.
9. Becker, E.W., Ehrfeld, W., Hagmann P., Manner A. and Münchmeyer D. (1986) Fabrication of microstructures with high aspect and great structural heights by synchrotron radiation lithography, galvanoforming and plastic moulding (LIGA process), *Journal of Microelectronic Engineering* **4**, 35-36.
10. Elwenspoek, M. and Wiegerink, R. (2000) *Mechanical Microsensors*, Springer-Verlag.
11. Kovacs, G.T.A. (1998) *Micromachined Transducer Sourcebook*, WCB McGraw-Hill.
12. Menz, W. and Bley, P. (1992) *Mikrosystemtechnik für Ingenieure*, VCH Verlagsgesellschaft mbH.
13. Muller, R.S. (1999) *Microactuators*, IEEE, A.P. Pisano ISBN: 0780334418.
14. Trimmer, S.T. (1996) *Micromechanics and MEMS: Classic and Seminal Papers from 1990*, IEEE Press, W.S. Trimmer ISBN: 0879422459.
15. Fan, L.S., Tai, Y.C. and Muller, R.S., (1988) IC-processed electrostatic micromotors, *IEEE International Electronic Devices*, 666-669.
16. Tai, Y.C., Fan, L.S. and Muller, R.S. (1989) IC-processed micromotors: design, technology and testing, *Proceedings IEEE Microelectromechanical Systems* 1-6.
17. Krygowski, T.W., Siegowski, J.J., Rodgers, M.S., Montague, S., Allen, J.J., Jakubczak, J.F. and Miller, S.L. (1999) Infrastructure, technology and applications of microelectromechanical systems (MEMS), Sensors Expo., Cleveland (USA).
18. Sniegowski, J.J. (1996) Chemical-mechanical polishing: enhancing the manufacturability of MEMS, *SPIE Micromachining and Microfabrication Process Technology*, Austin, TX (USA), SPIE Vol. 2879, 104-115.
19. Sniegowski, J.J., de Boer, M.P. (2000) IC-compatible polysilicon surface micromachining, *Annu. Rev. Mater. Sci.* **30**, 299-333.
20. Michalicek, M.A. (2000) Introduction to microelectromechanical system, MEMS Short Course, Air Force Research Laboratory, Space Vehicles Directorate, New Mexico.
21. Smith, J.H., Montague, S. and Sniegowski, J.J. (1995) Material and processing issues for the monolithic integration of microelectronics with surface-micromachined polysilicon sensors and actuators, *SPIE Conference on Micromachining and Microfabrication*, Vol. 2639, 64-73.

22. Howe, R. (1995) Polysilicon integrated microsystems : technologies and applications, *Proceedings Transducers'95*, 43-46.
23. Cohn, M.B., Böhringer, K.F., Noworolski, J.M., Singh, A., Keller, C.G., Goldberg, K.Y. and Howe, T.R. (1998) Microassembly technologies for MEMS, *SPIE Conference on Micromachined Devices and Components IV*, SPIE Vol. 3514, Santa Clara (US) 2-16.
24. Dario, P., Valleggi, R., Carrozza, M.C., Montesi, M.C. and Cocco, M. (1992) Microactuators for microrobots: a critical survey, *Journal of Micromechanics and Microengineering*, 2, 141-157.
25. Trimmer, W.S.N. (1989) Microrobots and micromechanical systems, *Sensors & Actuators*, 19, 267-287.
26. Fujita, H. and Gabriel, K.J. (1991) New opportunities for microactuators, *International Conference on Solid-State Sensors and Actuators*, San Francisco, (US) 14-20.
27. Fujita, H. (1997) A decade of MEMS and its future, *IEEE Proceedings on Micoelectromechanical Systems*, Nagoya (Japan), 1- 8.
28. Mehregany, M., Bart, S.F., Tavrow, L.S., Lang, J.H., Senturia, S.D. and Schlecht, M.F. (1990) A study of three microfabricated variable-capacitance motors, *Proceedings of the 5th International Conference on Solid-State Sensors and Actuators*, Vol.2, 173-179.
29. Trimmer, W. and Jebens, R. (1989) An operational harmonic electrostatic motor, *IEEE Proceedings on Microelectromechanical Systems*, 13-16.
30. Jacobsen, S.C., Price, R.H., Wood, J.E., Rytting, T.H. and Rafaelof, M. (1989) The wobble motor : an electrostatic, planetary armature, microactuator, *IEEE Proceedings on Microelectromechanical Systems*, 17-24.
31. Tang, W.C., Nguyen, T.C. and Howe, R.T. Laterally driven polysilicon resonant microstructures, *IEEE Proceedings on Microelectromechanical Systems*, 53-59.
32. Sniegowski, J.J. (1996) Multi-level polysilicon surface-micromachining technology : applications and issues, *ASME International Mechanical Engineering Congress*, Atlanta, GA (USA), Vol. 52, 751-759.
33. Horsley, D.A., Cohn, M.B., Singh, A., Horowitz, R. and Pisano, A. (1998) Design and fabrication of an angular microactuator for magnetic disk drives, *Journal of Microelectromechanical Systems* 7, 141-148.
34. Minotti, P., Langlet, P., Bourbon, G. and Masuzawa, T. (1998) Design and characterization of high-torque/low-speed silicon-based electrostatic micromotors using stator/rotor contact interactions, *Japanese Journal of Applied Physics*, 37, 359-361.
35. Minotti, P. and Le Moal, P. (2002) Evolutions récentes des lois de design des microactionneurs électrostatiques, *Traité EGEM Microactionneurs électroactifs*, Hermès-Lavoisier, 109-147.
36. Le Moal, P., Minotti, P., Bourbon, G. and Joseph, E. (2001) On-chip investigation of torque/speed characteristics on new high-torque micrometer-size polysilicon electrostatic actuators, *Japanese Journal of Applied Physics*, 40, 596-599.
37. Walter, V., Le Moal, P., Minotti, P., Joseph, E and Bourbon, G. (2002) Investigation of output mechanical power limits on high-torque electrostatic actuators using high-frequency CMOS camera combined with image processing software, *Japanese Journal of Applied Physics*, 41, 424-427.
38. Minotti, P., Langlet, P., Bourbon, G., Masuzawa, T. and P. Le Moal (1998) Direct-drive electrostatic micromotors using flexible polysilicon rotors, *Journal of Intelligent Material Systems and Structures*, 9, 829-836.

376

39. Minotti, P., Bourbon, G., Langlet, P. and Masuzawa, T. (1998) Arrayed electrostatic scratch drive actuators : toward visible effects up to the human scale, *Journal of Intelligent Material Systems and Structures*, 9, 837-846.
40. Minotti, P., Bourbon, G., Langlet, P. and Masuzawa, T. (1998) New cylindrical electrostatic micromotors using tubular combination of arrayed direct-drive actuators, *Japanese Journal of Applied Physics*, 37, 622-625.
41. Minotti, P., Le Moal, P., Joseph, E and Bourbon, G. (2001) Toward standard method for MEMS material measurement through on-chip electrostatic probing of micrometer size polysilicon tensile specimens, *Japanese Journal of Applied Physics*, 40, 120-122.
42. Berlin, A.A. (1997) MEMS based active control of macro-scale objects, Semiannual Technical Progress Report n° DABT63-95-C-0025, Defense Advanced Research Project Agency.
43. Temesvary, V., Wu, S., Hsiech, W.H., Tai, Y.C. and Miu, D.K. (1995) Design, fabrication and testing of silicon microgimbals for super-compact rigid disk drives, *Journal of Microelectromechanical Systems*, 4, 18-26.
44. Fujita, H. (1996) Future of actuators and microsystems, *Sensors & Actuators A*, 56, 105-111.
45. Fujita, H., Ataka, M. and Konishi, S. (1996) Group work of distributed microactuators, *Robotica*, 14, 487-492.
46. Ataka, M., Omokada, A., Takeshima, N. and Fujita, H. (1993) Polyimide bimorph actuators for a ciliary motion system, *Journal of Microelectromechanical Systems* 2, 146-150.
47. Konishi, S. and Fujita, H. (1994) A conveyence system using air flow based on the concept of distributed micro motion systems, *Journal of Microelectromechanical Systems* 3, 54-58.
48. Böhringer, K.F., Donald, B.R., Mihailovich, R. and Macdonald, N.C. (1994) A theory of manipulation and control for microfabricated actuator arrays, *Proceedings on Microelectromechanical Systems*, Oiso (Japan), 102-107.
49. Bourbon, G., Minotti, P., Hélin, P. and Fujita, H. (1999) Toward smart surfaces using high-density arrays of silicon-based mechanical oscillators, *Journal of Intelligent Material Systems and Structures*, 10, 534-540.
50. Böhringer, K.F., Donald, B.R., Mihailovich, R. and Macdonald, N.C. (1994) Sensorless manipulation using massively parallel microfabricated actuator arrays, *Proceedings of IEEE International Conference on Robotics and Automation*, San Diego (US), 826-833.
51. Suh, J.W., Darling, R.B., Böhringer, K.F., Donald, B.R., Baltes, H. and Kovacs, G.T.A. (1999) CMOS integrated ciliary actuator array as a general purpose micromanipulation tool for small objects, *Journal of Microelectromechanical Systems* 8, 483-496.
52. Whitehead, L.A. and Bolleman, B.J. (1995) Microstructured elastomeric electromechanical film transducer, *Journal of Acoustical Society of America*, 103, 389-395.
53. Collet, M. and Minotti, P. (2001) Toward acoustical impedance control of vibrating walls using silicon-based active skin, *Proceedings of the 5th SIAM Conference on Control and its Applications*, San Diego (US), P. 241.
54. Neumann, J.J. and Gabriel, K.J. (2002) CMOS-MEMS membrane for audio-frequency acoustic actuation, *Sensors & Actuators A*, 95, 175-182.
55. McMichael, J.M. and Francis, M.S. (1997) Micro air vehicles: toward a new dimension

in flight, http://www.darpa.mil/tto/mav

56. Fearing, R.S., Chiang, K.H., Dickinson, M.H., Pick, D.L., Sitti, M. and Yan, J. (2000) Wing transmission for a micromechanical flying insect, *Proceedings of IEEE International Conference on Robotics and Automation,* San Francisco (US), 1509-1516.

57. Pornsin-sirirak, T.N., Lee, S.W., Nassef, H., Grasmeyer, J., Tai, Y.C., Ho, C.M. and Keennon, M. (2000), *13th IEEE International Conference on Microelectromechanical Systems,* Miyazaki (Japan), 799-804.

58. Azuma, A. (1992) The biokinetics of flying and swimming, *Springer Verlag-Tokyo.*

POWER AMPLIFIERS FOR PIEZOELECTRIC ACTUATORS

H. JANOCHA, CH. STIEBEL, TH. WÜRTZ
Laboratory for Process Automation (LPA)
Saarland University
66121 Saarbrücken, Germany

Abstract

In the year 2000 diesel injection valves with piezoelectric actuators for automobiles have been used in series production for the first time. The introduction of the piezo technology in a mass product is expected to increase the acceptance of piezoelectric actuators for other applications by leaps and bounds. In light of this the efficient electric operation of piezoelectric actuators is gaining interest. This paper presents different, not only commercially available amplifier concepts but also totally new power amplifiers for piezoelectric actuators and their application-relevant features.

Keywords: piezo actuator, power amplifier, analogue amplifier, switching amplifier, hybrid amplifier

1. Introduction

Electrically speaking piezo actuators are foremost capacitive loads with capacitances C_A ranging between some hundreds of picofarads for piezoelectric disks or bending transducers and some tens of microfarads for large multi-layer stack transducers. To attain the maximum possible displacement, a voltage must be applied to the actuator, ranging from 150 V for low-voltage actuators to 1000 V for high-voltage actuators. Due to the actuator's capacitive character no power is required to maintain a certain position without a change in the mechanical load. This is the main advantage of piezo actuators compared to many other types of actuators, apart from their high dynamics (into to the range of kilohertz) and their high positioning precision (on the scale of nanometers). A change in the actuator's voltage v_A is possible by supplying the charge q_A, resulting in the current flow

$$i_A = \frac{dq_A}{dt} = C_A \frac{dv_A}{dt} .$$

(1)

379

A. Preumont (ed.), Responsive Systems for Active Vibration Control, 379–391.

With an applied voltage signal of $v_A = V_0 + \hat{V}_A \cos(2\pi f t)$ with $\hat{V}_A \leq \hat{V}_0$ according to Figure 1 a maximum current is required which is to be supplied by the driving amplifier:

$$\hat{I}_A = 2\pi f C_A \hat{V}_A. \tag{2}$$

2. Analogue amplifiers

Until the middle of the 1990s piezo actuators were mainly used for micro-positioning tasks. For such tasks the most important aspect is high precision, and the driving amplifier is above all expected to show an output signal of the highest possible quality. Analogue amplifiers are best suited to fulfil these expectations. They are the only type capable of continuously charging or discharging a capacitive load such as a piezo actuator. Their main disadvantage – the high losses in the power stage under highly-dynamic conditions – does not come into play in a quasi-static positioning task.

Figure 1 shows the basic structure of a power stage of an analogue amplifier for the control of piezoelectric actuators. The supply voltage V_B must exceed the voltage $V_0 + \hat{V}_A$ across the actuator. To control the actuator current i_A two transistors are necessary. The upper one serves to charge the piezo, the lower one to discharge it. The resulting voltage and current curves are also shown in Figure 1. An operation with one of the two transistors always being in an off-state is called class-C operation. In this kind of operation there is no current flow through the power stage transistors in idle condition (constant voltage) and thus no electrical losses occur. In energetic terms class-C operation is the most favourable of the analogue amplifiers.

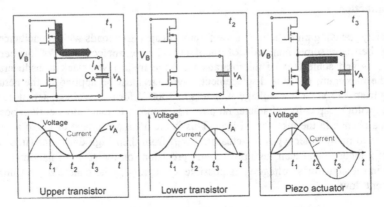

Figure 1. Voltage and current waveforms of a class-C Amplifier
t_1: Charging the actuator with maximum current
t_2: Maintaining the actuator voltage
t_3: Discharging the actuator with maximum current

By calculating the time average of the current-voltage product, the total power dissipation for both power stage transistors in periodic operation is

$$P_v = 2 C_A \hat{V}_A V_B f \cdot$$ (3)

Thus under full operating conditions, i.e. if $2\hat{V}_A = V_B$, $V_0 = \hat{V}_A$ and $\hat{I}_A = \hat{I}_{A\,max}$, twice the energy transmitted to the actuator, $\frac{1}{2}C_A V_B^2$, is converted into heat in the analogue power stage during each cycle of charging and discharging. In continuous operation of highly-dynamic large-signal applications, such as the generation of an oscillation, a power requirement in the range of some hundreds of watts is quickly attained and must be provided by the supply network and is mostly converted into heat in the heat sinks of the power stage. It is above all due to the dimensions of the heat sinks required for this purpose that a miniaturisation of such high-energy electronics can not very likely be achieved.

To safely and reliably limit the charging current of the actuator in case of a desired rectangular actuator voltage signal, comprehensive measures must be taken. If designed accordingly, pulse currents can be permitted that are considerably higher than the current $\hat{I}_{A\,max}$, necessary for continuous sinusoidal operation under full load. With such high pulse currents rectangular voltage pulses with high slew rates are possible. If the rate of pulse repetition is sufficiently low, pulse power on the order of several kilowatts can be dissipated safely with relatively small heat sinks.

A disadvantage of class-C operation are the distortions in the current signal at the zero crossing when the one transistor stops conducting and the other one just starts conducting. In class-A operation the distortions are minimized by always keeping both transistors in conducting operation (Figure 2). Figure 2 also shows that in class-A operation the quiescent current must be $\hat{I}_{A\,max}/2$ which leads to high power losses of $V_B \hat{I}_{A\,max}/2$. For this reason this operation is only suitable for a low power range, when mainly high quality of the output signal is required, as is the case with precision positioning.

For a better comparison of the different concepts with respect to the efficiency of the actuator system as a whole (amplifier, piezo actuator and mechanical load), the following energy flows are taken into consideration which appear during the electric drive of piezo actuators. Here all energy flows are with respect to the maximum energy content of the piezo actuator ($\frac{1}{2}C_A V_B^2$), i.e. the electrical energy required for the actuator's maximum displacement amplitude, corresponding to 100% in Figure 3.

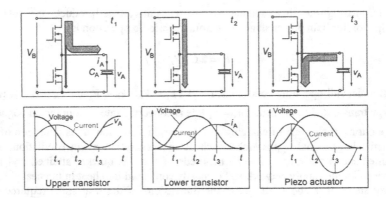

Figure 2. Voltage and current waveforms of a class-A Amplifier
t_1: Charging the actuator with maximum current
t_2: Maintaining the actuator voltage
t_3: Discharging the actuator with maximum current

Approximately 30% of the stored field energy is converted into mechanical work in the connected load and into heat as dielectric and hysteresis losses in the actuator. For the continuous adjustment of the charging and discharging current, analogue amplifiers use transistors in the current path of the actuator capacitance C_A functioning as controllable resistors. Thus, due to the dissipative charging procedure, only 50% of the supplied energy is released to the capacitive load; 100% of the required field energy is converted into heat in the analogue power stage.

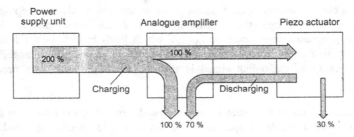

Figure 3. Energy flows for a piezoelectric actuator with analogue amplifier

In analogue power amplifiers a recovery of the energy stored in the actuator (70% of the field energy) is not possible during discharging, so this energy is also converted into heat. As a consequence, for every operating cycle (charging and subsequent discharging of the actuator) in an analogue power stage, twice the required field energy has to be fed from the power supply unit.

3. Switching amplifiers

One of the most important conditions for a possible miniaturisation of power electronics is to lower the power losses, requiring the driving electronics to be able to charge the actuator with a minimum of losses and to recover the energy stored in the actuator while discharging it. Both features can be fulfilled with switching amplifiers.

Figure 4 shows a possible circuit for such an amplifier. In contrast to the analogue amplifier the two power transistors are used here as switches only, which leads to a drastic reduction of transistor power losses. Furthermore, an additional element is necessary, a choke. It fulfils two main tasks: during phases I and III it serves to limit the actuator current without any losses, in phase IV it enables the feed back of the energy stored in the actuator (energy recovery). After turning on the respective power transistors, in phase II as well as in phase IV the energy stored in the choke L leads to a further charging or discharging of the actuator. This "uncontrollable" part of the charging current makes controlling such an amplifier more difficult.

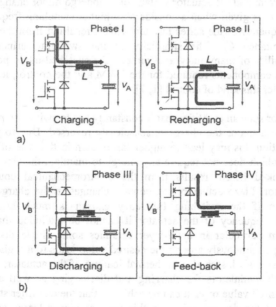

Figure 4. Bi-directional switching amplifier
a) charging cycle b) discharging cycle

As the dimensions of the choke have to be chosen according to the actuator's maximum charging current, in a switching amplifier a strongly elevated pulse current is not possible, as is the case with the analogue amplifier. So even in large-signal operation the dynamics in the switching amplifier remain basically lower than in the analogue amplifier. The high yield of energy of a switching amplifier comes to fruition only when the highly dynamic large-signal operation is permanently required, i.e. when an analogue drive would produce very high continuous losses (e.g. piezoelectric oscillator).

There are different approaches to the control of such a switching amplifier for piezoelectric actuators: The easiest one is fixed-frequency pulse-width modulation (PWM). In this case normally a transistor is switched with a fixed switching frequency and the duration of the closed-circuit condition is varied by means of the difference between actual and desired value. In the case of a purely capacitive load this procedure needs to be modified. In the easiest case, in every switching cycle first the upper switching transistor turns on, then at a given moment turns off again, and eventually the lower transistor turns on for the remaining time of the switching cycle. By varying the switching moment a control is possible as to whether the actuator is charged, discharged or whether the actual voltage value of the actuator is maintained. When maintaining the value, both the upper and the lower transistors are switched on supplying and releasing the same charge to the actuator on average, and the actuator voltage ripples around the desired value together with the switching frequency. The precondition for this procedure is that the choke current and the actuator voltage only undergo minor changes within one switching cycle. This is given when a considerably higher switching frequency is used than the natural frequency of the series resonant circuit resulting from the choke L and the actuator capacitance C_A. The simplicity of the switching principle and the commercial availability of many integrated circuits for switched-mode power supplies (ICs) containing all components required for the PWM and the control, lead to the fact that many users prefer this kind of control [1, 7].

It is characteristic for such an amplifier that it constantly has to switch. In particular with DC voltages on the actuator the charge is senselessly reversed. Due to the switching losses in DC operation this may lead to higher losses than in the case of the analogue class-C amplifier, which does not require any energy to maintain the actuator's charge. Another aspect linked hereto results from the electromechanical coupling in the piezoelectric actuator. Piezo actuators can convert changes in the charge into motion very quickly. Even if the switching frequency chosen is much higher than the mechanical natural frequency of the actuator it cannot be excluded that due to the switching ripples in particular in multi-layer ceramics some single disks are set in motion. So rippling continuously around the desired value leads to a higher mechanical load and due to hysteresis losses also to thermal loading of the actuator. By using two separate pulse-width modulators for charging and discharging as well as a tolerance band around the desired value of v_A it can be achieved that the amplifier stops switching when it has attained the desired value. In this way its power losses are reduced in particular in DC operation and the actuator is less loaded. A disadvantage is given by the fact that the introduction of a tolerance band also has a negative effect on the control accuracy.

A drawback that cannot be remedied with this method is the difficult design of the LC circuit. Due to great efforts for their design switching amplifiers with energy recovery only make sense when the driving power according to equation (3) is high. High driving powers occur mainly in dynamic operation of actuators. The natural frequency of the *LC* circuit on the one hand has to be high with respect to the maximum signal frequency (sensible for example 1 kHz), on the other hand considerably lower than the switching frequency in order to avoid a choke saturation by the PWM. For this reason switching frequencies of several hundreds of kilohertz are required quickly, which may become a problem, considering voltages of 150 to 1000 V and powers of several hundreds of watts, in particular as the upper transistor must be driven potential free.

Other approaches work with an actuator model and the mathematical description of the power stage and calculate the switching time required for every switching cycle and the respective choke current. In this way only the quantity of charge required for achieving a certain actuator voltage is transferred. For time-critical calculations a signal processor or a field-programmable gate-array (FPGA) [3] is employed. In [5] a circuit is presented which makes it possible to charge a piezoelectric actuator from 0 V to a certain, but fixed output voltage in one single switching cycle. The switching time required is calculated on the basis of the system differential equation and the energy balance. The switching time is realised by means of a simple timer circuit.

In Figure 5 a realistic efficiency of 95% is indicated for the switching amplifier during the charging cycle. Then the losses to be covered by the power supply unit only amount to 39% of the field energy; similar to Figure 3 the unavoidable losses in the actuator itself and the mechanical power in the connected load are supposed to amount to 30%. However, the losses caused by charging and discharging the actuator are reduced from 170% in analogue operation to only 9% in switching operation. In highly dynamic operation the losses lead to an unacceptably great warming of the actuator. Major investigations are carried out into this problem.

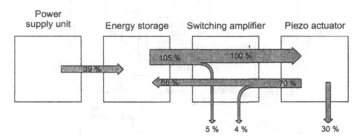

Figure 5. Energy flows for a piezoelectric actuator with switching amplifier and energy recovery

The values of the distortion factor, the deviation from the desired value and the control speed of such a switching power stage are worse than in an analogue amplifier. Furthermore, there are ripples on the output voltage, and the curve of the charging current is discontinuous, which all in all leads to a higher mechanical load of the piezo actuator if compared to analogue operation.

386

4. Switching amplifiers with "intelligent" switching

Continuing the last idea consistently leads to a control concept which enables the transmission of only that amount of energy in every switching cycle which is required to attain a certain actuator voltage [2]. In this way unnecessary switching cycles which only load the actuator and power semiconductors can be avoided. By the explicit evaluation of the differential equation of the power stage in any phase (Figure 6) the exact space of time Δt can be calculated for which the MOSFET M_1 or M_2 must be closed, in order to attain any desired value within one switching cycle starting from any actual value. The result is a non-linear equation in dependence of the actual value of the actuator voltage and the desired value.

In order not to limit the dynamics of the switching amplifier by long calculating times, instead of the computing network a semiconductor memory is used, where the corresponding switching times are stored for all pairs of desired and actual values (Figure 6). During operation this memory is directly driven by speedy A/D converters which directly record the system values. By modifying the switching table an over-current limitation can be implemented without current measuring. The whole control logic of the circuit is contained in a programmable logic device (CPLD).

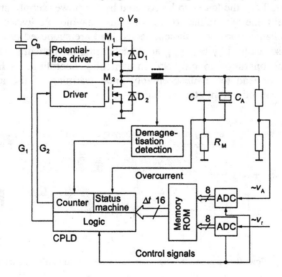

Figure 6. Circuit diagram of a switching amplifier with „intelligent" switching cycle

Figure 7 shows some test results which have been achieved with this concept. Compared to conventional switching amplifiers the quality of the output signal is excellent (Figure 7a). Only with input signals of higher frequencies (Figure 7b) the amplifier's switching mode of working becomes clearly apparent. The output voltage is always trailing the input voltage by one switching cycle, and a new switching cycle can only be started when the choke is demagnetised. The rectangular 700 Hz signal (Figure 7c) shows further features of the amplifier: the current limiting provides a gradual increase of the actuator voltage; with the last step the exact desired value is achieved, and the amplifier stops switching. As the actuator's capacitance is never overloaded, no additional switching cycles are necessary to correct the actuator voltage. This is a major advantage of this concept which could not be attained in this form with a simple pulse-width modulation. Figure 7b also shows that the charging current flows discontinuously with relatively high peak values up to 12 A. This is a characteristic resulting from the amplifier's switching mode of working, and it occurs in a more or less marked form in all switching amplifiers.

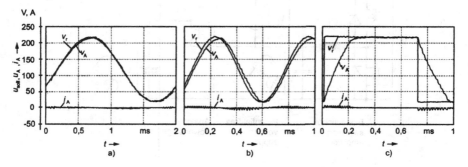

Figure 7. Measured output signals of a switching amplifier with „intelligent" switching cycle
a) 500 Hz sinusoidal signal b) 1.4 kHz sinusoidal signal c) 700 Hz rectangular signal ([2])

5. Hybrid amplifiers

The discontinuities in the actuator's charging current i_A as represented in Figure 7b are partly transformed into undesired high-frequency motion by the piezo ceramics. This leads to an increased load compared to an analogue drive with monotone frequencies with the same voltage amplitude and continuous charging current [6]. An ideal amplifier would be one that combines the high efficiency of the switching amplifier with the continuous current and voltage waveforms of an analogue amplifier. Figure 8 shows the basic circuit of such a "hybrid" amplifier.

As is the case with analogue amplifiers the piezo actuator is driven with high precision by an analogue power stage. But this stage is not supplied with the maximum output voltage, but only with the lower voltage $V_E = 20$ to 40 V. For this reasons the power losses remain low although the whole actuator current i_A flows through the analogue power stage being controlled and free from current ripples. In order to enable a charging of the actuator to attain the required high voltage $\hat{V}_A = 150$ to 1000 V despite the low supply voltage V_E of the power stage, the reference potential of the analogue power stage can be adjusted without losses to any voltage in the range of 0 V and $\hat{V}_A$ by means of a bi-directional switching amplifier. This potential adjustment requires potential free

388

feeding of the supply voltage and the whole control circuit of the analogue power stage as well as all input signals to the power stage. An analogue control is only possible as long as the reference potential of the analogue power stage follows exactly the voltage on the actuator, which places great demands on the signal quality of the switching amplifier (amplifier with "intelligent switching"). When all these conditions are fulfilled, the hybrid amplifier is the ideal combination of a switching amplifier with its high yield of energy and an analogue amplifier with its high-quality signal. Furthermore, the actuator is not loaded with high current peaks due to discontinuous charging currents.

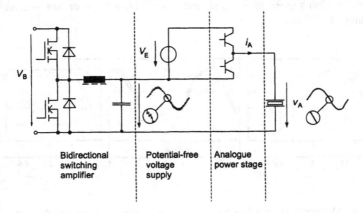

Figure 8. Basic circuit diagram of a hybrid amplifier

From the viewpoint of the load (piezo actuator) the hybrid amplifier has a pure analogue output stage that delivers an output voltage free of ripples whereas the main part of the driving energy is transmitted from the switching part with little losses. Figure 9 illustrates how the two energy flow diagrams from Figures 3 and 5 are combined to form the energy flow diagram of the hybrid amplifier.

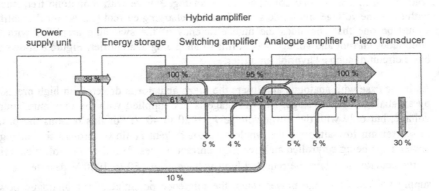

Figure 9. Energy flows for a piezoelectric actuator with a hybrid amplifier

In comparison with Figure 5 an analogue amplifier of low power has been added between switching amplifier and piezo actuator. When charging and discharging the piezo actuator, each time this analogue amplifier converts 5% of the energy into heat. So compared to the analogue amplifier (170% of losses) the power balance is no longer improved to 9%, but only to 19% of losses in the power path of electronics (9% in the switching amplifier and 10% in the analogue amplifier). However, this slightly higher power dissipation enables the achievement of an output signal which almost attains the same quality of a purely analogue amplifier (switching ripple on the load below 1% of the nominal voltage), and it is capable of transferring the main part of the field energy with little losses.

6. Comparison of the amplifier types

The analogue and switching amplifiers described in chapters 2 and 3 are available commercially from several manufacturers, whereas those introduced in chapters 4 and 5 are the results of research projects at the University of Saarland which were advanced to a prototype level by $D*ASS$ [8].

TABLE I. Comparison of different types of amplifiers for piezo actuators

Criterion	Analogue Class-A Amplifier	Analogue Class-C Amplifier	Switching Amplifier	Hybrid Amplifier
Losses in the power transistors	very high even in idle condition	high while driving	very low	low
Feed-back of stored field energy	not possible	not possible	possible	possible (for the most part)
Ripples on the output signal	extremely low	very low	high	low
Relation pulse/continuous current[1]	typical 3.14 (π)	up to 100	1	1
Dynamics in small-signal operation[2]	extremely high	very high	low	high
Load on the actuator[3]	very low	very low	high	low
Electromagnetic compatibility	very good	very good	poor, active disturbance	poor, active disturbance
Load range[4] ($C_A/C_{A\,Nom}$)	100	100	≈ 5	100

[1] important for maximum slew rate of single rectangular pulses for a given construction volume (continuous current)

[2] below current limitation

[3] load from parts of actuator current not resulting from input signal (e.g. current ripples, discontinuous charging current)

[4] range of variation of load capacitance around the nominal value without being forced to change the actuator's control parameters.

Table I provides a survey of the characteristic features of the different amplifiers. In many cases the construction volume available is an important selection criterion. With analogue amplifiers the construction volume is for example determined by the cooling required, i.e. by the continuous output current at a given maximum voltage. In class-C operation this continuous output current can be exceeded by up to one-hundred fold for a short term and with a low repetition rate (pulse operation). But this is not the case with switching amplifiers. Here the element that determines the volume is the choke, as it always has to be designed for the maximum pulse current which can correspond to the continuous current because of the few losses.

In practice this means that in case of a low continuous output power and some short rectangular pulses the construction volume of a class-C amplifier can possibly be smaller than that of a switching amplifier with energy recovery, as the choke of the switching amplifier has always to be designed for the maximum pulse current even if this is only required for certain periods. Concerning the losses the situation may be similar despite the energy recovery, if it is taken into consideration that the switching amplifier causes quiescent losses of some watts due to its "higher intelligence" (current consumption of the CPLD, etc.) compared to the class-C amplifier whose quiescent losses can be very, very low.

In general it can be stressed that there is no universal amplifier that is equally well suited for all possible applications of piezo actuators. Figure 10 reflects the authors' experience over several years and is intended to aid in the selection of a suitable amplifier. In difficult cases or when optimal solutions have to be found, advice by an expert is indispensable. This service is for example provided by *D*ASS* [8] whose products and services have long included the knowledge presented in this paper.

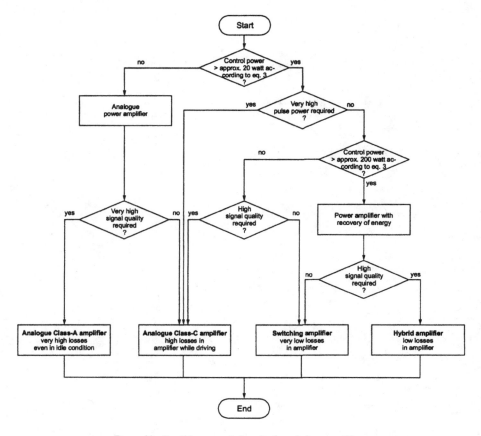

Figure 10. Possible approach for selection of piezo amplifier

7. References

1. Clingman, D.J.: *Drive Electronics for Large Piezo Actuators*; Proc. SPIE Vol. 3044; Industrial and Commercial Applications of Smart Structures Technologies; San Diego; USA; 1997; pp. 459-467
2. Janocha, H.; Stiebel, C.: *New Approach to a Switching Amplifier for Piezoelectric Actuators*, Proc. 6th Internat. Conf. on New Actuators; Bremen; Germany ; 1998; pp. 189-192
3. Kasper, R.; Schröder, J.; Wagner, A.: *Schnellschaltendes Hydraulikventil mit piezoelektrischem Stellantrieb*; O+P Ölhydraulik und Pneumatik; 1997; no. 41; pp. 694-698
4. N.N.: *Der neue Dieselspezialist, mit Piezohydraulik zur Serienfertigung*, Nutzfahrzeug-Dieseltechnik für Nordamerika; Siemens Automobiltechnik; http://www.siemens.de; 1999
5. Schmeißer, F.: *Piezoelektrische Aktuatoren: Kräftig und schnell*; Elektronik; 1984; no. 8; pp. 92-96
6. Stiebel, C.; Würtz, T.; Janocha, H.: *Analogverstärker mit Energierückgewinnung zum Ansteuern von piezoelektrischen Aktoren*, SPS/IPC/DRIVES; Elektrische Automatisierungstechnik - Systeme und Komponenten MESAGO, 1999. (Nürnberg), pp. 693-702
7. Zvonar, G.A.; Douglas, K.L.: *Nonlinear Electronic Control of an Electrostrictive Actuator*, Proc. SPIE Vol. 3044; Industrial and Commercial Applications of Smart Structures Technologies; San Diego; USA; 1997; pp. 448-458
8. *D*ASS* mbH: www.dass.de

Index